Finite Elements
and Approximation

FINITE ELEMENTS AND APPROXIMATION

O. C. ZIENKIEWICZ
K. MORGAN

University of Wales, Swansea, United Kingdom

A Wiley-Interscience Publication
John Wiley & Sons
New York   Chichester   Brisbane   Toronto   Singapore

Library of Congress Cataloging in Publication Data:

Zienkiewicz, O. C.
    Finite elements and approximation.

    "A Wiley-Interscience publication."
    Includes index.
    1. Approximation theory.   2. Finite element method.
I. Morgan, K. (Kenneth), 1945–   II. Title.

QA297.5.Z53   1982   515.3'53   82-16051
ISBN 0-471-98240-7

Printed in the United States of America

10 9 8 7 6 5 4 3 2 1

To Helen and Elizabeth

PREFACE _____

Today the finite element method is a powerful tool for the approximate solution of differential equations governing diverse physical phenomena. Its use in industry and research is extensive, and indeed it could be said that without it (and its handmaiden the computer) many problems would be incapable of solution. Despite this extensive use, comprehension of the principles involved is often lacking in the user who has been trained in a standard undergraduate course (and indeed in many postgraduate courses). It is the object of this book to address such an audience and to form the background text for an undergraduate or early postgraduate course on the subject. For some years now at the authors' institution a large part of this book has formed the basis of a course given to students of civil engineering, and we find that the principles are readily absorbed. When writing the text the authors kept in mind a wider audience of engineers and physicists, and the coverage is therefore suitable for a broad range of students.

It is now about 25 years since the phrase "finite element method" was coined. At the time its inspiration was the field of structural analysis, and analogies with such a discrete process were used for the solution of continuum problems. As the understanding of the basic process grew, its roots in other mathematical approximation methods (such as those due to Rayleigh, Ritz, and Galerkin) became obvious, and the generality opened up made the field an attractive one for mathematicians. Unfortunately, much of their work is couched in a language that others find difficult to follow. Therefore, in this book we attempt a presentation which, though reasonably rigorous, should be readily understood by those with a basic knowledge of calculus.

Many alternative numerical approximation processes existed before the advent of the finite element method. Here boundary solution techniques and finite difference methods have established their own useful existence—and proponents of these have at times crossed swords with those advocating finite element methods in claiming particular superiority. Today some of us see the essential unity of all approximation processes used in the solution of problems defined by differential equations, and in this book we stress this throughout. We endeavor to show that a "generalized finite element method" can be defined embracing all the alternative variants, thus leaving scope for choosing the "optimal approximation" to the user. For this reason the book begins with a chapter on finite difference methods—probably the most obvious (and oldest) of the approximation procedures.

We have endeavored to provide a sufficient number of illustrative examples as well as exercises to make this a suitable teaching text (or a self-study book). Any suggestions from the reader on detailed improvement of these will be welcome.

Finally, we should like to thank Dr. Don Kelly for contributing the major part of Chapter 8 on error estimates and the secretaries of the Civil Engineering Department at Swansea who typed the manuscript.

O. C. ZIENKIEWICZ
K. MORGAN

Swansea, Wales, United Kingdom
September 1982

CONTENTS _____

Finite Elements
and Approximation

_____

Continuum Boundary Value Problems and the Need for Numerical Discretization. Finite Difference Methods

1.1. INTRODUCTION

While searching for a quantitative description of physical phenomena, the engineer or the physicist establishes generally a system of ordinary or partial differential equations valid in a certain region (or domain) and imposes on this system suitable boundary and initial conditions. At this stage the mathematical model is complete, and for practical applications "merely" a solution for a particular set of numerical data is needed. Here, however, come the major difficulties, as only the very simplest forms of equations, within geometrically trivial boundaries, are capable of being solved exactly with available mathematical methods. Ordinary differential equations with constant coefficients are one of the few examples for which standard solution procedures are available—and even here, with a large number of dependent variables, considerable difficulties are encountered.

To overcome such difficulties and to enlist the aid of the most powerful tool developed in this century—the digital computer—it is necessary to recast the problem in a purely algebraic form, involving only the basic arithmetic operations. To achieve this, various forms of *discretization* of the continuum problem defined by the differential equations can be used. In such a discretization the infinite set of numbers representing the unknown function or functions is replaced by a finite number of unknown parameters, and this process, in general, requires some form of approximation.

Of the various forms of discretization which are possible, one of the simplest is the *finite difference process*. In this chapter we describe some of the essentials of this process to set the stage, but the remainder of this book is concerned with various *trial function* approximations falling under the general classification of *finite element methods*. The reader will find later that even the finite difference process can be included as a subclass of this more general category.

Before proceeding further we shall focus our attention on some particular problems which will serve as a basis for later examples. It is clearly impossible to deal in detail in a book of this length with a wide range of physical problems, each requiring an introduction to its background. It is our hope, however, that the few examples chosen will serve to introduce the general principles of approximation, which the readers can then apply to their own particular special cases.

1.2.   SOME EXAMPLES OF CONTINUUM PROBLEMS

Consider the example of Fig. 1.1a in which a problem of heat flow in a two-dimensional domain Ω is presented. If the heat flowing in the direction of the x and y axes per unit length and in unit time is denoted by q_x and q_y respectively, the difference D between outflow and inflow for an element of size $dx\, dy$ is given as

$$D = dy\left(q_x + \frac{\partial q_x}{\partial x}dx - q_x\right) + dx\left(q_y + \frac{\partial q_y}{\partial y}dy - q_y\right) \qquad (1.1)$$

For conservation of heat, this quantity must be equal to the sum of the heat generated in the element in unit time, say, $Q\, dx\, dy$, where Q may vary with position and time, and the heat released in unit time due to the temperature change, namely, $-\rho c(\partial \phi/\partial t)\, dx\, dy$, where c is the specific heat, ρ is the density and $\phi(x, y, t)$ is the temperature distribution. Clearly, this requirement of equality leads to the differential relationship

$$\frac{\partial q_x}{\partial x} + \frac{\partial q_y}{\partial y} - Q + \rho c\frac{\partial \phi}{\partial t} = 0 \qquad (1.2)$$

which has to be satisfied throughout the problem domain Ω.

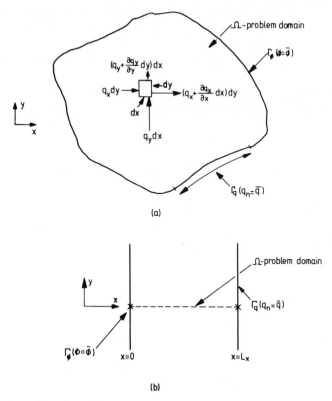

(a)

(b)

FIGURE 1.1. Examples of continuum problems. (*a*) Two-dimensional heat conduction. (*b*) One-dimensional heat conduction.

Introducing now a physical law governing the heat flow in an isotropic medium,[1] we can write, for the flow component in any direction n,

$$q_n = -k\frac{\partial\phi}{\partial n} \qquad (1.3)$$

where k is a property of the medium known as the conductivity. Specifically, in the x and y directions we can then write for an isotropic material

$$q_x = -k\frac{\partial\phi}{\partial x}$$

$$q_y = -k\frac{\partial\phi}{\partial y} \qquad (1.4)$$

Relationships (1.2) and (1.4) define a system of differential equations governing the problem at hand, and which now requires solution for the three dependent variables q_x, q_y, and ϕ.

Such a solution needs the specification of *initial conditions* at time, say, $t = t_0$ (e.g., the distribution of temperature may be given everywhere in Ω at this time) and of *boundary conditions* on the surface or boundary Γ of the problem. Typically two different kinds of boundary condition may be involved.

In the first condition, say applicable on a portion Γ_ϕ of the boundary, the values of the temperature are specified as $\bar{\phi}(x, y, t)$, so we have

$$\phi - \bar{\phi} = 0 \qquad \text{on } \Gamma_\phi \tag{1.5}$$

A boundary condition of this form is frequently referred to as being a Dirichlet boundary condition.

In the second condition, applicable on the remainder Γ_q of the boundary, the values of the heat outflow in the direction n normal to the boundary are prescribed as $\bar{q}(x, y, t)$. Then we can write

$$q_n - \bar{q} = 0 \qquad \text{on } \Gamma_q \tag{1.6a}$$

or, alternatively,

$$-k\frac{\partial \phi}{\partial n} - \bar{q} = 0 \qquad \text{on } \Gamma_q \tag{1.6b}$$

This type of boundary condition is often called a Neumann boundary condition.

The problem now is completely defined by Eq. (1.2), (1.4), (1.5), and (1.6), and numbers representing the distribution of ϕ, q_x, and q_y at all times can, in principle, be obtained by the solution of this set of equations.

This problem may be expressed in an alternative form by using Eq. (1.4) to eliminate the quantities q_x and q_y from Eq. (1.2), and now a higher order differential equation in a single independent variable results. Performing this elimination produces the equation

$$\frac{\partial}{\partial x}\left(k\frac{\partial \phi}{\partial x}\right) + \frac{\partial}{\partial y}\left(k\frac{\partial \phi}{\partial y}\right) + Q - \rho c\frac{\partial \phi}{\partial t} = 0 \tag{1.7}$$

which once again requires the specification of initial and boundary conditions.

In the above we have been concerned with a problem defined in time and space domains, with the former requiring the specification of initial conditions. The independent variables here were x, y, and t. If *steady-state* conditions are assumed (i.e., the problem is invariant with time and so $\partial/\partial t = 0$), the governing equation (1.2) or (1.7) simplifies. In the latter case we have

$$\frac{\partial}{\partial x}\left(k\frac{\partial \phi}{\partial x}\right) + \frac{\partial}{\partial y}\left(k\frac{\partial \phi}{\partial y}\right) + Q = 0 \tag{1.8}$$

which for solution requires only the imposition of boundary conditions of the form (1.5) and (1.6). Such boundary value problems will be the subject of discussion of the major part of this book, but in Chapter 7 we shall return to time-dependent equations and consider possible methods for their solution.

While we have written here the governing equations for a two-dimensional situation, this could have easily been extended to three dimensions to deal with more general problems. On the other hand, in some problems only a one-dimensional variation occurs; in Fig. 1.1*b*, for instance, we consider the heat flow through a slab in which conditions do not vary with *y*. Then, from Eq. (1.8), we have for steady state an ordinary differential equation

$$\frac{d}{dx}\left(k\frac{d\phi}{dx}\right) + Q = 0 \tag{1.9}$$

and the problem "domain" is now simply the range $0 \leqslant x \leqslant L_x$.

Such an ordinary differential equation can be solved analytically, but we shall use it and similar equations extensively to illustrate the application of discretization procedures. This will enable us to demonstrate the accuracy of approximate methods by comparing their results with the exact solutions.

The problem of heat flow just described is typical of many other physical situations and indeed can be identified with problems such as the following.

1. **Irrotational ideal fluid flow.** If we put $k = 1$, $Q = 0$, then Eq. (1.8) reduces to a simple Laplacian form;

$$\frac{\partial^2\phi}{\partial x^2} + \frac{\partial^2\phi}{\partial y^2} \equiv \nabla^2\phi = 0 \tag{1.10}$$

   which is the equation governing the distribution of the potential in irrotational ideal fluid flow.

2. **Flow of fluid through porous media.** Here we take $Q = 0$ and identify k as the medium permeability. The hydraulic head ϕ then satisfies Eq. (1.8).

3. **Small deformation of membranes under a lateral load.** With $k = 1$ and Q defined to be the ratio of the lateral load intensity to the in-plane tension of the membrane, Eq. (1.8) is the equation governing the transverse membrane deflection ϕ.

Other applications will occur to the reader familiar with different physical and engineering problems, and from time to time we shall introduce in this book different applications of the above differential equation and indeed other systems of differential equations.

Although at such times the full exploration of the origin and derivation of such equations may not always be apparent to all readers, we hope that the procedures of mathematical discretization adopted to produce a solution will be clear in each case.

1.3.   FINITE DIFFERENCES IN ONE DIMENSION

Suppose we are faced with a simple one-dimensional boundary value problem, that is, we wish to determine a function $\phi(x)$ which satisfies a given differential equation in the region $0 \leqslant x \leqslant L_x$, together with appropriate boundary conditions at $x = 0$ and $x = L_x$. As we have just seen, a typical example of this type of problem would be that of calculating the temperature distribution $\phi(x)$ through a slab of thickness L_x, of thermal conductivity k, with the faces $x = 0$ and $x = L_x$ maintained at given temperatures $\bar{\phi}_0$ and $\bar{\phi}_{L_x}$, respectively, and with heat generation at a rate $Q(x)$ per unit length in the slab. The governing differential equation for this problem is given by Eq. (1.9), which reduces to the equation

$$k \frac{d^2\phi}{dx^2} = -Q(x) \tag{1.11}$$

if we make the assumption that the material thermal conductivity is constant. The associated boundary conditions are of the type given in Eq. (1.5) and can be written as

$$\phi(0) = \bar{\phi}_0, \qquad \phi(L_x) = \bar{\phi}_{L_x} \tag{1.12}$$

To solve this problem by the finite difference method we begin by *differencing* the independent variable x, that is, we construct a set (or *grid* or *mesh*) of $L + 1$ discrete, equally spaced grid points x_l ($l = 0, 1, 2, \ldots, L$) on the range $0 \leqslant x \leqslant L_x$ (see Fig. 1.2) with $x_0 = 0$, $x_L = L_x$, and $x_{l+1} - x_l = \Delta x$.

The next step is to replace those terms in the differential equation that involve differentiation by terms involving algebraic operations only. This process, of necessity, involves an approximation and can be accomplished by making use of the finite difference approximations to function derivatives. The manner in which such approximations can be made are now discussed.

1.3.1.   The Finite Difference Approximation of Derivatives

Using Taylor's theorem with remainder we can write, exactly,

$$\phi(x_{l+1}) = \phi(x_l + \Delta x) = \phi(x_l) + \Delta x \frac{d\phi}{dx}\bigg|_{x=x_l} + \frac{\Delta x^2}{2} \frac{d^2\phi}{dx^2}\bigg|_{x=x_l+\theta_1 \Delta x}$$

$$\tag{1.13}$$

where θ_1 is some number in the range $0 \leqslant \theta_1 \leqslant 1$. Using the subscript l to

Typical mesh point

FIGURE 1.2. Construction of a finite difference mesh over the interval $0 \leqslant x \leqslant L_x$.

denote an evaluation at $x = x_l$, this can be written

$$\phi_{l+1} = \phi_l + \frac{\Delta x}{2} \frac{d\phi}{dx}\bigg|_l + \frac{\Delta x^2}{2} \frac{d^2\phi}{dx^2}\bigg|_{l+\theta_1} \tag{1.14}$$

and therefore

$$\frac{d\phi}{dx}\bigg|_l = \frac{\phi_{l+1} - \phi_l}{\Delta x} - \frac{\Delta x}{2} \frac{d^2\phi}{dx^2}\bigg|_{l+\theta_1} \tag{1.15}$$

This leads to the so-called *forward difference* approximation of the first derivative of a function in which

$$\frac{d\phi}{dx}\bigg|_l \approx \frac{\phi_{l+1} - \phi_l}{\Delta x} \tag{1.16}$$

The error E in this approximation can be seen to be given by

$$E = -\frac{\Delta x}{2} \frac{d^2\phi}{dx^2}\bigg|_{l+\theta_1} \tag{1.17}$$

and as E is equal to a constant multiplied by Δx, we say that this error is $O(\Delta x)$. This is known as the *order* of the error.

The exact magnitude of the error cannot be obtained from this expression, as the actual value of θ_1 is not given by Taylor's theorem, but it follows that

$$|E| \leqslant \frac{\Delta x}{2} \max_{[x_l, x_{l+1}]} \left|\frac{d^2\phi}{dx^2}\right| \tag{1.18}$$

Figure 1.3 shows a graphical interpretation of the approximation that we have derived mathematically. The first derivative of $\phi(x)$ at $x = x_l$ is the slope of the tangent to the curve $y = \phi(x)$ at this point, that is, the slope of the line AB. The forward difference approximation is the slope of the line AC, and it can be seen that the slope of this line approaches that of the line AB as the mesh spacing Δx gets smaller.

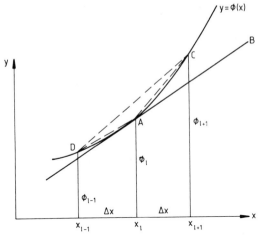

FIGURE 1.3. A graphical interpretation of some finite difference approximations to $d\phi/dx|_l$. Forward difference—slope of AC; backward difference—slope of DA; central difference—slope of DC.

In a similar manner we can use Taylor's theorem to obtain

$$\phi_{l-1} = \phi_l - \Delta x \frac{d\phi}{dx}\bigg|_l + \frac{\Delta x^2}{2} \frac{d^2\phi}{dx^2}\bigg|_{l-\theta_2} \tag{1.19}$$

where $0 \leqslant \theta_2 \leqslant 1$. Rewriting this expression in the form

$$\frac{d\phi}{dx}\bigg|_l = \frac{\phi_l - \phi_{l-1}}{\Delta x} + \frac{\Delta x}{2} \frac{d^2\phi}{dx^2}\bigg|_{l-\theta_2} \tag{1.20}$$

we can produce the *backward difference* approximation

$$\frac{d\phi}{dx}\bigg|_l \simeq \frac{\phi_l - \phi_{l-1}}{\Delta x} \tag{1.21}$$

The error E in this approximation is again $O(\Delta x)$, and now

$$|E| \leqslant \frac{\Delta x}{2} \max_{[x_{l-1},\, x_l]} \left| \frac{d^2\phi}{dx^2} \right| \tag{1.22}$$

The graphical representation of the backward difference approximation can be seen in Figure 1.3; the slope of the line AB is now approximated by the slope of the line AD.

In both the forward and the backward difference approximations the error is of the same order, that is, $O(\Delta x)$. However, if we replace the expansions of

Eqs. (1.14) and (1.19) by

$$\phi_{l+1} = \phi_l + \Delta x \frac{d\phi}{dx}\bigg|_l + \frac{\Delta x^2}{2}\frac{d^2\phi}{dx^2}\bigg|_l + \frac{\Delta x^3}{6}\frac{d^3\phi}{dx^3}\bigg|_{l+\theta_3}, \qquad 0 \leqslant \theta_3 \leqslant 1 \quad (1.23a)$$

$$\phi_{l-1} = \phi_l - \Delta x \frac{d\phi}{dx}\bigg|_l + \frac{\Delta x^2}{2}\frac{d^2\phi}{dx^2}\bigg|_l - \frac{\Delta x^3}{6}\frac{d^3\phi}{dx^3}\bigg|_{l-\theta_4}, \qquad 0 \leqslant \theta_4 \leqslant 1 \quad (1.23b)$$

then a more accurate representation for the first derivative can be obtained by subtracting Eq. (1.23b) from Eq. (1.23a). The resulting equation

$$\phi_{l+1} - \phi_{l-1} = 2\,\Delta x \frac{d\phi}{dx}\bigg|_l + \frac{\Delta x^3}{6}\left(\frac{d^3\phi}{dx^3}\bigg|_{l+\theta_3} + \frac{d^3\phi}{dx^3}\bigg|_{l-\theta_4}\right) \qquad (1.24)$$

can be used to derive the *central difference* approximation

$$\frac{d\phi}{dx}\bigg|_l \simeq \frac{\phi_{l+1} - \phi_{l-1}}{2\,\Delta x} \qquad (1.25)$$

and the error E in this approximation satisfies

$$|E| \leqslant \frac{\Delta x^2}{6} \max_{[x_{l-1},\, x_{l+1}]} \left|\frac{d^3\phi}{dx^3}\right| \qquad (1.26)$$

As the error here is $O(\Delta x^2)$, this should now be a better representation than either the forward or the backward difference approximation. This can again be seen in Fig. 1.3, where the graphical interpretation is that we are now approximating to the slope of the line AB by the slope of the line DC. Again, adding the Taylor expansions

$$\phi_{l+1} = \phi_l + \Delta x \frac{d\phi}{dx}\bigg|_l + \frac{\Delta x^2}{2}\frac{d^2\phi}{dx^2}\bigg|_l + \frac{\Delta x^3}{6}\frac{d^3\phi}{dx^3}\bigg|_l + \frac{\Delta x^4}{24}\frac{d^4\phi}{dx^4}\bigg|_{l+\theta_5},$$

$$0 \leqslant \theta_5 \leqslant 1 \quad (1.27a)$$

$$\phi_{l-1} = \phi_l - \Delta x \frac{d\phi}{dx}\bigg|_l + \frac{\Delta x^2}{2}\frac{d^2\phi}{dx^2}\bigg|_l - \frac{\Delta x^3}{6}\frac{d^3\phi}{dx^3}\bigg|_l + \frac{\Delta x^4}{24}\frac{d^4\phi}{dx^4}\bigg|_{l-\theta_6},$$

$$0 \leqslant \theta_6 \leqslant 1 \quad (1.27b)$$

we find that the terms involving the first and third derivatives disappear. The result is that

$$\frac{d^2\phi}{dx^2}\bigg|_l = \frac{\phi_{l+1} - 2\phi_l + \phi_{l-1}}{\Delta x^2} - \frac{\Delta x^2}{24}\left(\frac{d^4\phi}{dx^4}\bigg|_{l+\theta_5} + \frac{d^4\phi}{dx^4}\bigg|_{l-\theta_6}\right) \quad (1.28)$$

and so we can approximate the second derivative by

$$\frac{d^2\phi}{dx^2}\bigg|_l \approx \frac{\phi_{l+1} - 2\phi_l + \phi_{l-1}}{\Delta x^2} \quad (1.29)$$

The error E in this approximation is $O(\Delta x^2)$ and satisfies

$$|E| \leqslant \frac{\Delta x^2}{12} \max_{[x_{l-1}, x_{l+1}]} \left|\frac{d^4\phi}{dx^4}\right| \quad (1.30)$$

These approximations to first and second derivatives are sufficient for our present purposes, but approximations (of increasing complexity) to higher order derivatives can be obtained in a similar manner if so required. This point is briefly considered further in Section 1.10.

1.3.2. Solution of a Differential Equation by the Finite Difference Method

If we evaluate Eq. (1.11) at a typical grid point x_l, we obtain, exactly,

$$k\frac{d^2\phi}{dx^2}\bigg|_l = -Q_l \quad (1.31)$$

and using the approximation of Eq. (1.29) for the second derivative produces the equation

$$k\frac{\phi_{l+1} - 2\phi_l + \phi_{l-1}}{\Delta x^2} = -Q_l \quad (1.32)$$

An equation of this form arises at each of the interior grid points x_l ($l = 1, 2, \ldots, L - 1$) on the finite difference mesh. Writing down these equations separately gives [changing the sign and inserting the boundary conditions of

Eq. (1.12)]

$$-\phi_2 + 2\phi_1 = \frac{\Delta x^2 Q_1}{k} + \bar{\phi}_0$$

$$-\phi_3 + 2\phi_2 - \phi_1 = \frac{\Delta x^2 Q_2}{k}$$

$$-\phi_4 + 2\phi_3 - \phi_2 = \frac{\Delta x^2 Q_3}{k}$$

$$\vdots \qquad\qquad \vdots \qquad\qquad (1.33)$$

$$-\phi_{L-1} + 2\phi_{L-2} - \phi_{L-3} = \frac{\Delta x^2 Q_{L-2}}{k}$$

$$2\phi_{L-1} - \phi_{L-2} = \frac{\Delta x^2 Q_{L-1}}{k} + \bar{\phi}_{L_x}$$

If ϕ is a column vector whose transpose is $(\phi_1, \phi_2, \phi_3, \ldots, \phi_{L-2}, \phi_{L-1})$, then this set of equations may be written as a single matrix equation,

$$\mathbf{K}\phi = \mathbf{f} \qquad\qquad (1.34)$$

where

$$\mathbf{K} = \begin{bmatrix} 2 & -1 & 0 & 0 & 0 & & & \\ -1 & 2 & -1 & 0 & 0 & & & \\ 0 & -1 & 2 & -1 & 0 & & & \\ 0 & 0 & -1 & 2 & -1 & & & \\ & & & & \ddots & \ddots & & \\ & & & & & \ddots & \ddots & \\ & & & & -1 & 2 & -1 \\ & & & & 0 & -1 & 2 \end{bmatrix} \qquad (1.35)$$

$$\mathbf{f} = \begin{bmatrix} \dfrac{\Delta x^2}{k} Q_1 + \bar{\phi}_0 \\[2ex] \dfrac{\Delta x^2}{k} Q_2 \\[1ex] \vdots \\[1ex] \dfrac{\Delta x^2}{k} Q_{L-1} + \bar{\phi}_{L_x} \end{bmatrix} \qquad (1.36)$$

Thus the original problem of determining an unknown continuous function $\phi(x)$ has been replaced by the problem of solving a matrix equation for the discrete set of values $\phi_1, \phi_2, \phi_3, \ldots, \phi_{L-1}$.

The finite difference method will therefore give information about the function values at the mesh points, but it gives us no information about the function values between these points. Indeed we have only approximated to the governing equation at a discrete number of points and not throughout the region.

The solution to Eq. (1.34) can be efficiently computed by noting that the matrix **K** is symmetric, positive definite and tridiagonal, and then using an inversion algorithm specifically designed for such an equation system.[2] It must be remembered that the resulting solution ϕ only approximates to the exact solution of the problem as originally posed because of the approximation involved in replacing Eq. (1.31) by Eq. (1.32). However, the fact that the error in the approximation is $O(\Delta x^2)$ indicates that reducing the mesh spacing should reduce the error involved and produce a more accurate solution.

The practical use of the finite difference method will now be illustrated by applying the general theory outlined above to a particular simple example.

Example 1.1

It is required to obtain the solution of the equation $d^2\phi/dx^2 - \phi = 0$ which satisfies the boundary conditions $\phi = 0$ at $x = 0$ and $\phi = 1$ at $x = 1$. A mesh spacing $\Delta x = \frac{1}{3}$ is chosen, as shown in Fig. 1.4, and the solution may then be found by the finite difference method.

The only unknowns are ϕ_1 and ϕ_2, the values of the solution at the points $x = \frac{1}{3}$ and $x = \frac{2}{3}$, respectively, while the given boundary conditions imply that $\phi_0 = 0$ and $\phi_3 = 1$.

The equation evaluated at a general grid point x_l is

$$\left. \frac{d^2\phi}{dx^2} \right|_l - \phi_l = 0$$

which can be expressed in finite difference form, using the approximation of Eq. (1.29), as

$$\phi_{l+1} - 2\phi_l + \phi_{l-1} - \Delta x^2 \phi_l = 0$$

Using this equation for $l = 1$ and $l = 2$, that is, the two interior points, and

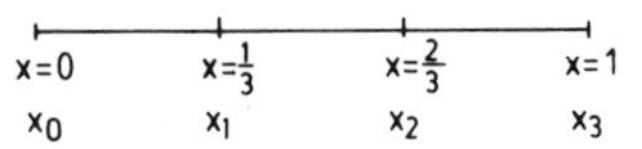

FIGURE 1.4. Finite difference mesh adopted for the solution of Examples 1.1 and 1.2.

inserting the known conditions, gives

$$-\phi_2 + 2\tfrac{1}{9}\phi_1 = 0$$

$$2\tfrac{2}{9}\phi_2 - \phi_1 = 1$$

with solution

$$\phi_1 = 0.2893, \qquad \phi_2 = 0.6107$$

The reader can readily obtain the analytical solution for this simple example and the above results can then be compared with the exact grid point values of 0.2889 and 0.6102, respectively.

If the calculation is repeated with $\Delta x = \tfrac{1}{6}$, the solution of the resulting equation system gives values of 0.2890 and 0.6104 for the finite difference approximations to the value of ϕ at $x = \tfrac{1}{3}$ and $x = \tfrac{2}{3}$, respectively. The improvement in the accuracy of the finite difference approximations with a decrease in mesh spacing is then apparent.

EXERCISES

1.1. Solve the equation $d^2\phi/dx^2 + \phi = 0$ subject to the conditions $\phi = 1$ at $x = 0$ and $\phi = 0$ at $x = 1$, using a mesh point spacing $\Delta x = 0.25$. Compare the resulting finite difference solution with the exact solution.

1.2. The distribution of bending moment M in a beam subjected to loading by a distributed load $w(x)$ per unit length satisfies the equation $d^2M/dx^2 = w(x)$. A beam of unit length is simply supported (i.e., $M = 0$) at both ends and carries a load $w(x) = \sin \pi x$ per unit length. Calculate the distribution of the bending moment by the finite difference method using a mesh point spacing $\Delta x = 0.25$.

1.3. The equation governing the variation in temperature T in a viscous fluid flowing between two parallel plates ($y = 0$ and $y = 2H$) is given to be

$$\frac{d^2T}{dy^2} = -\frac{4U^2\mu}{H^4k}(H - y)^2$$

where μ, k, and U are the viscosity, thermal conductivity, and maximum velocity of the fluid respectively. If $\mu = 0.1$, $k = 0.08$, $H = 3.0$, and $U = 3.0$, calculate the temperature distribution when one plate is held at $T = 0$ and the other at $T = 5$, using the finite difference method and a mesh point spacing $\Delta y = 0.5H$.

1.4. A cable at tension T is held fixed at its ends $x = 0$ and $x = 1$ and rests on an elastic foundation of stiffness k. When the cable is loaded transversely with load w per unit length, the deflection ϕ of the cable satisfies the equation $d^2\phi/dx^2 - k\phi/T = -w/T$. Solve this equation by the finite difference method for the case $k/T = 1, w/T = 1$, using a mesh point spacing $\Delta x = 0.1$ and a suitable computer program. Compare your solutions with the exact answer to this problem.

1.5. On a finite difference mesh with point spacing Δx, the first derivative is to be replaced by the approximation $d\phi/dx|_l \approx a\phi_{l+1} + b\phi_l$ where a and b are constants. Show that if it is required that this approximation be exact whenever ϕ is a linear function of x, then $a = 1/\Delta x, b = -1/\Delta x$, that is, the forward difference method. If the same requirement is applied to the approximation $d\phi/dx|_l \approx a\phi_l + b\phi_{l-1}$, show that the result is the backward difference method. (*Hint*: First move the origin to $x = x_l$ using $X = x - x_l$, and then require equality for both $\phi = 1$ and $\phi = X$.)

1.6. Prove that if the approximation $d^2\phi/dx^2|_l \approx a\phi_{l+1} + b\phi_l + c\phi_{l-1}$ is exact whenever ϕ is a quadratic function of x, then a, b, and c are such that the approximation is that of Eq. (1.29). Show that this approximation is also exact if ϕ is a cubic.

1.7. Construct an approximation

$$\left.\frac{d\phi}{dx}\right|_l \approx a\phi_{l+2} + b\phi_{l+1} + c\phi_l + d\phi_{l-1} + e\phi_{l-2}$$

by requiring that the approximation be exact if ϕ is a quartic function of x. What is the order of the error in this approximation?

1.4. DERIVATIVE BOUNDARY CONDITIONS

Frequently in real problems one (or more) of the associated boundary conditions may be expressed in terms of a derivative; for example, returning to our heat flow problem of Section 1.3, suppose that we now assume that the surface $x = L_x$ of the slab is subjected to a condition of prescribed heat flux $\bar{q}$ across the surface (i.e., a condition of the form of Eq. (1.6) is to be applied). Using Eq. (1.6), the appropriate condition at $x = L_x$ is now not that the temperature itself is specified at this point, but that the gradient of the temperature is specified, namely,

$$-k\frac{d\phi}{dx} = \bar{q} \quad \text{at } x = L_x \tag{1.37}$$

Then if we repeat the work of the previous section and write down the finite difference equation at each interior point, we obtain

$$-\phi_2 + 2\phi_1 = \frac{\Delta x^2 Q_1}{k} + \bar{\phi}_0$$

$$-\phi_3 + 2\phi_2 - \phi_1 = \frac{\Delta x^2 Q_2}{k}$$

$$-\phi_4 + 2\phi_3 - \phi_2 = \frac{\Delta x^2 Q_3}{k}$$

$$\vdots \qquad \vdots \tag{1.38}$$

$$-\phi_{L-1} + 2\phi_{L-2} - \phi_{L-3} = \frac{\Delta x^2 Q_{L-2}}{k}$$

$$-\phi_L + 2\phi_{L-1} - \phi_{L-2} = \frac{\Delta x^2 Q_{L-1}}{k}$$

which, since ϕ_L is now unknown, is a set of $L - 1$ equations in the L unknowns $\phi_1, \phi_2, \phi_3, \ldots, \phi_{L-1}, \phi_L$. The missing equation has to be provided by the boundary condition of Eq. (1.37), which can be written as

$$\frac{d\phi}{dx}\bigg|_L = -\frac{\bar{q}}{k} \tag{1.39}$$

If the derivative is replaced by the backward difference approximation of Eq. (1.21), then this condition becomes

$$\frac{\phi_L - \phi_{L-1}}{\Delta x} = -\frac{\bar{q}}{k} \tag{1.40}$$

which together with Eq. (1.38) produces a complete set of L equations in the unknowns $\phi_1, \phi_2, \ldots, \phi_L$.

Example 1.2

Return again to the equation considered in Example 1.1, but subjected now to the boundary conditions $\phi = 0$ at $x = 0$ and $d\phi/dx = 1$ at $x = 1$. If the finite difference mesh shown in Fig. 1.4 is used, the unknowns are ϕ_1, ϕ_2, and ϕ_3, while the boundary conditions give $\phi_0 = 0$, $d\phi/dx|_3 = 1$.

The finite difference approximation of the governing equation at x_1 and at x_2 becomes

$$-\phi_2 + 2\tfrac{1}{9}\phi_1 = 0$$

$$-\phi_3 + 2\tfrac{1}{9}\phi_2 - \phi_1 = 0$$

and using a backward difference representation of the derivative boundary condition at x_3 produces

$$\frac{\phi_3 - \phi_2}{1/3} = 1$$

The solution of this set of equations can be found to be

$$\phi_1 = 0.2477, \qquad \phi_2 = 0.5229, \qquad \phi_3 = 0.8563$$

and the corresponding exact solutions in this case are 0.2200, 0.4648, and 0.7616.

There is an inconsistency in the above analysis in that we have represented the differential equation to within an error which is $O(\Delta x^2)$, whereas our backward difference approximation to the derivative means that we have represented the boundary condition to within an error which is $O(\Delta x)$. This can be rectified by handling the derivative boundary condition in a different manner. First, we introduce a fictitious mesh point $x_{L+1}(= x_L + \Delta x)$ with associated "temperature" ϕ_{L+1}. This "temperature" has no physical significance as the point x_{L+1} lies outside the slab that is being analyzed. We can then write down the finite difference representation of the governing equation at each point x_l $(l = 1, 2, \dots, L)$ and obtain

$$-\phi_2 + 2\phi_1 = \frac{\Delta x^2 Q_1}{k} + \bar{\phi}_0$$

$$-\phi_3 + 2\phi_2 - \phi_1 = \frac{\Delta x^2 Q_2}{k}$$

$$-\phi_4 + 2\phi_3 - \phi_2 = \frac{\Delta x^2 Q_3}{k} \tag{1.41}$$

$$\vdots \qquad\qquad \vdots$$

$$-\phi_L + 2\phi_{L-1} - \phi_{L-2} = \frac{\Delta x^2 Q_{L-1}}{k}$$

$$-\phi_{L+1} + 2\phi_L - \phi_{L-1} = \frac{\Delta x^2 Q_L}{k}$$

which is a set of L equations for the $L + 1$ unknown $\phi_1, \phi_2, \ldots, \phi_L, \phi_{L+1}$. Again the missing equation is provided by the boundary condition at $x = L_x$, but instead of using the backward difference representation, used previously in Eq. (1.40), the central difference approximation is now applied. This means that we write

$$\left. \frac{d\phi}{dx} \right|_L \approx \frac{\phi_{L+1} - \phi_{L-1}}{2\,\Delta x} \tag{1.42}$$

and the appropriate form of the boundary condition of Eq. (1.37) is then that

$$\phi_{L+1} - \phi_{L-1} = -\frac{2\bar{q}\,\Delta x}{k} \tag{1.43}$$

In this way the governing equation and the boundary condition can be represented to the same order of approximation.

The reader will observe that the resulting equation system, when cast in the matrix form of Eq. (1.34), loses its symmetry, and this is a point of some computational importance.

Example 1.3

The problem solved in Example 1.2 will now be reconsidered and a central difference approximation used for the derivative boundary condition at $x = 1$. With the same mesh spacing $\Delta x = \frac{1}{3}$, this will necessitate the introduction of the fictitious mesh point x_4, with corresponding "solution" value ϕ_4, as shown in Fig. 1.5.

The finite difference form of the differential equation at x_1, x_2, x_3 gives

$$-\phi_2 + 2\tfrac{1}{9}\phi_1 = 0$$

$$-\phi_3 + 2\tfrac{1}{9}\phi_2 - \phi_1 = 0$$

$$-\phi_4 + 2\tfrac{1}{9}\phi_3 - \phi_2 = 0$$

where use has been made of the boundary condition

$$\phi_0 = 0$$

Central differencing of the derivative boundary condition at x_3 produces the additional relationship

$$\frac{\phi_4 - \phi_2}{2/3} = 1$$

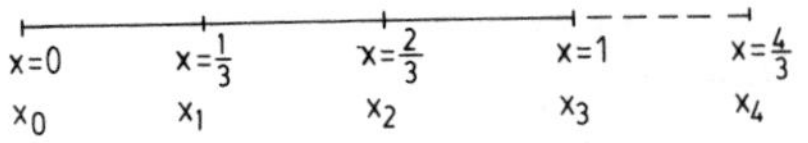

FIGURE 1.5. Finite difference mesh adopted for the solution of Example 1.3.

The solution of this equation set is

$$\phi_1 = 0.2168, \qquad \phi_2 = 0.4576, \qquad \phi_3 = 0.7493$$

which can be seen to be considerably more accurate than the solution calculated in Example 1.2 using the backward difference representation of the derivative boundary condition.

EXERCISES

1.8. Solve the equation $d^2\phi/dx^2 + \phi = x$ subject to the boundary conditions $\phi = 0$ at $x = 0$ and $d\phi/dx + \phi = 0$ at $x = 1$ using a suitable mesh point spacing.

1.9. Repeat the problem of Exercise 1.3, but applying in this case the boundary conditions that one plate is held at temperature $T = 0$ while there is no heat flow (i.e., $dT/dy = 0$) across the other plate.

1.10. Consider one-dimensional steady heat flow in a rod of length 10 cm and diameter 1 cm. One end of the rod is held at a temperature of 50°C, while heat is input at the other end at a rate of 200 W/cm². If $k = 75$ W/cm. °C and if heat is being generated within the rod at a rate of $150T$ W/cm² per unit length, where T is the temperature, calculate the temperature distribution using the finite difference method with a mesh point spacing $\Delta x = 1.0$ cm.

1.5. NONLINEAR PROBLEMS

The mathematical modeling of physical problems frequently produces governing differential equations and/or boundary conditions that are nonlinear in character. Whereas analytical methods of solution for linear equations normally fail to cope with nonlinear differential equations, the finite difference method can be applied without modification to both linear and nonlinear problems. We have seen that application of the finite difference method to linear boundary value problems requires the solution of a set of linear equations of the form of Eq. (1.34). When the boundary value problem is nonlinear, application of the finite difference method produces a set of nonlinear algebraic equations.

Returning to our heat conduction example of Section 1.3, we can consider the physically realistic problem where the thermal conductivity k is a given function of the temperature ϕ and the governing equation is the nonlinear equation

$$\frac{d}{dx}\left[k(\phi)\frac{d\phi}{dx}\right] = -Q(x) \qquad (1.44)$$

Now, using a central difference approximation, we can write

$$\left.\frac{d\psi}{dx}\right|_{l} = \frac{\psi_{l+1/2} - \psi_{l-1/2}}{\Delta x} \qquad (1.45)$$

where the subscript $l + \frac{1}{2}$ indicates an evaluation at the point midway between x_l and x_{l+1}, and the subscript $l - \frac{1}{2}$ is defined similarly. Thus if we take $\psi = k(\phi)d\phi/dx$, the original differential equation may be replaced by the finite difference approximation

$$k(\phi_{l+1/2})\left.\frac{d\phi}{dx}\right|_{l+1/2} - k(\phi_{l-1/2})\left.\frac{d\phi}{dx}\right|_{l-1/2} = -\Delta x\, Q_l \qquad (1.46)$$

and changing the sign and replacing the appropriate derivatives by central difference approximations, produces

$$-k(\phi_{l+1/2})[\phi_{l+1} - \phi_l] + k(\phi_{l-1/2})[\phi_l - \phi_{l-1}] = \Delta x^2 Q_l \qquad (1.47)$$

Thus application of the finite difference method to the original nonlinear differential equation has produced the set of nonlinear algebraic equations

$$-k(\phi_{l+1/2})\phi_{l+1} + \left[k(\phi_{l+1/2}) + k(\phi_{l-1/2})\right]\phi_l - k(\phi_{l-1/2})\phi_{l-1} = \Delta x^2 Q_l,$$

$$l = 1, 2, \ldots, L - 1 \qquad (1.48)$$

It should be noted that this equation reduces to Eq. (1.32) when k is constant. Defining the column vectors ϕ and $\mathbf{f}$ as previously, the solution can be obtained by solving this set of nonlinear equations, which may be conveniently expressed in the form

$$\mathbf{K}(\phi)\phi = \mathbf{f} \qquad (1.49)$$

This solution can be achieved by many standard, albeit iterative and costly, methods. We shall not discuss in this text details of such procedures,[3] but we shall restrict discussion to one of the simplest and most obvious methods which is that of simple iteration in which the system of equations (1.49) is solved repeatedly with successively improved values of $\mathbf{K}(\phi)$.

If we start from some initial guess

$$\phi = \phi_0 \qquad (1.50)$$

and evaluate the matrix

$$\mathbf{K}(\boldsymbol{\phi}_0) = \mathbf{K}_0 \tag{1.51}$$

an improved approximation for $\boldsymbol{\phi}(= \boldsymbol{\phi}_1)$ can be obtained as

$$\boldsymbol{\phi}_1 = \mathbf{K}_0^{-1}\mathbf{f} \tag{1.52}$$

This process can be obviously continued writing

$$\boldsymbol{\phi}_n = \mathbf{K}_{n-1}^{-1}\mathbf{f} \tag{1.53}$$

and proceeding until the difference between $\boldsymbol{\phi}_n$ and $\boldsymbol{\phi}_{n-1}$ is within a suitable tolerance. This approach to the solution of nonlinear systems can best be illustrated by considering its application to a particular problem.

Example 1.4

Solve the equation

$$\frac{d}{dx}\left[k\frac{d\phi}{dx}\right] = -10x$$

with $\phi = 0$ when $x = 0$, $\phi = 0$ when $x = 1$, and where $k = 1 + 0.1\phi$. Here, as in Example 1.1, the value $\Delta x = 1/3$ is chosen, and the finite difference equation at $x = x_l$ can be written using Eq. (1.48) as

$$-k_{l+1/2}\phi_{l+1} + \left(k_{l+1/2} + k_{l-1/2}\right)\phi_l - k_{l-1/2}\phi_{l-1} = 10x_l\Delta x^2$$

Using this form for $l = 1, 2$ and inserting the boundary conditions $\phi_0 = \phi_3 = 0$ produces

$$-k_{3/2}\phi_2 + \left(k_{3/2} + k_{1/2}\right)\phi_1 = 10x_1\Delta x^2$$

$$\left(k_{5/2} + k_{3/2}\right)\phi_2 - k_{3/2}\phi_1 = 10x_2\Delta x^2$$

But, by definition,

$$k_{3/2} = 1 + 0.1\phi_{3/2}$$

which introduces the value of ϕ at a point midway between the mesh points x_1

and x_2. One method of obtaining this value is to use the approximation

$$\phi_{3/2} \simeq \frac{(\phi_1 + \phi_2)}{2}$$

and then

$$k_{3/2} \simeq 1 + 0.1 \frac{(\phi_1 + \phi_2)}{2}$$

With $k_{1/2}$, $k_{5/2}$ evaluated similarly, the above set of nonlinear equations only involves the two unknown values ϕ_1 and ϕ_2, and the iteration procedure outlined above may be started. Now

$$\mathbf{K}(\phi) = \begin{bmatrix} 2 + 0.05(2\phi_1 + \phi_2) & -1 - 0.05(\phi_1 + \phi_2) \\ -1 - 0.05(\phi_1 + \phi_2) & 2 + 0.05(\phi_1 + 2\phi_2) \end{bmatrix}$$

$$\phi = \begin{bmatrix} \phi_1 \\ \phi_2 \end{bmatrix}$$

$$\mathbf{f} = \begin{bmatrix} \frac{10}{27} \\ \frac{20}{27} \end{bmatrix}$$

and so, starting from the initial guess $\phi_1 = \phi_2 = 0$,

$$\mathbf{K}_0 = \begin{bmatrix} 2 & -1 \\ -1 & 2 \end{bmatrix}$$

and the solution is

$$\phi_1 = \begin{bmatrix} 0.49383 \\ 0.61728 \end{bmatrix}$$

The new matrix $\mathbf{K}_1$ is then obtained by insertion of these values into the matrix $\mathbf{K}$ above to produce

$$\mathbf{K}_1 = \begin{bmatrix} 2.08025 & -1.05556 \\ -1.05556 & 2.08642 \end{bmatrix}$$

and the solution is now

$$\phi_2 = \begin{bmatrix} 0.48190 \\ 0.59883 \end{bmatrix}$$

Repeating the above process gives after four cycles

$$\phi_4 = \begin{bmatrix} 0.48220 \\ 0.59931 \end{bmatrix}$$

which as the reader may confirm by direct integration of the equation, agrees with the exact answer at the grid points to four places of decimals.

EXERCISES

1.11. Solve the equation $e^{-\phi} d^2\phi/dx^2 = 1$, subject to the boundary conditions $\phi = 0$ at $x = 0$ and at $x = 1$, using a mesh point spacing $\Delta x = 0.25$. This equation is typical of problems of heat conduction in chemically reactive materials for which $Q \propto e^{\phi}$, where ϕ is the temperature. Redo the problem with the right-hand side of the governing equation equal to 10.

1.12. The surfaces $x = 0$ and $x = 1$ of a slab of unit conductivity are maintained at temperatures of 10 and 80, respectively. Using a finite difference mesh with mesh point spacing $\frac{1}{3}$, obtain the steady-state distribution ϕ, of temperature if heat is being generated within the slab at a rate $0.1 \phi^2$ per unit length.

1.13. It is required to solve the equation $\phi(d\phi/dx) = d^2\phi/dx^2$ subject to the conditions $\phi = -2$ at $x = 0$ and $\phi = -1$ at $x = 1$. When the finite difference method is used, devise two alternative representations, say $\mathbf{K}^1(\phi)$ and $\mathbf{K}^2(\phi)$, for the matrix $\mathbf{K}(\phi)$ of Eq. (1.49) and obtain expressions for the entries of $\mathbf{K}^1(\phi)$ and $\mathbf{K}^2(\phi)$. Solve both equation sets, using simple iteration and a mesh point spacing $\Delta x = \frac{1}{3}$, and compare the resulting solutions with the exact solution $\phi = -2/(1 + x)$.

1.6. FINITE DIFFERENCES IN MORE THAN ONE DIMENSION

The problem of approximating to differential equations in two or more independent variables is obviously a little more involved, although the principles used are identical to those employed in one dimension. Let us consider once again the heat conduction problem given in the steady state by Eq. (1.8).

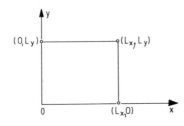

FIGURE 1.6. The two-dimensional rectangular region over which the problem of steady heat conduction is to be solved.

For simplicity we shall take the region Ω to be a rectangle, as shown in Fig. 1.6, and, with a constant conductivity k, the governing equation becomes

$$k\left(\frac{\partial^2 \phi}{\partial x^2} + \frac{\partial^2 \phi}{\partial y^2} \right) = -Q(x, y) \qquad (1.54)$$

If the sides of the rectangle are held at a constant temperature $\bar{\phi}$, then the associated boundary conditions can be written as

$$\phi(0, y) = \phi(L_x, y) = \phi(x,0) = \phi(x, L_y) = \bar{\phi} \qquad (1.55)$$

The origin of the (x, y) coordinate system has been placed at one corner of the rectangle, with the axes lying along the sides of the rectangle, as shown in Fig. 1.6.

To apply the finite difference method in this situation, we proceed exactly as for the one-dimensional case. To this end we construct the set of equally spaced grid points x_l $(l = 0, 1, 2, \ldots, L)$ on the range $0 \leqslant x \leqslant L_x$ with $x_0 = 0$, $x_L = L_x$, $x_{l+1} - x_l = \Delta x$, and also the set of equally spaced grid points y_m $(m = 0, 1, 2, \ldots, M)$ on the range $0 \leqslant y \leqslant L_y$ with $y_0 = 0$, $y_M = L_y$, $y_{m+1} - y_m = \Delta y$. The region in which the solution is required is then covered by a rectangular finite difference grid by drawing a line parallel to OY through each point x_l and a line parallel to OX through each point y_m (see Fig. 1.7). A typical grid point then has coordinates of the form (x_l, y_m). The finite difference method is now to be applied to Eq. (1.54), which means that we will replace terms involving partial derivatives by corresponding finite difference approximations.

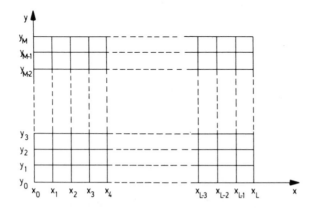

FIGURE 1.7.   Finite difference grid covering the rectangular region of Fig. 1.6.

1.6.1. Finite Difference Approximations to Partial Derivatives

Using Taylor's theorem for a function of two variables, it is possible to write, exactly,

$$\phi(x_{l+1}, y_m) = \phi(x_l + \Delta x, y_m)$$

$$= \phi(x_l, y_m) + \Delta x \left.\frac{\partial \phi}{\partial x}\right|_{x_l, y_m} + \frac{\Delta x^2}{2} \left.\frac{\partial^2 \phi}{\partial x^2}\right|_{x_l + \theta_1 \Delta x, y_m},$$

$$0 \leqslant \theta_1 \leqslant 1 \quad (1.56)$$

or using the subscript l, m to denote evaluation at (x_l, y_m),

$$\phi_{l+1, m} = \phi_{l, m} + \Delta x \left.\frac{\partial \phi}{\partial x}\right|_{l, m} + \frac{\Delta x^2}{2} \left.\frac{\partial^2 \phi}{\partial x^2}\right|_{l+\theta_1, m} \quad (1.57)$$

By repeating the process outlined in Section 1.3.1, the following results can then be obtained:

1. A forward difference approximation for $\partial \phi / \partial x$, namely,

$$\left.\frac{\partial \phi}{\partial x}\right|_{l, m} \simeq \frac{\phi_{l+1, m} - \phi_{l, m}}{\Delta x} \quad (1.58)$$

   with an error which is $O(\Delta x)$.

2. A backward difference approximation for $\partial \phi / \partial x$, namely,

$$\left.\frac{\partial \phi}{\partial x}\right|_{l, m} \simeq \frac{\phi_{l, m} - \phi_{l-1, m}}{\Delta x} \quad (1.59)$$

   with an error which is $O(\Delta x)$.

3. A central difference approximation for $\partial \phi / \partial x$, namely,

$$\left.\frac{\partial \phi}{\partial x}\right|_{l, m} \simeq \frac{\phi_{l+1, m} - \phi_{l-1, m}}{2 \Delta x} \quad (1.60)$$

   with an error which is $O(\Delta x^2)$.

4. A finite difference approximation for $\partial^2 \phi / \partial x^2$, namely,

$$\left.\frac{\partial^2 \phi}{\partial x^2}\right|_{l, m} \simeq \frac{\phi_{l+1, m} - 2\phi_{l, m} + \phi_{l-1, m}}{\Delta x^2} \quad (1.61)$$

   with an error which is $O(\Delta x^2)$.

In addition, by considering the Taylor expansion of $\phi(x_l, y_m + \Delta y)$, similar expressions can be obtained for finite difference approximations to $\partial \phi / \partial y$ and $\partial^2 \phi / \partial y^2$.

1.6.2. Solution of a Partial Differential Equation by the Finite Difference Method

Evaluating Eq. (1.54) at a typical mesh point (x_l, y_m) produces, exactly,

$$k\left(\left.\frac{\partial^2\phi}{\partial x^2}\right|_{l,m} + \left.\frac{\partial^2\phi}{\partial y^2}\right|_{l,m}\right) = -Q_{l,m} \tag{1.62}$$

and using the approximation of Eq (1.61) for the second derivatives, gives

$$k\left[\frac{\phi_{l+1,m} - 2\phi_{l,m} + \phi_{l-1,m}}{\Delta x^2} + \frac{\phi_{l,m+1} - 2\phi_{l,m} + \phi_{l,m-1}}{\Delta y^2}\right] = -Q_{l,m} \tag{1.63}$$

approximately. The given boundary conditions mean that

$$\phi_{0,m} = \phi_{L,m} = \bar{\phi}, \quad m = 0,1,2,\dots, M$$

$$\phi_{l,0} = \phi_{l,M} = \bar{\phi}, \quad l = 0,1,2,\dots, L \tag{1.64}$$

Thus writing down Eq. (1.63) at each of the interior mesh points (i.e., $l = 1,2,3,\dots, L-1$; $m = 1,2,3,\dots, M-1$) and inserting the boundary conditions, produces a set of $(L-1) \times (M-1)$ linear equations in the $(L-1) \times (M-1)$ unknowns $\phi_{1,1}, \phi_{1,2}, \dots, \phi_{L-1,M-1}$. If we define

$$\phi_l^T = (\phi_{l,1}, \phi_{l,2}, \phi_{l,3}, \dots, \phi_{l,M-1}) \tag{1.65}$$

and

$$\phi^T = (\phi_1, \phi_2, \phi_3, \dots, \phi_{L-1}) \tag{1.66}$$

then this equation set can be expressed in the standard form of Eq. (1.34). The matrix $\mathbf{K}$ in this case is a banded symmetric positive definite matrix which can be conveniently written as

$$\mathbf{K} = \begin{bmatrix} \bar{\mathbf{K}} & -\mathbf{I} & \mathbf{0} & \mathbf{0} & & & \\ -\mathbf{I} & \bar{\mathbf{K}} & -\mathbf{I} & \mathbf{0} & & & \\ \mathbf{0} & -\mathbf{I} & \bar{\mathbf{K}} & -\mathbf{I} & & & \\ & & \cdot & \cdot & \cdot & & \\ & & & \cdot & \cdot & \cdot & \\ & & & & \cdot & \cdot & \cdot \\ & & & & -\mathbf{I} & \bar{\mathbf{K}} & -\mathbf{I} \\ & & & & \mathbf{0} & -\mathbf{I} & \bar{\mathbf{K}} \end{bmatrix} \tag{1.67}$$

where $\mathbf{I}$ denotes the unit $(M-1) \times (M-1)$ matrix and $\bar{\mathbf{K}}$ is the $(M-1) \times$

$(M - 1)$ tridiagonal matrix defined by

$$
\overline{\mathbf{K}} =
\begin{bmatrix}
4 & -1 & 0 & 0 & & & & \\
-1 & 4 & -1 & 0 & & & & \\
0 & -1 & 4 & -1 & & & & \\
& & & \cdot & \cdot & \cdot & & \\
& & & & \cdot & \cdot & \cdot & \\
& & & & & \cdot & \cdot & \cdot \\
& & & & -1 & 4 & -1 \\
& & & & 0 & -1 & 4
\end{bmatrix}
\tag{1.68}
$$

This set of equations can be solved by standard methods[4] to produce the approximations to the values of the solution at the mesh points.

The above approach has to be modified along the lines described in Section 1.4 if one or more of the boundary conditions are given in terms of the derivative of the required function.

Example 1.5

In problems of elastic torsion of prismatic bars the equation

$$
\frac{\partial^2 \phi}{\partial x^2} + \frac{\partial^2 \phi}{\partial y^2} = -2G\theta
$$

is encountered, where G is the elastic shear modulus and θ the angle of twist for each section. Here ϕ is a stress function such that $\phi = 0$ on the boundaries.[5]

The twisting moment T is given by

$$
T = 2 \int \int_{\Omega} \phi \, dx \, dy
$$

and the shear stress in a direction n in the section can be obtained from

$$
\tau = \frac{\partial \phi}{\partial n}
$$

It is required to determine the value of T and the maximum shear stress for a given value of $G\theta$ and the rectangular section shown in Fig. 1.8a.

By symmetry considerations, the solution need only be obtained for a quarter of the section, as shown in Fig. 1.8b, and a mesh with $\Delta x = \Delta y = 1$ will be used.

We note that the value of ϕ will be proportional to the constant $G\theta$, and for simplicity we take $G\theta = 1$. The typical form of an approximate difference equation is now, using Eq. (1.63),

$$
\phi_{l+1, m} + \phi_{l-1, m} + \phi_{l, m+1} + \phi_{l, m-1} - 4\phi_{l, m} = -2
$$

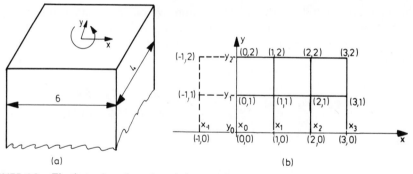

FIGURE 1.8. Elastic torsion of a prismatic bar. (*a*) Bar of rectangular cross section. (*b*) Finite difference mesh used.

The use of symmetry requires that $\partial \phi / \partial x = 0$ along the y axis shown in Fig. 1.8*b*, and that similarly $\partial \phi / \partial y = 0$ along the x axis. These slope conditions can be imposed as shown in Eq. (1.42) and then at a point such as (x_0, y_1), the resulting equation becomes $\phi_{-1,1} = \phi_{1,1}$.

Inserting the zero values at the boundaries $x = 3$ and $y = 2$, we can write the equation system for the six unknown values of ϕ in a matrix form as:

$$
\begin{bmatrix}
4 & -2 & 0 & -2 & 0 & 0 \\
-1 & 4 & -1 & 0 & -2 & 0 \\
0 & -1 & 4 & 0 & 0 & -2 \\
-1 & 0 & 0 & 4 & -2 & 0 \\
0 & -1 & 0 & -1 & 4 & -1 \\
0 & 0 & -1 & 0 & -1 & 4
\end{bmatrix}
\begin{bmatrix}
\phi_{0,0} \\
\phi_{1,0} \\
\phi_{2,0} \\
\phi_{0,1} \\
\phi_{1,1} \\
\phi_{2,1}
\end{bmatrix}
=
\begin{bmatrix}
2 \\
2 \\
2 \\
2 \\
2 \\
2
\end{bmatrix}
$$

The reader can verify by back substitution that

$$\phi^T = [3.1370, 2.8866, 1.9971, 2.3873, 2.2062, 1.5508]$$

is the solution to these equations. By using a form of the trapezoidal rule* in two-dimensions, the integral determining the torque can be evaluated approximately with the result that $T = 65.41$, which may be compared with the exact solution of $T = 76.4$. Similarly, the maximum slope[5] will be formed at the point $(0, 2)$, and so a possible approximation to the absolute value of the maximum shear stress is $|\tau_{max}| = 2.3873$, where the backward difference approximation for the derivative $\partial \phi / \partial n$ has been used. This again may be compared with the exact maximum value of the stress given by $|\tau_{max}| = 2.96$.

The approximation obtained by the use of a backward difference formula has already been shown to be less accurate than the approximation used for

---

*This and other methods of numerical integration are discussed in some detail in Chapter 5.

the main differential equation. We can improve our approximation to τ_{max} by using three values of ϕ on the centerline section as follows. Denoting the point (x_0, y_2) as A, (x_0, y_1) as B, and (x_0, y_0) as C, we can write, by Taylor's theorem,

$$\phi_B = \phi_A - \Delta y \left.\frac{\partial \phi}{\partial y}\right|_A + \frac{1}{2}\Delta y^2 \left.\frac{\partial^2 \phi}{\partial y^2}\right|_A - \frac{\Delta y^3}{6}\left.\frac{\partial^3 \phi}{\partial y^3}\right|_D$$

and

$$\phi_C = \phi_A - 2\Delta y \left.\frac{\partial \phi}{\partial y}\right|_A + \frac{1}{2}(2\Delta y)^2 \left.\frac{\partial^2 \phi}{\partial y^2}\right|_A - \frac{(2\Delta y)^3}{6}\left.\frac{\partial^3 \phi}{\partial y^3}\right|_E$$

where D lies in AB and E lies in AC.

It is now possible to eliminate $\partial^2 \phi/\partial y^2|_A$ between these two equations and produce the result

$$\left.\frac{\partial \phi}{\partial y}\right|_A = \frac{\phi_C - 4\phi_B + 3\phi_A}{2\Delta y} + O(\Delta y^2)$$

Using the first term on the right-hand side as an approximation, the error made is then of the same order as that made in approximating the governing equation, and inserting the appropriate values gives $|\tau_{max}| = 3.21$. This result is closer to the exact value than that obtained above by use of the backward difference approximation.

EXERCISES

1.14. The temperature along the sides of a square plate is maintained at the values shown in the diagram. Obtain the internal steady temperature distribution using the square mesh illustrated.

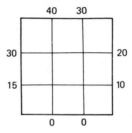

1.15. The temperature along three sides of a square plate is maintained at the values shown in the diagram, while the remaining side is thermally insulated. Obtain the steady temperature distribution using the square mesh illustrated.

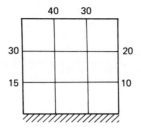

1.16. Solve by the finite difference method the problem of heat transfer through the hollow conduit shown in the diagram. The mesh point spacing is given by $\Delta x = \Delta y = 1$. The inner boundary is held at a temperature $\phi = 100$ while the outer boundary is subject to a convection condition of the form $k \partial\phi/\partial n = -\alpha\phi$. Consider the case $\alpha = k = 1$.

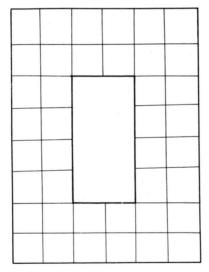

1.17. Obtain a finite difference form of Laplace's equation

$$\frac{\partial^2\phi}{\partial r^2} + \frac{1}{r}\frac{\partial\phi}{\partial r} + \frac{1}{r^2}\frac{\partial^2\phi}{\partial\theta^2} = 0$$

with respect to cylindrical polar coordinates (r, θ). The temperature along the sides of a sector of an annulus is maintained at the values shown in the diagram. Obtain the temperature distribution using the mesh illustrated in which $\Delta r = \frac{1}{4}$ and $\Delta\theta = \pi/18$.

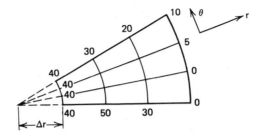

1.18. The potential ϕ in groundwater flow under a certain coffer dam satisfies the equation

$$\frac{\partial^2\phi}{\partial x^2} + \frac{\partial^2\phi}{\partial y^2} = 0$$

Determine the potential by analyzing the region illustrated in the figure and imposing the boundary conditions shown. Take a finite difference mesh in which $\Delta x = 2$ and $\Delta y = 1.5$.

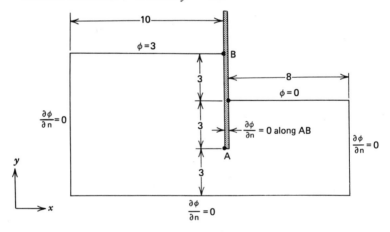

1.19. In Example 1.5 an alternative finite difference approximation to $\partial\phi/\partial y$ was developed in which the error was $O(\Delta y^2)$. Repeat Example 1.2 using this approximation to represent the derivative boundary condition at $x = 1$, and compare the results produced with those obtained in Examples 1.2 and 1.3.

1.7. PROBLEMS INVOLVING IRREGULARLY SHAPED REGIONS

The examples considered in the last section have involved boundaries made up of straight lines which have intersected the rectangular mesh at points which are also mesh points. However, when the boundary is curved, this will no

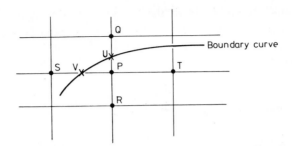

FIGURE 1.9. Finite difference grid in the neighborhood of a curved boundary.

longer be the case, and the previously derived finite difference formulas for derivatives will have to be modified in the neighborhood of the boundary.

Suppose that the boundary curve in a certain problem intersects the rectangular finite difference grid as shown in Fig. 1.9. Then $PT = PS = \Delta x$ and $PQ = PR = \Delta y$. Let $PU = \lambda \, \Delta y$ and $PV = \mu \, \Delta x (0 < \lambda, \mu < 1)$ where U and V are the points where the boundary cuts the lines PQ and PS, respectively. Using Taylor's theorem, we can write

$$\phi_T = \phi_P + \Delta x \left. \frac{\partial \phi}{\partial x} \right|_P + \frac{\Delta x^2}{2} \left. \frac{\partial^2 \phi}{\partial x^2} \right|_P + \frac{\Delta x^3}{6} \left. \frac{\partial^3 \phi}{\partial x^3} \right|_{P_1} \tag{1.69}$$

and

$$\phi_V = \phi_P - \mu \, \Delta x \left. \frac{\partial \phi}{\partial x} \right|_P + \frac{\mu^2 \, \Delta x^2}{2} \left. \frac{\partial^2 \phi}{\partial x^2} \right|_P - \frac{\mu^3 \, \Delta x^3}{6} \left. \frac{\partial^3 \phi}{\partial x^3} \right|_{P_2} \tag{1.70}$$

where P_1 is some point on PT and P_2 is some point on PV.

From these two equations we obtain the approximations

$$\left. \frac{\partial \phi}{\partial x} \right|_P \simeq \frac{\mu^2 \phi_T - \phi_V - (\mu^2 - 1)\phi_P}{\mu(\mu - 1) \, \Delta x} \tag{1.71}$$

with error $O(\Delta x^2)$ and

$$\left. \frac{\partial^2 \phi}{\partial x^2} \right|_P \simeq \frac{2[\mu \phi_T + \phi_V - (\mu + 1)\phi_P]}{\mu(\mu + 1) \, \Delta x^2} \tag{1.72}$$

with error $O(\Delta x)$ that is, the accuracy for the approximation of the second derivative in the neighborhood of the boundary is not as good as the approximation at an ordinary mesh point. Similarly, approximations can be obtained for $\partial \phi / \partial y |_P$ and $\partial^2 \phi / \partial y^2 |_P$. Thus if we are trying to solve the heat conduction equation

$$k \left[\frac{\partial^2 \phi}{\partial x^2} + \frac{\partial^2 \phi}{\partial y^2} \right] = -Q(x, y) \tag{1.73}$$

inside a curved boundary along which the value of the function $\phi(x, y)$ is given, we can apply the approximation of Eq. (1.63) at points which are not adjacent to the boundary, while at points such as the point P in Fig. 1.9 the appropriate finite difference approximation would then be

$$\frac{\mu \phi_T + \phi_V - (\mu + 1) \phi_P}{\mu(\mu + 1) \Delta x^2} + \frac{\phi_U + \lambda \phi_R - (\lambda + 1) \phi_P}{\lambda(\lambda + 1) \Delta y^2} = -\frac{Q_P}{2k} \tag{1.74}$$

where the values ϕ_U and ϕ_V are given by the boundary condition.

The situation becomes more difficult if the problem boundary conditions give information on the derivative of the function along a curved boundary, for example, if the solution of the heat conduction equation (1.73) is required subject to the condition that a certain portion Γ_q of a curved boundary is subjected to a prescribed heat flux $\bar{q}$. As we have seen, the appropriate form of the boundary condition is then

$$-k \left. \frac{\partial \phi}{\partial n} \right|_{\Gamma_q} = \bar{q} \tag{1.75}$$

where n is the direction of the outward normal to Γ_q. The method of handling this problem by the finite difference method will not be considered in this book, and the interested reader should refer to other works.[6]

1.8. NONLINEAR PROBLEMS IN MORE THAN ONE DIMENSION

In one dimension it has already been shown how the finite difference method can be applied equally well to both linear and nonlinear boundary value problems, and the same is true in more than one dimension, for example, if we wish to solve the problem of steady heat conduction across a rectangular block which has a temperature-dependent thermal conductivity, then we need to solve the nonlinear equation

$$\frac{\partial}{\partial x} \left[k(\phi) \frac{\partial \phi}{\partial x} \right] + \frac{\partial}{\partial y} \left[k(\phi) \frac{\partial \phi}{\partial y} \right] = -Q(x, y) \tag{1.76}$$

Application of the finite difference method, using the approximations developed in Section 1.5, produces the nonlinear equation set

$$
\left[-k_{l+1/2,\,m}\phi_{l+1,\,m} + \left(k_{l+1/2,\,m} + k_{l-1/2,\,m}\right)\phi_{l,\,m} - k_{l-1/2,\,m}\phi_{l-1,\,m}\right]/\Delta x^2
$$

$$
+ \left[-k_{l,\,m+1/2}\phi_{l,\,m+1} + \left(k_{l,\,m+1/2} + k_{l,\,m-1/2}\right)\phi_{l,\,m} \right. \tag{1.77}
$$

$$
\left. -k_{l,\,m-1/2}\phi_{l,\,m-1}\right]/\Delta y^2
$$

$$
= Q_{l,\,m}, \qquad l = 1, 2, \ldots, L - 1 \quad m = 1, 2, \ldots, M - 1
$$

Inserting the associated conditions, this set can be solved by an appropriate iterative technique, although clearly the computation effort required will now be substantially greater.

1.9. APPROXIMATION AND CONVERGENCE

The one- and two-dimensional solutions to ordinary and partial differential equations derived in this chapter by numerical procedures of the finite difference type have, we hope, illustrated the possibilities of discretization. The seemingly intractable (or at least mathematically difficult) problem of the differential equation has been replaced by a purely algebraic problem in which a number of simultaneous equations have to be solved. This solution may not be readily obtainable by hand calculation if the number of equations is large, but it can always be obtained on a computer with some investment of time and cost. Obviously this cost can be reduced by optimizing the algorithms used in such numerical solutions, and much effort has gone into this aspect of the problem.[7]

The important thing is to realize that the possibility of solution now exists, even though this involves an approximation. We have already shown that the error in the finite difference approximation decreases with some power of the mesh size. To apply the process in practice to a situation in which we have no exact solution, it is necessary to study the *convergence* of the method as the mesh size is refined in an attempt to estimate the magnitude of the errors occurring in the approximations produced.

If for instance the error in an approximation is of $O(\Delta x^2)$, then the results of two solutions on meshes with grid spacing Δx_1 and Δx_2 can be extrapolated as follows.

Let ϕ_l^1 correspond to the finite difference solution on the first mesh and ϕ_l^2 to the solution on the second mesh. If ϕ_l^e is the exact solution at x_l, then, even

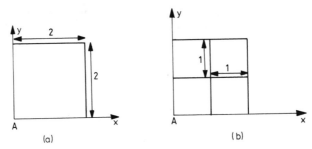

FIGURE 1.10. Application of the Richardson extrapolation to the elastic torsion problem of Example 1.6. The finite difference meshes used on a quarter section have (*a*) $\Delta x_1 = \Delta y_1 = 2$ and (*b*) $\Delta x_2 = \Delta y_2 = 1$.

though we do not know the error magnitude, we can write

$$\frac{\phi_I^e - \phi_I^1}{\phi_I^e - \phi_I^2} = \left(\frac{\Delta x_1}{\Delta x_2}\right)^2 \tag{1.78}$$

and ϕ_I^e, the exact solution, can be determined. This relationship is not exactly true as "order" of error and its actual magnitude are not the same thing, but improved results can often be obtained in this way. The procedure is known as Richardson extrapolation.

Example 1.6

Consider the problem of torsion of a bar of square section with side length 4. Again we must solve the torsion equation of Example 1.5 for the stress function ϕ subject to $\phi = 0$ on the boundaries, and by symmetry only a quarter section need be analyzed. With $\Delta x_1 = \Delta y_1 = 2$ (Fig. 1.10a) the finite difference approximation produces $\phi_A^1 = 2G\theta$ where A denotes the center of the section. Using $\Delta x_1 = \Delta y_2 = 1$ (Fig. 1.10b), gives $\phi_A^2 = 2.25G\theta$. Application of the extrapolation procedure of Eq. (1.78) then gives $\phi_A^e = 2.333G\theta$, which compares well with the exact value of $\phi_A = .2.357G\theta$ for this problem.

1.10.   CONCLUDING REMARKS

The scope of this chapter has been limited to consideration of the solution of boundary value problems by the finite difference process. We have restricted our attention to problems involving first- and second-order derivatives, although the reader will by now have recognized the possibility of extending the processes to arbitrary differential equations involving higher order derivatives. For instance the fourth-order ordinary differential can be approximated by using twice expressions such as those given in Eq. (1.29). For a general finite difference mesh as shown in Fig. 1.2, we can write

$$\frac{d^4\phi}{dx^4}\bigg|_l = \frac{d^2}{dx^2}\left(\frac{d^2\phi}{dx^2}\right)\bigg|_l \simeq \frac{d^2\phi/dx^2|_{l+1} - 2d^2\phi/dx^2|_l + d^2\phi/dx^2|_{l-1}}{\Delta x^2}$$

$$(1.79)$$

and inserting

$$\frac{d^2\phi}{dx^2}\bigg|_{l+1} \simeq \frac{\phi_{l+2} - 2\phi_{l+1} + \phi_l}{\Delta x^2} \qquad (1.80)$$

and so on, we obtain a five-point approximation

$$\frac{d^4\phi}{dx^4}\bigg|_l \simeq \frac{\phi_{l+2} - 4\phi_{l+1} + 6\phi_l - 4\phi_{l-1} + \phi_{l-2}}{\Delta x^4} \qquad (1.81)$$

Such approximations and others allow the finite difference process to be extended widely, and many successful applications have been reported in a variety of fields. However, the difficulties of approximating to curved boundaries and gradient boundary conditions remain, and in later sections of this book we shall indicate how such difficulties can be removed by the use of mapping or irregular grids. However, alternative, more general, procedures, which we discuss in the next chapters, will also be seen to eliminate easily such difficulties.

EXERCISES

1.20. The equation governing the deflection of a beam resting on an elastic foundation of stiffness k is $EI\, d^4\phi/dx^4 + k\phi = w$, where EI is the constant flexural rigidity of the beam, and w is the load per unit length carried by the beam. If the beam is of unit length and is clamped at both ends so that $\phi = d\phi/dx = 0$ at $x = 0$ and $x = 1$, determine the deflection when $w/EI = k/EI = 1$, using a finite difference method and a mesh spacing $\Delta x = \frac{1}{3}$. Repeat the process with $\Delta x = \frac{1}{6}$, and then use the Richardson extrapolation. Compare your results at $x = \frac{1}{3}$ and $x = \frac{2}{3}$ with the exact solution.

1.21. The small normal deflection ϕ of a uniform thin elastic plate Ω, of uniform flexural rigidity D, simply supported along its edges Γ, and subjected to unit uniform transverse load per unit area, is governed by the differential equation

$$\frac{\partial^4 \phi}{\partial x^4} + 2\frac{\partial^4 \phi}{\partial x^2 \, \partial y^2} + \frac{\partial^4 \phi}{\partial y^4} = 1/D$$

in Ω and the boundary conditions $\phi = \partial^2\phi/\partial n^2 = 0$ on Γ. Obtain the deflection for the rectangular plate $|x| \leqslant 3$, $|y| \leqslant 2$, using the finite difference method and a mesh spacing $\Delta x = \Delta y = \frac{1}{2}$.

1.22. On a finite difference mesh, with spacing Δx, the fourth derivative is to be replaced by the approximation

$$\left.\frac{d^4 \phi}{dx^4}\right|_l \approx a\phi_{l+2} + b\phi_{l+1} + c\phi_l + d\phi_{l-1} + e\phi_{l-2}$$

where a, b, c, d, and e are constants. Show that if it is required that this approximation be exact whenever ϕ is a quartic function of x, then the constants are such that the approximation is as in Eq. (1.81). What is the order of the error in this approximation?

1.23. Determine the steady-state internal temperature distribution ϕ for the semicircular region shown in the figure. Use the finite difference mesh and boundary temperatures indicated. Investigate the problems which would be encountered in the finite difference approach if the boundary condition along the curved surface were replaced by the condition $\partial\phi/\partial n = 0$.

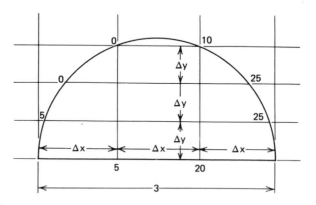

REFERENCES

[1] See, for example, H. S. Carslaw and J. C. Jaeger, *Conduction of Heat in Solids*, 2nd ed., Clarendon, Oxford, 1959.

[2] Details of this and of more general matrix inversion procedures are given in D. Potter, *Computational Physics*, Wiley-Interscience, New York, 1973.

[3] For details of more general methods the reader should consult O. C. Zienkiewicz, *The Finite Element Method*, 3rd ed., McGraw-Hill, New York, 1977, and the references given therein.

[4] See, for example, Ref. 2.

[5] A detailed analysis of this problem can be found in I. S. Sokolnikoff, *Mathematical Theory of Elasticity*, McGraw-Hill, New York, 1956.

[6] This method is described fully in S. H. Crandall, *Engineering Analysis*, McGraw-Hill, New York, 1956.

[7] For further details of this subject see, R. Wait, *The Numerical Solution of Algebraic Equations*, Wiley-Interscience, New York, 1979.

SUGGESTED FURTHER READING

I. Fried, *Numerical Solution of Differential Equations*, Academic, New York, 1979.

L. A. Hageman and D. M. Young, *Applied Iterative Methods*, Academic, New York, 1981.

A. R. Mitchell and D. F. Griffiths, *The Finite Difference Method in Partial Differential Equations*, Wiley, Chichester, 1980.

G. D. Smith, *Numerical Solution of Partial Differential Equations*, Oxford University Press, Oxford, 1971.

CHAPTER TWO _____

Weighted Residual Methods: Use of Continuous Trial Functions

2.1. INTRODUCTION — APPROXIMATION BY TRIAL FUNCTIONS

We show in this chapter that the key to the problem of determining numerically the solution of differential equations lies in the ability to develop accurate function approximation methods. In the finite difference method we have concentrated on defining the value of the unknown function $\phi(x)$ at a finite number of values of x. Alternative methods for determining numerically the solution to differential equations can however be developed by making the process of function approximation more systematic and general.

Suppose that we wish to approximate a given function ϕ in some region Ω bounded by a closed curve Γ. In problems involving differential equations it is required to find the solution satisfying certain boundary conditions, and we shall therefore attempt initially to construct approximations which are exactly equal to prescribed values of ϕ on the boundary curve Γ. If any function ψ can be found which takes on the same values as ϕ on Γ, that is, $\psi|_\Gamma = \phi|_\Gamma$ and if a set of independent *trial functions* $\{N_m; m = 1, 2, 3, \dots\}$ is introduced such that $N_m|_\Gamma = 0$ for all m, then at all points in Ω we can approximate to ϕ by

$$\phi \simeq \hat{\phi} = \psi + \sum_{m=1}^{M} a_m N_m \qquad (2.1)$$

where a_m $(m = 1, 2, \dots, M)$ are some parameters which are computed so as to

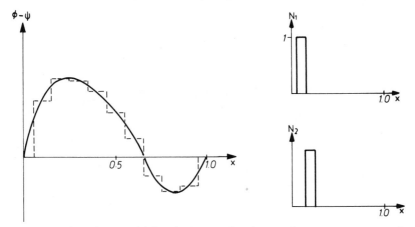

FIGURE 2.1. Discontinuous trial functions possessing the completeness property over $0 \leqslant x \leqslant 1$.

obtain a good "fit." Trial functions of this type are frequently referred to as *shape* or *basis* functions, and we shall occasionally use this alternative nomenclature. The manner in which ψ and the trial function set are defined then automatically ensures that this approximation has the property that $\hat{\phi}|_\Gamma = \phi|_\Gamma$, whatever the values of the parameters a_m. The trial function set should clearly be chosen so as to ensure that improvement in the approximation occurs with increase in the number M of trial functions used. One obvious condition for this convergence of the approximation is that the trial function set be such that the combination $\psi + \sum_{m=1}^{M} a_m N_m$ can adequately represent any function ϕ, satisfying $\phi|_\Gamma = \psi|_\Gamma$, as $M \to \infty$. This is the so-called *completeness* requirement, and on occasions it is evident by inspection. This is illustrated by the example of Fig. 2.1 in which Ω is the one-dimensional region $0 \leqslant x \leqslant 1$ and the chosen functions N_m are of a discontinuous form, shown to have the value unity on a suitable interval and the value zero elsewhere. Provided that no interior of the domain is omitted, it is clear that any well-behaved single-valued function can be approximated as closely as desired by dividing the total domain into ever smaller intervals. Such discontinuous functions are the subject of further attention in Chapter 3, but we shall not consider again the question of completeness for other trial function sets which will be used in this chapter.

To illustrate this general approach to function approximation, we shall begin by considering some obvious methods of determining the parameters a_m used in the representation of Eq. (2.1).

2.1.1. Point Fitting of Functions

In this method the parameters a_m are chosen by requiring that the approximation $\hat{\phi}$ be exactly equal to the function ϕ at M distinct, arbitrarily chosen, points in Ω. This requirement leads to a system of linear equations which must

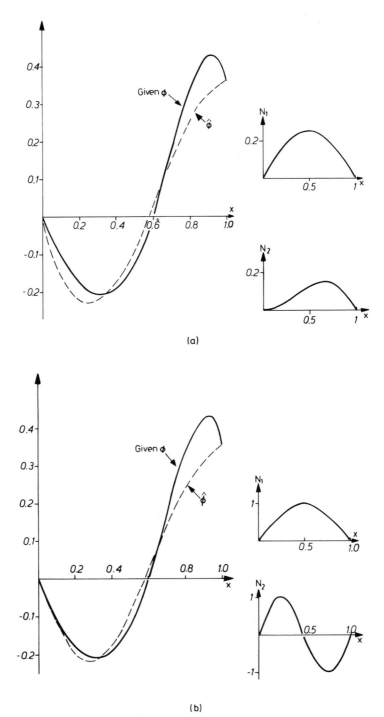

(a)

(b)

FIGURE 2.2. Trial function approximation of a given function ϕ using point fitting and the first two terms of the trial function sets (a) $\langle N_m = x^m(1 - x); \quad m = 1, 2, \dots \rangle$ and (b) $\langle N_m = \sin m\pi x; \quad m = 1, 2, \dots \rangle$.

be solved for the parameter set $\{a_m; m = 1, 2, \ldots, M\}$. In Fig. 2.2 it is shown how an arbitrary function in the domain $0 \leqslant x \leqslant 1$ can be approximated by using the first two members of the following trial function sets, namely, in Fig. 2.2*a*,

$$\{N_m = x^m(1 - x); m = 1, 2, \ldots\}$$

and in Fig. 2.2*b*,

$$\{N_m = \sin m\pi x; m = 1, 2, \ldots\}$$

In this case the function ψ was simply chosen as the linear function taking on the prescribed values of ϕ at $x = 0$ and $x = 1$, while the method adopted for choosing the parameters was to require that $\hat{\phi} = \phi$ at $x = \frac{1}{3}$ and at $x = \frac{2}{3}$.

2.1.2. Fourier Sine Series

Using the theory of Fourier series[1] it is possible to develop an approximation to an arbitrary function $\phi(x)$, over a range of values $0 \leqslant x \leqslant L_x$, in the form of Eq. (2.1). The theory requires that the function $\phi(x)$ have only a finite number of discontinuities and only a finite number of local maxima and minima in the range of interest, requirements which are satisfied by most functions encountered during the solution of physical problems. Assuming that ϕ is a continuous function, and defining ψ again as the linear function taking on the prescribed values of ϕ at $x = 0$ and $x = L_x$, then the infinite series

$$\psi + \sum_{m=1}^{\infty} a_m \sin \frac{m\pi x}{L_x} \tag{2.2}$$

where the coefficients are given by

$$a_m = \frac{2}{L_x} \int_0^{L_x} (\phi - \psi) \sin \frac{m\pi x}{L_x} dx \tag{2.3}$$

converges everywhere in the interval to the value of the function $\phi(x)$. Terminating the series (2.2) after a finite number M of terms enables an approximation to be produced:

$$\phi(x) \simeq \hat{\phi}(x) = \psi(x) + \sum_{m=1}^{M} a_m \sin \frac{m\pi x}{L_x}, \qquad 0 \leqslant x \leqslant L_x \tag{2.4}$$

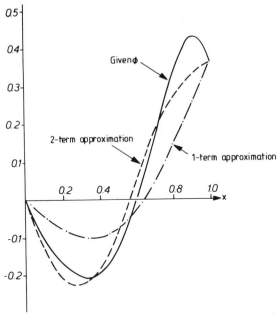

FIGURE 2.3. Truncated Fourier series used to approximate a given function.

The theory of Fourier series can be used to prove the completeness of this trial function set and to show that this approximation improves with increasing M. The use of this method for function fitting is illustrated in Fig. 2.3, where the results of using one and then two terms to construct an approximation to a given function over the range $0 \leqslant x \leqslant 1$ are displayed.

2.2. WEIGHTED RESIDUAL APPROXIMATIONS

We shall now attempt to develop a general method for determining the constants in approximations of the form of Eq. (2.1), and the particular methods introduced in the preceding section will be shown to be just special cases of this general method. We begin by introducing the error, or *residual*, R_Ω in the approximation which is defined by

$$R_\Omega = \phi - \hat{\phi} \qquad (2.5)$$

and it should be noted that R_Ω is a function of position in Ω. In an attempt to reduce this residual in some overall manner over the whole domain Ω, we could require that an appropriate number of integrals of the error over Ω, weighted in different ways, be zero, that is, we attempt to make

$$\int_\Omega W_l(\phi - \hat{\phi})\, d\Omega \equiv \int_\Omega W_l R_\Omega \, d\Omega = 0; \qquad l = 1, 2, \ldots, M \qquad (2.6)$$

Here $\langle W_l; l = 1, 2, 3, \dots \rangle$ is a set of independent *weighting functions*. The general convergence requirement that $\hat{\phi} \to \phi$ as $M \to \infty$ can then be cast in an alternative form by requiring that Eq. (2.6) be satisfied for all l as $M \to \infty$. The reader can readily verify that this can only be true if $R_\Omega \to 0$ at all points of the domain as required.

Replacing $\hat{\phi}$ in Eq. (2.6) by Eq. (2.1), we note that the *weighted residual statement* of Eq. (2.6) leads to a set of simultaneous linear equations for the unknown coefficients a_m, which can be written quite generally as

$$\mathbf{Ka} = \mathbf{f} \tag{2.7}$$

where

$$\mathbf{a}^T = (a_1, a_2, a_3, \dots, a_M) \tag{2.8a}$$

$$K_{lm} = \int_\Omega W_l N_m \, d\Omega, \qquad 1 \leqslant l, m \leqslant M \tag{2.8b}$$

$$f_l = \int_\Omega W_l(\phi - \psi) \, d\Omega, \qquad 1 \leqslant l \leqslant M \tag{2.8c}$$

Thus when the function ϕ to be approximated is given, Eq. (2.7) can be solved to obtain the coefficients in the approximation (2.1), having first determined the function ψ and chosen suitable trial and weighting function sets.

Various forms of weighting function sets $\langle W_l; l = 1, 2, \dots \rangle$ can be used in practice,[2] each leading to a different *weighted residual approximation* method. We list some of the commonly adopted choices below, in the context of a one-dimensional example.

2.2.1. Point Collocation

Here the members W_l of the weighting function set are given by

$$W_l = \delta(x - x_l) \tag{2.9a}$$

where $\delta(x - x_l)$ is the Dirac delta function defined to have the properties

$$\delta(x - x_l) = 0, \qquad x \neq x_l$$

$$\delta(x - x_l) = \infty, \qquad x = x_l$$

$$\int_{x < x_l}^{x > x_l} G(x) \, \delta(x - x_l) \, dx = G(x_l) \tag{2.9b}$$

Choosing this form of weighting is thus, by Eq. (2.6), equivalent to making the residual R_Ω equal to zero (i.e., $\phi = \hat{\phi}$) at a number of chosen points x_l, and the matrix $\mathbf{K}$ and vector $\mathbf{f}$ of Eq. (2.8) then have typical elements

$$K_{lm} = N_m|_{x=x_l}, \qquad f_l = [\phi - \psi]_{x=x_l} \qquad (2.10)$$

This is of course the basis of the point fitting method used to construct an approximation in Section 2.1.1.

2.2.2. Subdomain Collocation

If the weighting functions are chosen to satisfy

$$W_l = \begin{cases} 1, & x_l < x < x_{l+1} \\ 0, & x < x_l, x > x_{l+1} \end{cases} \qquad (2.11)$$

the weighted residual statement of Eq. (2.6) simply requires that the integrated error over M subregions of the domain should each be zero. Now the approximation is completed by solving Eq. (2.7), where

$$K_{lm} = \int_{x_l}^{x_{l+1}} N_m \, dx$$

$$f_l = \int_{x_l}^{x_{l+1}} (\phi - \psi) \, dx \qquad (2.12)$$

2.2.3. The Galerkin Method

In this, the most popular weighted residual method, we make the obvious choice of taking the trial functions themselves as the weighting functions, that is,

$$W_l = N_l \qquad (2.13)$$

instead of looking for a new set of functions. In this case the matrix $\mathbf{K}$ and the vector $\mathbf{f}$ of Eq. (2.7) have typical elements

$$K_{lm} = \int_{\Omega} N_l N_m \, dx$$

$$f_l = \int_{\Omega} N_l (\phi - \psi) \, dx \qquad (2.14)$$

and we notice the computational advantages of the method in that the matrix $\mathbf{K}$ is symmetric. The method was first used by Galerkin, and his name is now usually attached to it.

It is informative to note here that using the Galerkin method with the trial function set $\{N_m = \sin(m\pi x/L_x); \; m = 1, 2, 3, \dots \}$ to approximate a function

$\phi(x)$ over the range $0 \leqslant x \leqslant L_x$, by Eq. (2.14), leads to typical coefficients

$$K_{lm} = \int_0^{L_x} \sin \frac{l\pi x}{L_x} \sin \frac{m\pi x}{L_x} dx$$

$$f_l = \int_0^{L_x} (\phi - \psi) \sin \frac{l\pi x}{L_x} dx \tag{2.15}$$

Performing the integration shows that

$$K_{lm} = \begin{cases} L_x/2, & l = m \\ 0, & l \neq m \end{cases} \tag{2.16}$$

and the equation system (2.7) therefore has the particularly simple form of a diagonal system in this case, resulting immediately in the solution

$$a_m = \frac{2}{L_x} \int_0^{L_x} (\phi - \psi) \sin \frac{m\pi x}{L_x} dx; \qquad m = 1, 2, \ldots, M \tag{2.17}$$

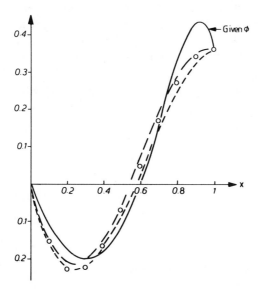

FIGURE 2.4. Approximation of a given function using the point collocation, the subdomain collocation, and the Galerkin methods. In each case the first two terms of the trial function set $\{N_m = x^m(1 - x); m = 1, 2, \ldots\}$ are used.

This can be seen to be identical to the expression used for determining the coefficients in Eq. (2.3) in Section 2.1.2, and so the truncated Fourier sine series representation of a function can be regarded as a Galerkin weighted residual approximation. The particular simplicity of the equations produced by the Galerkin approximation in this case was due to the *orthogonality* property of the trial functions that were being used, resulting in

$$\int_{\Omega} N_l N_m \, d\Omega = 0, \qquad l \neq m \tag{2.18}$$

and this characteristic of elements of a trial function set is particularly useful.

Figure 2.4 shows a comparison between the results obtained when three different weighted residual methods are used to approximate a given function $\phi(x)$.

2.2.4. Other Weightings

Obviously there exists a large number of other possible choices for the weighting function set, and the behavior of many different types of weighting has been explored. One such obvious choice is the use of the set $\{W_l = x^{l-1};\ l = 1, 2, \dots\}$, which is sometimes called the method of moments, requiring as it does that the area under the error curve and its various moments about the origin be zero. We shall not consider any of the other possibilities, as the methods so far described will prove sufficient for our purposes in this text. It is worth noting at this point, however, that the method of least squares, although not normally viewed as a weighted residual method, can be shown to belong to this class of methods. The standard least-squares approach is to attempt to minimize the sum of the squares of the residual, or error, at each point in the domain Ω. Here this requires minimization of

$$I(a_1, a_2, a_3, \dots, a_M) = \int_{\Omega} (\phi - \hat{\phi})^2 \, d\Omega \tag{2.19}$$

so that we attempt to make

$$\frac{\partial I}{\partial a_l} = 0, \qquad l = 1, 2, \dots, M \tag{2.20}$$

Carrying out the differentiation, and noting from Eq. (2.1) that

$$\frac{\partial \hat{\phi}}{\partial a_l} = N_l \tag{2.21}$$

it can be seen that I is a minimum when

$$\int_{\Omega} (\phi - \hat{\phi}) N_l \, d\Omega = 0 \tag{2.22}$$

This is exactly the same form as the standard weighted residual statement of

Eq. (2.6) with $W_l = N_l$, and so the least-squares method is equivalent to the Galerkin method in this case.

Example 2.1

An experiment has been performed to determine the deflection $u(x, y)$ of a loaded square plate of unit side length which is held fixed along its edges. The plate is covered by a square grid of spacing $\Delta x = \Delta y = 0.25$, and the deflections of the grid points have been measured and are as shown in Fig. 2.5a. By

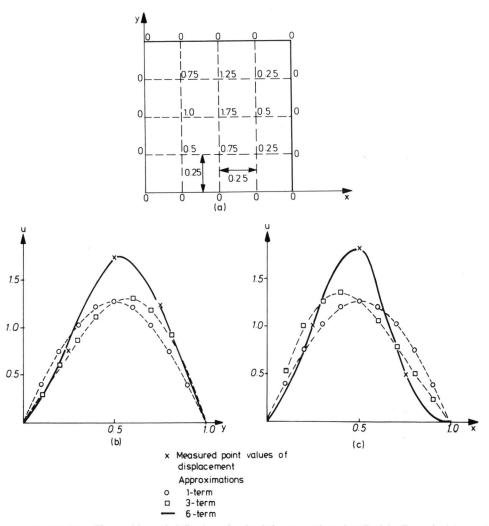

FIGURE 2.5. The problem of deflection of a loaded square plate described in Example 2.1. (*a*) Observed point deflections and behavior of the Galerkin approximations on the lines (*b*) $x = 0.5$ and (*c*) $y = 0.5$.

using the techniques of the previous sections we can fit a smooth curve to this set of experimental data to obtain information on the form of the displacement field between the grid points. We note that the trivial function $\psi = 0$ satisfies the boundary conditions of zero displacement along the edges of the plate, and as a suitable set of trial functions vanishing on the plate boundary we choose $\langle \sin l\pi x \sin m\pi y; l, m = 1, 2, \dots \rangle$.

Then, from Eq. (2.1), a six-term approximation $\hat{u}$ to the displacement field can be written as

$$\hat{u} = \sum_{\substack{l, m=1 \\ l+m \leqslant 4}}^{3} a_{lm} \sin l\pi x \sin m\pi y$$

where the constants a_{lm} can be determined by, for example, the Galerkin statement

$$\int_{\Omega} (u - \hat{u}) \sin l\pi x \sin m\pi y \, dx \, dy = 0$$

Performing the integration produces the result

$$a_{lm} = 4 \int_0^1 \int_0^1 u \sin l\pi x \sin m\pi y \, dx \, dy$$

with this simple form resulting from the orthogonality of the trial functions over the region Ω. A numerical integration,* using the two-dimensional extension of the trapezoidal rule, over the grid shown in Figure 2.5a, enables the values of the constants to be found with the result that

$$\hat{u} = 1.275 \sin \pi x \sin \pi y - 0.169 \sin \pi x \sin 2\pi y + 0.258 \sin 2\pi x \sin \pi y$$

$$- 0.130 \sin \pi x \sin 3\pi y - 0.063 \sin 2\pi x \sin 2\pi y - 0.307 \sin 3\pi x \sin \pi y$$

The behavior of the one-, three- and six-term Galerkin weighted residual approximations along the lines $x = 0.5$ and $y = 0.5$ is plotted and compared with the point values of the displacement function u in Fig. 2.5b and c.

EXERCISES

2.1. Use a suitable polynomial trial function set to approximate the function $\phi = 1 + \sin(\pi x/2)$ over the range $0 \leqslant x \leqslant 1$. Use both the point collocation and the Galerkin methods and investigate numerically the convergence of successive approximations to the given function.

---

*Numerical integration procedures are discussed in some detail in Chapter 5.

2.2. An experiment on one-dimensional steady heat conduction produces the following readings for the temperature at various points:

Distance	0	0.2	0.4	0.6	0.8	1.0
Temperature	20	30	50	65	40	30

Fit a smooth curve to this set of data by using the Galerkin method and any suitable trial function set.

2.3. Return to Example 2.1 and examine the behavior of a six-term expansion where the constants are determined by point collocation applied at the six interior grid points.

2.3. APPROXIMATION TO THE SOLUTIONS OF DIFFERENTIAL EQUATIONS AND THE USE OF TRIAL FUNCTION — WEIGHTED RESIDUAL FORMS. BOUNDARY CONDITIONS SATISFIED BY CHOICE OF TRIAL FUNCTIONS

Let us consider a differential equation of the type we have discussed in Chapter 1, which we will now write quite generally as

$$A(\phi) = \mathcal{L}\phi + p = 0 \quad \text{in } \Omega \tag{2.23}$$

where $\mathcal{L}$ is an appropriate *linear* differential operator and p is independent of ϕ. A particular example could be, for instance, Eq. (1.8) of Chapter 1 representing two-dimensional linear heat conduction which can be written in the above notation by defining

$$\mathcal{L}\phi = \frac{\partial}{\partial x}\left(k\frac{\partial \phi}{\partial x}\right) + \frac{\partial}{\partial y}\left(k\frac{\partial \phi}{\partial y}\right)$$

$$p = Q \tag{2.24}$$

where k and Q are independent of ϕ. The discussion here will be concerned with the general equation (2.23), and not with a particular equation such as Eq. (2.24), so as to demonstrate the general applicability of the procedures to be described.

We are required to solve Eq. (2.23) subject to its associated boundary conditions, which again we write in general form as

$$B(\phi) = \mathfrak{M}\phi + r = 0 \quad \text{on } \Gamma \tag{2.25}$$

where $\mathfrak{M}$ is an appropriate linear operator and r is independent of ϕ. For instance, the Dirichlet and Neumann boundary conditions of Eqs. (1.5) and

(1.6) can be expressed in the general form of Eq. (2.25) by taking

$$\mathfrak{M}\phi = \phi, \qquad r = -\bar{\phi} \qquad \text{on } \Gamma_{\phi}$$

$$\mathfrak{M}\phi = -k\frac{\partial\phi}{\partial n}, \qquad r = -\bar{q} \qquad \text{on } \Gamma_{q} \qquad (2.26)$$

Following the techniques introduced in the previous sections of this chapter, we attempt to construct an approximation $\hat{\phi}$ to the actual solution ϕ by using, as in Eq. (2.1), an expansion

$$\phi \simeq \hat{\phi} = \psi + \sum_{m=1}^{M} a_m N_m \qquad (2.27)$$

Now the function ψ and the trial functions N_m are chosen such that

$$\left.\begin{array}{l} \mathfrak{M}\psi = -r \\ \mathfrak{M}N_m = 0, \qquad m = 1, 2, \ldots \end{array}\right\} \text{ on } \Gamma \qquad (2.28)$$

and then $\hat{\phi}$ automatically satisfies the boundary conditions of Eq. (2.25) for all values of the coefficients a_m. The approximation of Eq. (2.27) can now be used to approximate the derivatives of ϕ by straightforward differentiation. Thus, provided that the functions N_m are continuous in the problem domain Ω and that all their derivatives exist, we can write

$$\phi \simeq \hat{\phi} = \psi + \sum_{m=1}^{M} a_m N_m$$

$$\frac{\partial\phi}{\partial x} \simeq \frac{\partial\hat{\phi}}{\partial x} = \frac{\partial\psi}{\partial x} + \sum_{m=1}^{M} a_m \frac{\partial N_m}{\partial x} \qquad (2.29)$$

$$\frac{\partial^2\phi}{\partial x^2} \simeq \frac{\partial^2\hat{\phi}}{\partial x^2} = \frac{\partial^2\psi}{\partial x^2} + \sum_{m=1}^{M} a_m \frac{\partial^2 N_m}{\partial x^2}$$

and so on.

In this chapter we assume the shape functions to be continuously differentiable in this fashion, but in Chapter 3 we relax such restrictions.

As we have constructed our expansion so as to satisfy the boundary conditions, to approximate the required function ϕ we need only ensure that $\hat{\phi}$ approximately satisfies the differential equation (2.23). Substitution of $\hat{\phi}$ into this equation yields a residual R_{Ω}, which can be written

$$R_{\Omega} \equiv A(\hat{\phi}) \equiv \mathcal{L}\hat{\phi} + p \equiv \mathcal{L}\psi + \left(\sum_{m=1}^{M} a_m \mathcal{L}N_m\right) + p \qquad (2.30)$$

since $\mathcal{L}$ is a linear operator.

Immediately the procedures of Section 2.2 can be applied in an attempt to make $R_\Omega \approx 0$ everywhere in Ω, and hence make $\hat{\phi}$ approximately equal to the required function. Once again this can be accomplished by using the weighted residual process, choosing a set of weighting functions $\{W_l; l = 1, 2, \dots\}$ and requiring that

$$\int_\Omega W_l R_\Omega \, d\Omega \equiv \int_\Omega W_l \left\{ \mathcal{L}\psi + \left(\sum_{m=1}^M a_m \mathcal{L} N_m \right) + p \right\} d\Omega = 0 \qquad (2.31)$$

The total number of unknowns is M, and applying Eq. (2.31) for $l = 1, 2, 3, \dots, M$ means that we produce a set of M linear algebraic equations which can be written

$$\mathbf{Ka} = \mathbf{f} \qquad (2.32)$$

where now

$$K_{lm} = \int_\Omega W_l \mathcal{L} N_m \, d\Omega, \qquad\qquad 1 \leqslant l, m \leqslant M$$

$$(2.33)$$

$$f_l = -\int_\Omega W_l p \, d\Omega - \int_\Omega W_l \mathcal{L}\psi \, d\Omega, \qquad 1 \leqslant l \leqslant M$$

When these matrix coefficients have been evaluated, this set of equations can be solved to determine the unknowns a_m, $m = 1, 2, 3, \dots, M$ and so complete the process of determining an approximate solution to the given differential equation. Normally the matrix $\mathbf{K}$ will be full and will not exhibit the banded structure which was characteristic of the matrices produced by the finite difference method in the previous chapter.

The various choices of the weighting functions, explained in the preceding section, can once again be made here, and to illustrate the procedure in detail we consider again the solution of Examples 1.1 and 1.5.

Example 2.2

As in Example 1.1, we require the solution of the equation $d^2\phi/dx^2 - \phi = 0$ subject to the boundary conditions $\phi = 0$ at $x = 0$ and $\phi = 1$ at $x = 1$.

The given boundary conditions can be expressed in the general form of Eq. (2.25), provided that we take $\mathfrak{M}\phi = \phi$ and $r = 0$ at $x = 0$, $r = -1$ at $x = 1$. Then, by Eq. (2.28), ψ and the trial functions N_m must be chosen such that

$$\psi = N_m = 0 \quad \text{at } x = 0; \qquad \psi = 1, N_m = 0 \quad \text{at } x = 1.$$

The function $\psi = x$ satisfies the required conditions on ψ, and as trial functions, vanishing at $x = 0$ and at $x = 1$, we can take the set $\{N_m = \sin m\pi x; m = 1, 2, 3, \dots\}$.

The problem boundary conditions are then satisfied by the expansion

$$\hat{\phi} = x + \sum_{m=1}^{M} a_m \sin m\pi x$$

The equation to be solved may be expressed in the general form $\mathcal{L}\phi + p = 0$ if we identify $\mathcal{L}\phi = d^2\phi/dx^2 - \phi$ and $p = 0$, and we can write a set of M equations, using the form given in Eqs. (2.32) and (2.33), once the weighting functions are chosen. We shall take $M = 2$, so that two unknown parameters a_1 and a_2 are involved, and use here two alternatives, namely, point collocation, with $W_l = \delta(x - x_l)$, and the Galerkin method with $W_l = N_l$. The reader can verify directly that the two equations to be solved are now

$$K_{11}a_1 + K_{12}a_2 = f_1$$

$$K_{21}a_1 + K_{22}a_2 = f_2$$

where, for $l, m = 1, 2$,

$$K_{lm} = \int_0^1 (1 + m^2\pi^2) W_l \sin m\pi x \, dx$$

$$f_l = -\int_0^1 W_l x \, dx$$

For the collocation process $\left(\text{with } R_\Omega \text{ made equal to zero at } x_1 = \tfrac{1}{3}, x_2 = \tfrac{2}{3}\right)$ we get

$$K_{11} = (1 + \pi^2)\sin\frac{\pi}{3}, \qquad K_{12} = (1 + 4\pi^2)\sin\frac{2\pi}{3}$$

$$K_{21} = (1 + \pi^2)\sin\frac{2\pi}{3}, \qquad K_{22} = (1 + 4\pi^2)\sin\frac{4\pi}{3}$$

$$f_1 = -\tfrac{1}{3}, \qquad\qquad f_2 = -\tfrac{2}{3}$$

while for the Galerkin method

$$K_{11} = \tfrac{1}{2}(1 + \pi^2), \qquad K_{12} = 0$$

$$K_{21} = 0, \qquad\qquad K_{22} = \tfrac{1}{2}(1 + 4\pi^2)$$

$$f_1 = -\frac{1}{\pi}, \qquad\qquad f_2 = \frac{1}{2\pi}$$

The numerical solution of the equations gives parameters

$$a_1 = -0.053\,12, \qquad a_2 = 0.004\,754 \qquad \text{for point collocation}$$

$$a_1 = -0.058\,57, \qquad a_2 = 0.007\,864 \qquad \text{for Galerkin}$$

In Fig. 2.6 we show the numerical solutions obtained in this case, using $M = 1$ and $M = 2$, with the two weighting procedures. Comparison with the exact solution demonstrates how convergence occurs in this example.

It is also of interest to contrast the solutions just obtained with those produced by the finite difference method in Chapter 1. This is done in the

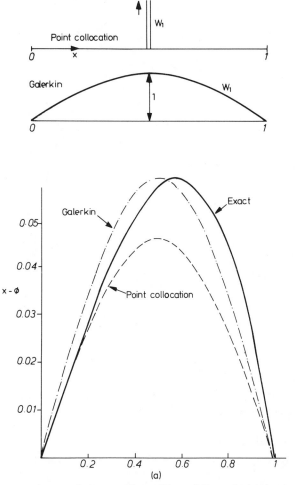

FIGURE 2.6. Approximate solutions to the problem of Example 2.2 produced by the point collocation and the Galerkin methods using (*a*) one-term and (*b*) two-term approximations.

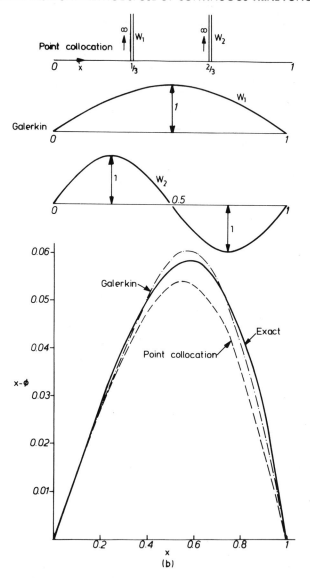

FIGURE 2.6. (*continued*).

following table, where we show the approximate values and the exact values at the finite difference mesh points $x = \frac{1}{3}$ and $x = \frac{2}{3}$.

x	Finite Difference $(\Delta x = \frac{1}{3})$	Point Collocation	Galerkin	Exact
$\frac{1}{3}$	0.2893	0.2914	0.2894	0.2889
$\frac{2}{3}$	0.6107	0.6165	0.6091	0.6102

It should be noted that again the Galerkin process used in conjunction with suitable trigonometric functions has resulted in a diagonal form of the matrix **K** due to the orthogonality of the shape functions. For this Galerkin method we can therefore easily determine all the coefficients in our approximation series directly as

$$a_m = \frac{-\int_0^1 x \sin m\pi x \, dx}{\int_0^1 (1 + m^2\pi^2)\sin^2 m\pi x \, dx} = \frac{2(-1)^m}{m\pi(1 + m^2\pi^2)}, \qquad m = 1, 2, 3, \ldots$$

and such a solution allows a simple estimation to be made of the magnitude of the error involved at any stage of the approximation.

We could easily have chosen alternative sets of weighting and trial functions. For instruction the reader should now try the set $\langle N_m = x^m(1 - x); \ m = 1, 2, \ldots \rangle$ to replace the trigonometric terms used previously, and compare the resulting approximation with that obtained above.

Example 2.3

The application of the weighted residual method to a problem in two dimensions can be illustrated by returning to the torsion problem described in Example 1.5. Here we require the solution of the equation $\partial^2\phi/\partial x^2 + \partial^2\phi/\partial y^2 = -2$ in the rectangular region $-3 \leqslant x \leqslant 3$, $-2 \leqslant y \leqslant 2$ subject to the condition that $\phi = 0$ on the boundaries, that is, on $x = \pm 3$ and on $y = \pm 2$.

The solution is obviously symmetric about the lines $x = 0$ and $y = 0$, and so we restrict ourselves to the use of even trial functions which also possess this symmetry property. For example, if the set $N_1 = \cos(\pi x/6)\cos(\pi y/4)$, $N_2 = \cos(3\pi x/6)\cos(\pi y/4)$, $N_3 = \cos(\pi x/6)\cos(3\pi y/4)$, and so on, is chosen, then a three-term approximation of the form

$$\hat{\phi} = a_1\cos\frac{\pi x}{6}\cos\frac{\pi y}{4} + a_2\cos\frac{3\pi x}{6}\cos\frac{\pi y}{4} + a_3\cos\frac{\pi x}{6}\cos\frac{3\pi y}{4} = \sum_{m=1}^{3} a_m N_m$$

automatically satisfies the required boundary conditions and is an even function. Again the equation to be solved may be written in the general form $\mathcal{L}\phi + p = 0$ if we identify $\mathcal{L}\phi = \partial^2\phi/\partial x^2 + \partial^2\phi/\partial y^2$ and $p = 2$. Once the weighting functions have been chosen, Eqs. (2.32) and (2.33) can be used to give directly the set of three equations which must be solved to obtain the values of a_1, a_2, and a_3. If the Galerkin method is adopted, the elements of the matrices **K** and **f** become

$$K_{lm} = -\int_{-3}^{3}\int_{-2}^{2}\left(\frac{\partial^2 N_m}{\partial x^2} + \frac{\partial^2 N_m}{\partial y^2}\right)N_l \, dy \, dx$$

$$f_l = \int_{-3}^{3}\int_{-2}^{2} 2N_l \, dy \, dx$$

and, because of the orthogonality of the trial functions over the rectangular region, this system reduces to diagonal form with immediate solution

$$a_1 = \frac{4608}{13\pi^4}, \qquad a_2 = \frac{-4608}{135\pi^4}, \qquad a_3 = \frac{-4608}{255\pi^4}$$

The approximation to the twisting moment T may then be calculated from

$$T = 2\int_{-3}^{3}\int_{-2}^{2} \hat{\phi}\, dy\, dx = 74.265$$

which can be compared with the exact value of 76.4, while the approximation to the maximum value of the shear stress in this case is

$$|\tau| = 3.02$$

as opposed to the exact value of 2.96.

As in Section 2.2, it is possible again to bring the method of least squares within the general weighted residual formulation of Eq. (2.31). For now we attempt to minimize

$$I(a_1, a_2, \ldots, a_M) = \int_{\Omega} R_{\Omega}^2\, d\Omega = \int_{\Omega}\left\{ \mathcal{L}\psi + \left(\sum_{m=1}^{M} a_m \mathcal{L}N_m \right) + p \right\}^2 d\Omega \qquad (2.34)$$

by requiring that

$$\frac{\partial I}{\partial a_l} = 0, \qquad l = 1, 2, \ldots, M \qquad (2.35)$$

which can be seen to lead to the equation set

$$\int_{\Omega} R_{\Omega} \frac{\partial R_{\Omega}}{\partial a_l}\, d\Omega = 0, \qquad l = 1, 2, \ldots, M \qquad (2.36)$$

This is of the general weighted residual form with weighting functions defined by

$$W_l = \frac{\partial R_{\Omega}}{\partial a_l} = \mathcal{L}N_l \qquad (2.37)$$

and we note that the least squares form is not equivalent to the Galerkin approximation in this case.

Before concluding this section the reader should observe that, in Examples 2.2 and 2.3 the Galerkin process resulted in a set of symmetric equations while the collocation process gave nonsymmetric coefficients. This particular feature of the Galerkin method has already been observed in Section 2.2 in the context of function approximation, and the reasons for its reappearance here will be demonstrated later. Note again, however, that this presents computational advantages in larger problems and is one of the reasons for the popularity of the Galerkin weighting choice.

EXERCISES

2.4. Return to Exercise 1.2 and obtain the bending moment distribution by the point collocation, Galerkin, and least-squares methods, using a suitable set of polynomial trial functions. Compare the answers with the exact solution and with those obtained by the finite difference method.

2.5. A certain problem of one-dimensional steady heat transfer with a distributed heat source is governed by the equation $d^2\phi/dx^2 + \phi + 1 = 0$ and the boundary conditions $\phi = 0$ at $x = 0$ and $d\phi/dx = -\phi$ at $x = 1$. Calculate an approximate solution by the Galerkin method and investigate the convergence properties of the method by comparing the results with the exact solution.

2.6. Return to Exercise 1.20 and find the deflection of the loaded beam on an elastic foundation by the point collocation and Galerkin methods. Compare the answers with the exact solution and with those produced by the finite difference method.

2.7. In a certain two-dimensional steady heat conduction problem in a square of side length 1, the temperature on the sides $x = \pm 1$ varies as $1 - y^2$, while the temperature on the sides $y = \pm 1$ varies as $1 - x^2$. Obtain an approximation to the distribution of temperature over the square using the Galerkin method.

2.8. Return to Exercise 1.21 and obtain approximations for the deflection of the loaded plate by using the point collocation and Galerkin methods.

2.4. SIMULTANEOUS APPROXIMATION TO THE SOLUTIONS OF DIFFERENTIAL EQUATIONS AND TO THE BOUNDARY CONDITIONS

In the previous section we have shown how we can solve approximately a differential equation by using an expansion in terms of trial functions, constructing an approximating function $\hat{\phi}$ which satisfies identically all the problem boundary conditions. In this section we shall relax this requirement, which obviously limited the choice of possible trial function forms.

If now we postulate that an expansion

$$\phi \simeq \hat{\phi} = \sum_{m=1}^{M} a_m N_m \tag{2.38}$$

does not satisfy a priori some or all of the problem boundary conditions, then the residual in the domain

$$R_\Omega = A(\hat{\phi}) = \mathcal{L}\hat{\phi} + p \qquad \text{in } \Omega \tag{2.39}$$

is supplemented by a boundary residual

$$R_\Gamma = B(\hat{\phi}) = \mathfrak{M}\hat{\phi} + r \qquad \text{on } \Gamma \tag{2.40}$$

We can attempt to reduce the weighted sum of the residual on the boundary and the residual on the domain by writing

$$\int_\Omega W_l R_\Omega \, d\Omega + \int_\Gamma \overline{W}_l R_\Gamma \, d\Gamma = 0 \tag{2.41}$$

where, in general, the weighting functions W_l and $\overline{W}_l$ can be chosen independently.

Clearly, if Eq. (2.41) is satisfied for a very large number of arbitrary functions W_l and $\overline{W}_l$, then the approximation $\hat{\phi}$ must approach the exact solution ϕ, provided that the expansion of Eq. (2.38) is capable of so doing. The position is unchanged if W_l and $\overline{W}_l$ are related in some manner.

Quite generally the system of equations which will have to be solved can again be written in the form

$$\mathbf{Ka} = \mathbf{f} \tag{2.42}$$

where now

$$K_{lm} = \int_\Omega W_l \mathcal{L} N_m \, d\Omega + \int_\Gamma \overline{W}_l \mathfrak{M} N_m \, d\Gamma$$

$$\tag{2.43}$$

$$f_l = -\int_\Omega W_l p \, d\Omega - \int_\Gamma \overline{W}_l r \, d\Gamma$$

To illustrate the application of this process we shall reconsider the solution of Example 2.2, but using now a trial expansion that does not satisfy, a priori, the boundary conditions.

Example 2.4

In Example 2.2 the equation $d^2\phi/dx^2 - \phi = 0$ with $\phi = 0$ at $x = 0$ and $\phi = 1$ at $x = 1$ was solved by constructing an approximation $\hat{\phi}$ which automatically satisfied the given conditions at $x = 0$ and at $x = 1$. If we approach this problem using the method just described we need not look for a function ψ satisfying the boundary conditions on ϕ, and we do not require that our trial functions should vanish on the boundaries. A possible trial function set is taken now simply as $\{N_m = x^{m-1}; \; m = 1, 2, 3, \dots \}$.

In this case the boundary curve Γ consists of the two points $x = 0$ and $x = 1$, so that the integration over the boundary in Eq. (2.41) reduces to two discrete residuals, and the appropriate form of the statement here becomes

$$\int_0^1 W_l R_\Omega \, dx + \left[\overline{W}_l R_\Gamma \right]_{x=0} + \left[\overline{W}_l R_\Gamma \right]_{x=1} = 0$$

In this example the weighting functions to be used will be defined by $W_l = N_l$, $\overline{W}_l = -N_l|_\Gamma$, and then the above statement requires that

$$\int_0^1 \left(\frac{d^2\hat{\phi}}{dx^2} - \hat{\phi} \right) N_l \, dx - \left[N_l \hat{\phi} \right]_{x=0} - \left[N_l (\hat{\phi} - 1) \right]_{x=1} = 0$$

Using a three-term expansion $\hat{\phi} = a_1 + a_2 x + a_3 x^2$ produces an equation of the form (2.42), where now

$$\mathbf{K} = \begin{bmatrix} 3 & \frac{3}{2} & -\frac{2}{3} \\ \frac{3}{2} & \frac{4}{3} & \frac{1}{4} \\ \frac{4}{3} & \frac{5}{4} & \frac{8}{15} \end{bmatrix}$$

$$\mathbf{f} = \begin{bmatrix} 1 \\ 1 \\ 1 \end{bmatrix}$$

It should be noted that $\mathbf{K}$ is not symmetric in this case, even though a Galerkin-type method has been applied.

The solution of this set is

$$a_1 = 0.068, \qquad a_2 = 0.632, \qquad a_3 = 0.226$$

The convergence of the approximation to the prescribed conditions at $x = 0$ and at $x = 1$ is shown in the following table, which compares the behavior of the one-, two-, and three-term approximations at these two points.

	One Term	Two Terms	Three Terms	Exact
$x = 0$	$\frac{1}{3}$	-0.095	0.068	0
$x = 1$	$\frac{1}{3}$	0.762	0.925	1

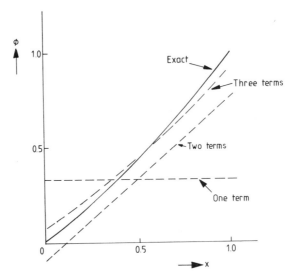

FIGURE 2.7. Comparison between the exact and the trial function solutions to the problem of Example 2.4.

The convergence of the approximating sequence over the whole range $0 \leqslant x \leqslant 1$ can be seen in Fig. 2.7, where these three approximations are compared with the exact solution.

In the one-dimensional example just illustrated the boundary integral in Eq. (2.41) has reduced to the determination of two point values—but the process is obviously valid when we have two or more independent variables in the domain. This is illustrated in the next example.

Example 2.5

We return to the solution of the torsion problem already considered in Examples 1.5 and 2.3, and which is defined by the equation $\partial^2\phi/\partial x^2 + \partial^2\phi/\partial y^2 = -2$ in the rectangular region $-3 \leqslant x \leqslant 3$, $-2 \leqslant y \leqslant 2$ subject to the condition that $\phi = 0$ on the boundaries. As previously, we restrict ourselves to even trial functions, but we now relax the requirement that our approximation should automatically satisfy all the boundary conditions. For example, if we choose the set

$$N_1 = (4 - y^2), \qquad N_2 = (4 - y^2)x^2, \qquad N_3 = (4 - y^2)y^2,$$

$$N_4 = (4 - y^2)x^2y^2, \qquad N_5 = (4 - y^2)x^4, \qquad \text{and so on}$$

and form a five-term approximation

$$\hat{\phi} = \left(4 - y^2\right)\left(a_1 + a_2 x^2 + a_3 y^2 + a_4 x^2 y^2 + a_5 x^4\right)$$

then $\hat{\phi}$ immediately satisfies the required conditions on $y = \pm 2$, but the condition that $\hat{\phi} = 0$ on $x = \pm 3$ is not automatically satisfied. The imposition of the boundary condition on the lines $x = \pm 3$ is then accomplished by incorporation of this condition into the weighted residual statement, as in Eq. (2.41), and the appropriate form of this statement for this problem is then

$$\int_{-3}^{3}\int_{-2}^{2}\left(\frac{\partial^2 \hat{\phi}}{\partial x^2} + \frac{\partial^2 \hat{\phi}}{\partial y^2} + 2\right) W_I \, dy \, dx + \int_{-2}^{2} \overline{W}_I \hat{\phi}|_{x=3} \, dy - \int_{2}^{-2} \overline{W}_I \hat{\phi}|_{x=-3} \, dy = 0$$

Using weighting functions $W_I = N_I$, $\overline{W}_I = N_I|_\Gamma$ and performing the integrations, produces an equation of form (2.42), where

$$\mathbf{K} = \begin{bmatrix} 6.66667 & -44 & 43.7333 & 80 & -813.6 \\ -12 & -333.6 & 105.6 & 355.2 & -5{,}748.68 \\ 11.7333 & -16 & 159.086 & 389.486 & -547.2 \\ 9.6 & -163.2 & 433.371 & 1971.57 & -3{,}932.43 \\ -237.6 & -3{,}156.69 & 432 & 1{,}473.74 & -46{,}239.43 \end{bmatrix}$$

$$\mathbf{f} = \begin{bmatrix} 12 \\ 36 \\ 16 \\ 48 \\ 194.4 \end{bmatrix}$$

The convergence of the approximation to the exact values of the twisting moment T and of the maximum shear stress τ is shown in the following table, where the results obtained by using the two-term and five-term approximations are displayed.

	Two Terms	Five Terms	Exact
T	58.94	73.60	76.4
$|\tau|$	3.52	3.33	2.96

The improvement in the accuracy of representation of the boundary condition $\hat{\phi} = 0$ on the lines $x = \pm 3$ as more terms are included is illustrated in Fig. 2.8, which shows the behavior of the two-term and the five-term approximations on these boundary lines.

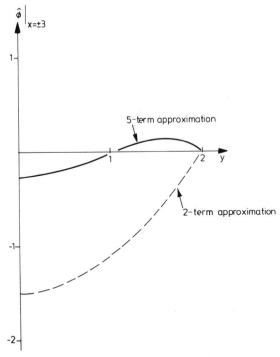

FIGURE 2.8. Behavior of the two-term and five-term approximations of Example 2.5 on the lines $x = \pm 3$.

EXERCISES

2.9. Use the weighted residual method to solve the problem of Exercise 1.1 by using first an approximation that satisfies the given boundary conditions and then an approximation in which the boundary conditions are not automatically satisfied. Compare the performance of these two methods with the behavior of the exact solution.

2.10. Obtain, by the weighted residual method, the distribution of the bending moment in the beam described in Exercise 1.2 by using an approximation which satisfies the condition at $x = 0$, but does not satisfy automatically the other condition at $x = 1$. Use boundary weighting functions $\overline{W}_l = \alpha N_l|_\Gamma$, where α is a constant, and compare the results for $\alpha = \pm 0.1, \pm 1, \pm 10, \pm 100$ in turn with those produced in Exercise 2.4 by an approximation automatically satisfying both boundary conditions.

2.11. Return to Exercise 2.7 and obtain the distribution of temperature by using an approximation which satisfies only the boundary conditions on the sides $x = \pm 1$. Demonstrate the convergence of the approximation to the required condition on the sides $y = \pm 1$.

2.12. In Exercise 1.21 the problem of determining the deflection of a loaded elastic plate was described. Return to this problem and determine an approximate solution which satisfies at the outset only the boundary condition of zero displacement at the edge of the plate. Demonstrate the improvement in the satisfaction of the other boundary condition at the plate edge as more terms are included in the approximation.

2.5. NATURAL BOUNDARY CONDITIONS

The examples of the previous section have shown that it is possible to evaluate the coefficients in the expansion of Eq. (2.38), and thus determine an approximate solution, without a priori satisfaction of the boundary conditions. Nevertheless the process was more complex, even in the two simple examples illustrated, as more parameters were needed to produce a similar approximation to that obtained by satisfying the boundary conditions at the outset, that is, for a given accuracy, a larger number of equations has to be solved.

Further, the formulation of Eq. (2.41) could require the evaluation of integrals involving derivatives of $\hat{\phi}$ along the boundaries which may present difficulties if these boundaries are of curved or otherwise complicated shape.

In this section we show, for certain equations and boundary conditions, how such boundary derivative evaluations can be made unnecessary and how the general form may be amended.

Again returning to the weighted residual statement of Eq. (2.41), it should be observed that the first term in this equation

$$\int_{\Omega} W_l R_\Omega \, d\Omega \equiv \int_{\Omega} W_l (\mathcal{L}\hat{\phi} + p) \, d\Omega \tag{2.44}$$

can frequently be rearranged to yield an expression of the form

$$\int_{\Omega} W_l \mathcal{L}\hat{\phi} \, d\Omega \equiv \int_{\Omega} (\mathcal{C}W_l)(\mathcal{D}\hat{\phi}) \, d\Omega + \int_{\Gamma} W_l \mathcal{E}\hat{\phi} \, d\Gamma \tag{2.45}$$

where $\mathcal{C}$, $\mathcal{D}$, and $\mathcal{E}$ are linear differential operators involving an order of differentiation lower than that of the original operator $\mathcal{L}$. The resulting expression is often termed the *weak form* of the weighted residual statement for the problem under consideration.

When relation (2.45) is inserted into Eq. (2.41), it may be possible to arrange for the last term of Eq. (2.45) to cancel with part of the last term of Eq. (2.41) by a suitable choice of the boundary weighting function $\overline{W}_l$, thus eliminating the integral involving $\hat{\phi}$ or its derivatives along the boundary. This will only be possible for certain boundary conditions that we term *natural*. In general, boundary conditions involving prescribed values of the function itself will not benefit from this treatment, while certain boundary conditions on derivatives will.

A further advantage of such a rearrangement is that certain symmetries of algebraic equations can be immediately observed, now that lower orders of differentiation of the trial functions N_m will generally arise. This particular feature will be of importance in the discussions of the next chapter.

To illustrate, and to gain understanding of, the process it is helpful at this stage to consider a particular example.

Example 2.6

We return here to a differential equation which has been studied previously, namely $d^2\phi/dx^2 - \phi = 0$, but which is now to be solved subject to the boundary conditions $\phi = 0$ at $x = 0$ and $d\phi/dx = 20$ at $x = 1$. Let us assume that we choose an approximation

$$\hat{\phi} = \psi + \sum_{m=1}^{M} a_m N_m$$

where ψ and the set N_m is such that the condition at $x = 0$ is automatically satisfied, for example $\psi = 0$; $\{N_m = x^m; m = 1, 2, \ldots, M\}$ could be a suitable choice here.

The weighted residual statement for both the differential equation and the yet unsatisfied boundary condition at $x = 1$ is then

$$\int_0^1 \left(\frac{d^2\hat{\phi}}{dx^2} - \hat{\phi} \right) W_l \, dx + \left[\left(\frac{d\hat{\phi}}{dx} - 20 \right) \overline{W}_l \right]_{x=1} = 0$$

If integration by parts is used on the first term, this expression may be rewritten as

$$-\int_0^1 \frac{d\hat{\phi}}{dx} \frac{dW_l}{dx} \, dx - \int_0^1 \hat{\phi} W_l \, dx + \left[W_l \frac{d\hat{\phi}}{dx} \right]_{x=1}$$

$$-\left[W_l \frac{d\hat{\phi}}{dx} \right]_{x=0} + \left[\overline{W}_l \left(\frac{d\hat{\phi}}{dx} - 20 \right) \right]_{x=1} = 0$$

and if we now choose W_l and $\overline{W}_l$ to be such that

$$\overline{W}_l|_{x=1} = -W_l|_{x=1}$$

and

$$W_l|_{x=0} = 0$$

we have simply, as the required statement,

$$\int_0^1 \frac{d\hat{\phi}}{dx} \frac{dW_l}{dx} dx + \int_0^1 \hat{\phi} W_l \, dx = 20 W_l|_{x=1}$$

Thus in this formulation there is no need to evaluate the derivative of $\hat{\phi}$ at $x = 1$, and the boundary condition to be applied at this point is a natural condition. Further, we observe that if the required condition at $x = 1$ in this case is $d\phi/dx = 0$ then no term involving a boundary evaluation is required. This indeed underlines the natural nature of the gradient boundary condition in such a formulation.

From this, our set of approximating equations arises in the standard form, once the trial function $\hat{\phi}$ is inserted. We have again

$$\mathbf{Ka} = \mathbf{f}$$

with

$$K_{lm} = \int_0^1 \frac{dW_l}{dx} \frac{dN_m}{dx} dx + \int_0^1 W_l N_m \, dx$$

$$f_l = 20 W_l|_{x=1}$$

and all of these can easily be evaluated. With two unknown parameters a_1 and a_2 and trial functions defined by $N_m = x^m$, the Galerkin method gives $a_1 = 11.7579$, $a_2 = 3.4582$. The corresponding values of $\hat{\phi}$ at $x = \frac{1}{2}$ and $x = 1$ are contrasted with the exact values in the following table.

	Two-term Galerkin	Exact
$x = 0.5$	6.7435	6.7540
$x = 1$	15.2161	15.2319

The convergence of $d\hat{\phi}/dx|_{x=1}$ to the required value of 20 is illustrated by the fact that with one term the approximation produces $d\hat{\phi}/dx|_{x=1} = 15$, and with two terms the result is $d\hat{\phi}/dx|_{x=1} = 18.67$.

Notice that the Galerkin method may be legitimately used here as the restriction $W_l|_{x=0} = 0$, imposed above during the derivation of the final weighted residual statement, is automatically satisfied by each member of the trial function set.

2.5.1. Natural Boundary Conditions for the Heat Conduction Equation

We can return to the heat conduction equation of Chapter 1 and discuss the question of natural boundary conditions more generally. We consider the solution of the equation

$$\frac{\partial}{\partial x}\left(k\frac{\partial \phi}{\partial x}\right) + \frac{\partial}{\partial y}\left(k\frac{\partial \phi}{\partial y}\right) + Q = 0 \qquad (2.46)$$

subject to the conditions

$$\phi = \bar{\phi} \qquad \text{on } \Gamma_\phi$$

$$k\frac{\partial \phi}{\partial n} = -\bar{q} \qquad \text{on } \Gamma_q \qquad (2.47)$$

A trial expansion

$$\hat{\phi} = \psi + \sum_{m=1}^{M} a_m N_m \qquad (2.48)$$

is used with ψ and the trial functions chosen so that $\phi = \bar{\phi}$ on Γ_ϕ, that is, $\psi = \bar{\phi}$ and $N_m = 0$, $m = 1, 2, \ldots, M$ on Γ_ϕ. Writing the weighted residual statement as

$$\int_\Omega \left[\frac{\partial}{\partial x}\left(k\frac{\partial \hat{\phi}}{\partial x}\right) + \frac{\partial}{\partial y}\left(k\frac{\partial \hat{\phi}}{\partial y}\right)\right] W_l \, dx \, dy + \int_\Omega W_l Q \, dx \, dy$$

$$+ \int_{\Gamma_q} \left(k\frac{\partial \hat{\phi}}{\partial n} + \bar{q}\right)\overline{W}_l \, d\Gamma = 0, \qquad l = 1, 2, \ldots, M \qquad (2.49)$$

ensures the satisfaction of both the differential equation and the gradient condition on Γ_q in the limit as $M \to \infty$. The first integral may be rewritten in the form of Eq. (2.45) by using the well-known Green's lemma[3] which states that, for suitably differentiable functions α and β, we can write the identities

$$\int_\Omega \alpha \frac{\partial \beta}{\partial x} dx \, dy = -\int_\Omega \frac{\partial \alpha}{\partial x}\beta \, dx \, dy + \int_\Gamma \alpha\beta n_x \, d\Gamma \qquad (2.50a)$$

$$\int_\Omega \alpha \frac{\partial \beta}{\partial y} dx \, dy = -\int_\Omega \frac{\partial \alpha}{\partial y}\beta \, dx \, dy + \int_\Gamma \alpha\beta n_y \, d\Gamma \qquad (2.50b)$$

where n_x and n_y are the direction cosines of the outward normal n to the closed curve Γ surrounding an area Ω in the (x, y) plane, and the integration around Γ is made in an anticlockwise direction. Using this identity, and noting that

$$n_x \frac{\partial \alpha}{\partial x} + n_y \frac{\partial \alpha}{\partial y} = \frac{\partial \alpha}{\partial n} \qquad (2.51)$$

we can transform our weighted residual statement to

$$-\int_{\Omega}\left(\frac{\partial W_l}{\partial x}k\frac{\partial\hat{\phi}}{\partial x}+\frac{\partial W_l}{\partial y}k\frac{\partial\hat{\phi}}{\partial y}\right)dx\,dy+\int_{\Omega}W_lQ\,dx\,dy$$

$$+\int_{\Gamma_\phi+\Gamma_q}k\frac{\partial\hat{\phi}}{\partial n}W_l\,d\Gamma+\int_{\Gamma_q}\left(k\frac{\partial\hat{\phi}}{\partial n}+\bar{q}\right)\overline{W}_l\,d\Gamma=0 \tag{2.52}$$

Limiting now the choice of the weighting functions so that

$$W_l=0\qquad\text{on }\Gamma_\phi \tag{2.53a}$$

and

$$\overline{W}_l=-W_l\text{ on }\Gamma_q \tag{2.53b}$$

we see that the term involving the weighted integral of the gradient of $\hat{\phi}$ on the boundary disappears and the approximating equation becomes

$$\int_{\Omega}\left(\frac{\partial W_l}{\partial x}k\frac{\partial\hat{\phi}}{\partial x}+\frac{\partial W_l}{\partial y}k\frac{\partial\hat{\phi}}{\partial y}\right)dx\,dy-\int_{\Omega}W_lQ\,dx\,dy+\int_{\Gamma_q}W_l\bar{q}\,d\Gamma=0$$

$$\tag{2.54}$$

Equation (2.54) leads, on substitution of the trial expansion of Eq. (2.48), to the usual form

$$\mathbf{Ka}=\mathbf{f} \tag{2.55}$$

with

$$K_{lm}=\int_{\Omega}\left(\frac{\partial W_l}{\partial x}k\frac{\partial N_m}{\partial x}+\frac{\partial W_l}{\partial y}k\frac{\partial N_m}{\partial y}\right)dx\,dy,\qquad 1\leqslant l,m\leqslant M$$

$$f_l=\int_{\Omega}W_lQ\,dx\,dy-\int_{\Gamma_q}W_l\bar{q}\,d\Gamma-\int_{\Omega}\left(\frac{\partial W_l}{\partial x}k\frac{\partial\psi}{\partial x}+\frac{\partial W_l}{\partial y}k\frac{\partial\psi}{\partial y}\right)dx\,dy, \tag{2.56}$$

$$1\leqslant l\leqslant M$$

Again it should be noted that, within the context of this formulation, a Galerkin choice of $W_l=N_l$ is admissible as the condition of Eq. (2.53a) is automatically satisfied by the trial functions N_m used in the expansion of Eq. (2.48). Further, it can be seen immediately that with such a Galerkin form a symmetric matrix $\mathbf{K}$ results since

$$K_{lm}=K_{ml} \tag{2.57}$$

We have observed this fact earlier in Example 2.3, but before Green's lemma was applied to the weighted residual statement this result was by no means evident.

We recall that the finite difference method applied to this problem, with central differencing of the derivative boundary condition, leads to a nonsymmetric matrix equation system. In conclusion, we have demonstrated here how the boundary condition

$$k \frac{\partial \phi}{\partial n} = -\bar{q} \tag{2.58}$$

is in some way natural for this problem as the formulation eliminates the need for an actual evaluation of $\partial \phi / \partial n$ on the boundaries and note that, if $\bar{q} = 0$, such boundaries do not enter explicitly into the formulation of Eq. (2.54).

Example 2.7

It is required to solve the problem of steady-state heat conduction in a material of unit thermal conductivity and occupying the square region $-1 \leqslant x \leqslant 1$, $-1 \leqslant y \leqslant 1$. The sides $y = \pm 1$ are maintained at zero temperature, while heat is supplied at the rate $\cos(\pi y / 2)$ per unit length on the sides $x = \pm 1$.

The governing equation for steady heat conduction with no heat generation follows from Eq. (2.46), and is

$$\frac{\partial^2 \phi}{\partial x^2} + \frac{\partial^2 \phi}{\partial y^2} = 0$$

and the boundary conditions

$$\phi = 0 \qquad \qquad \text{on } y = \pm 1, \text{ that is, on } \Gamma_\phi$$

$$\frac{\partial \phi}{\partial n} = \cos \frac{\pi y}{2} \qquad \text{on } x = \pm 1, \text{ that is, on } \Gamma_q$$

are of the form given in Eq. (2.47). If we use the trial function set $N_1 = 1 - y^2$, $N_2 = (1 - y^2)x^2$, $N_3 = (1 - y^2)y^2$, $N_4 = (1 - y^2)x^2y^2$, $N_5 = (1 - y^2)x^4$, and so on, then a five-term approximation

$$\hat{\phi} = (1 - y^2)(a_1 + a_2 x^2 + a_3 y^2 + a_4 x^2 y^2 + a_5 x^4)$$

immediately satisfies the conditions on Γ_ϕ. The required form for the weighted residual statement of Eq. (2.49) for use with this approximation is then

$$\int_{-1}^{1}\int_{-1}^{1}\left(\frac{\partial^2\hat{\phi}}{\partial x^2} + \frac{\partial^2\hat{\phi}}{\partial y^2}\right)W_l\,dx\,dy + \int_{\Gamma_q}\left(\frac{\partial\hat{\phi}}{\partial n} - \cos\frac{\pi y}{2}\right)\overline{W}_l\,d\Gamma = 0$$

from which, using Eqs. (2.50), we can produce the equation

$$\int_{-1}^{1}\int_{-1}^{1}\left(\frac{\partial\hat{\phi}}{\partial x}\frac{\partial W_l}{\partial x} + \frac{\partial\hat{\phi}}{\partial y}\frac{\partial W_l}{\partial y}\right)dx\,dy$$

$$-\int_{\Gamma_\phi+\Gamma_q}\frac{\partial\hat{\phi}}{\partial n}W_l\,d\Gamma - \int_{\Gamma_q}\left(\frac{\partial\hat{\phi}}{\partial n} - \cos\frac{\pi y}{2}\right)\overline{W}_l\,d\Gamma = 0$$

If we choose $W_l = N_l$, then the integral over Γ_ϕ disappears since $N_l = 0$ on this portion of the boundary.

It follows that the above statement can be written as

$$\int_{-1}^{1}\int_{-1}^{1}\left(\frac{\partial\hat{\phi}}{\partial x}\frac{\partial N_l}{\partial x} + \frac{\partial\hat{\phi}}{\partial y}\frac{\partial N_l}{\partial y}\right)dx\,dy$$

$$-\int_{\Gamma_q}\frac{\partial\hat{\phi}}{\partial n}N_l\,d\Gamma - \int_{\Gamma_q}\left(\frac{\partial\hat{\phi}}{\partial n} - \cos\frac{\pi y}{2}\right)\overline{W}_l\,d\Gamma = 0$$

The appearance of the gradient term involving $\hat{\phi}$ on Γ_q can now be removed by taking $\overline{W}_l = -N_l|_{\Gamma_q}$, as in Eq. (2.53b). This shows that the condition on Γ_q is a natural boundary condition for this formulation, and the result is that

$$\int_{-1}^{1}\int_{-1}^{1}\left(\frac{\partial\hat{\phi}}{\partial x}\frac{\partial N_l}{\partial x} + \frac{\partial\hat{\phi}}{\partial y}\frac{\partial N_l}{\partial y}\right)dx\,dy = \int_{\Gamma_q}N_l\cos\frac{\pi y}{2}\,dy$$

The substitution of our approximation for $\hat{\phi}$ will produce a matrix equation of the form (2.55), where now

$$K_{lm} = \int_{0}^{1}\int_{0}^{1}\left(\frac{\partial N_l}{\partial x}\frac{\partial N_m}{\partial x} + \frac{\partial N_l}{\partial y}\frac{\partial N_m}{\partial y}\right)dx\,dy, \qquad 1 \leqslant l, m \leqslant 5$$

$$f_l = \int_{0}^{1}N_l|_{x=1}\cos\frac{\pi y}{2}\,dy, \qquad\qquad 1 \leqslant l \leqslant 5$$

On evaluation of the matrix elements it is found that

$$
\mathbf{K} = \begin{bmatrix}
1.33333 & 0.444444 & 0.266667 & 0.0888889 & 0.266667 \\
 & 0.977778 & 0.0888889 & 0.154921 & 1.4381 \\
 & & 0.419048 & 0.139683 & 0.0533333 \\
 & \text{symmetric} & & 0.117672 & 0.16 \\
 & & & & 1.3672
\end{bmatrix}
$$

$$
\mathbf{f} = \begin{bmatrix}
0.516025 \\
0.516025 \\
0.0704803 \\
0.0704803 \\
0.516025
\end{bmatrix}
$$

and the matrix equation can be solved to yield the vector **a** of unknown coefficients. If this process is carried out, it is then possible to demonstrate the

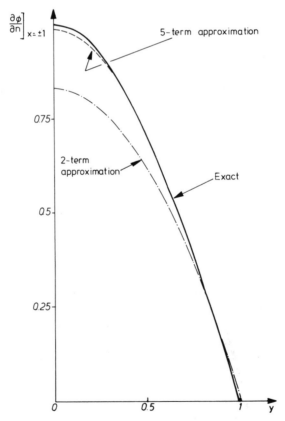

FIGURE 2.9. Comparison between the normal gradient of the exact and the trial function solutions along the lines $x = \pm 1$ for the problem of Example 2.7.

convergence of the approximation produced in this way to the natural boundary condition on the lines $x = \pm 1$ as shown in Fig. 2.9.

EXERCISES

2.13. Use the Galerkin method to recalculate the heat conduction problem of Exercise 1.9 and compare the solution with the exact solution to the problem.

2.14. Return to Exercise 1.10 and obtain the solution by the Galerkin weighted residual method. Demonstrate convergence to the exact solution when the number of terms involved in the approximation is increased.

2.15. The problem of determining the deflection of a loaded beam clamped at both ends and resting on an elastic foundation has been considered in Exercise 1.20. Identify the natural boundary conditions and solve by the Galerkin method the case of such a beam which is simply supported at its ends, using an approximation which does not satisfy automatically the natural boundary conditions. The appropriate boundary conditions for the simply supported beam are $\phi = d^2\phi/dx^2 = 0$ at the two ends.

2.16. Obtain the appropriate weighted residual statement for two-dimensional steady heat conduction in a region Ω subject to the boundary conditions $\phi = \bar{\phi}$ on Γ_ϕ and $k\,\partial\phi/\partial n = -\bar{q} + h(\phi - \phi_0)$ on Γ_q. Here ϕ is the temperature at any point, k is the thermal conductivity of the medium, and $\bar{\phi}$, $\bar{q}$, h, and ϕ_0 are given functions of position.

2.17. Solve the problem of steady-state heat conduction in a material occupying the square region $-1 \leqslant x, y \leqslant 1$ when the sides $y = \pm 1$ are maintained at a temperature of 100 while the sides $x = \pm 1$ are subjected to the condition $\partial\phi/\partial n = -1 - \phi$.

2.18. Show that in the problem of determining the deflection of a loaded thin elastic plate, as described in Exercise 1.21, the boundary condition $\partial^2\phi/\partial n^2 = 0$ on Γ is a natural boundary condition. Solve the problem and compare the answers with those obtained in Exercise 2.12.

2.6. BOUNDARY SOLUTION METHODS

In our attempt to construct approximate solutions to differential equations subject to appropriate boundary conditions we have in this chapter used trial function sets which are such that either

1. The boundary conditions are satisfied totally by the approximation $\hat{\phi}$, but the differential equation is not satisfied a priori in the domain Ω.

2.   The boundary conditions are not satisfied a priori, either in part or in total, by the approximation $\hat{\phi}$, and neither is the differential equation.

A third possibility obviously exists in which we choose trial functions such that the approximation $\hat{\phi}$ automatically satisfies the differential equation, but does not immediately satisfy the boundary conditions.[4] If the differential equation under consideration is linear, this can be accomplished by choosing trial functions which themselves are solutions of the differential equation. Choosing our trial functions in this way, we can take

$$\phi \simeq \hat{\phi} = \sum_{m=1}^{M} a_m N_m \qquad (2.59)$$

and the weighted residual statement of Eq. (2.41) reduces to

$$\int_{\Gamma} \overline{W}_l R_{\Gamma} \, d\Gamma = 0 \qquad (2.60)$$

since

$$R_{\Omega} = A(\hat{\phi}) = \sum_{m=1}^{M} a_m A(N_m) = 0 \qquad (2.61)$$

Only one set of weighting functions $\overline{W}_l$ need now be defined and, further, the set need only be defined on the boundary Γ rather than throughout the domain.

However, this set of trial functions is more difficult to choose. Certain general, if rather complicated, procedures are possible, but here we will restrict ourselves to an illustration of the solution procedure by considering the example of the Laplace differential equation in which the choice of the trial function set is particularly easy.

If we write any analytic function of the complex variable $z = x + iy$, then its real and imaginary parts must satisfy automatically the Laplace equation. We can demonstrate this by assuming that

$$f(z) = u + iv \qquad (2.62)$$

where u and v are real, is such an analytic function, and then we have

$$\frac{\partial^2 f}{\partial x^2} = f''$$

$$\frac{\partial^2 f}{\partial y^2} = i^2 f'' = .-f'' \qquad (2.63)$$

where $f'' = d^2f/dz^2$. Thus

$$\nabla^2 f = \nabla^2 u + i\nabla^2 v = 0 \tag{2.64}$$

and this can only be true if $\nabla^2 u = \nabla^2 v = 0$. From this result we can immediately use an analytic function such as

$$f(z) = z^n \tag{2.65}$$

to generate a set of functions satisfying the Laplace equation. This leads to the following set:

$$
\begin{array}{llll}
n = 1, & u = x, & v = y & \\
n = 2, & u = x^2 - y^2, & v = 2xy & \\
n = 3, & u = x^3 - 3xy^2, & v = 3x^2y - y^3 & (2.66) \\
n = 4, & u = x^4 - 6x^2y^2 + y^4, & v = 4x^3y - 4xy^3 &
\end{array}
$$

and so on.

Having generated a suitable set of functions, we can illustrate the application of the method by solving once again the torsion problem already considered in Examples 1.5, 2.3, and 2.5.

Example 2.8

The torsion problem as previously described is governed by the equation $\partial^2\phi/\partial x^2 + \partial^2\phi/\partial y^2 = -2$ in the region defined by $-3 \leqslant x \leqslant 3, -2 \leqslant y \leqslant 2$, and the required solution is to satisfy the condition that $\phi = 0$ on the boundaries. In the previous section we showed how it is possible to generate a series of functions which satisfy the Laplace equation. To enable us to use this set of functions in this example we first introduce a new variable θ defined by

$$\phi = \theta - \tfrac{1}{2}(x^2 + y^2)$$

The torsion problem can be reformulated in terms of the new variable θ, and it becomes the problem of solving the equation

$$\frac{\partial^2\theta}{\partial x^2} + \frac{\partial^2\theta}{\partial y^2} = 0, \qquad -3 \leqslant x \leqslant 3, -2 \leqslant y \leqslant 2$$

subject to $\theta = \tfrac{1}{2}(x^2 + y^2)$ on the boundaries. We now attempt to solve this problem by the boundary solution method described above. The required solution will be symmetric in x and y, and so from Eq. (2.66), we can use as trial functions the set $N_1 = 1$, $N_2 = x^2 - y^2$, $N_3 = x^4 - 6x^2y^2 + y^4$, and so on.

A three-term approximation would then be

$$\hat{\theta} = a_1 + a_2(x^2 - y^2) + a_3(x^4 - 6x^2y^2 + y^4)$$

which satisfies the governing equation in terms of θ exactly, for all values of a_1, a_2, a_3, since each trial function is a solution of the Laplace equation.

The weighted residual statement is thus of the form of Eq. (2.60), and for this example it may be written

$$\int_{\Gamma}[\hat{\theta} - \tfrac{1}{2}(x^2 + y^2)]\overline{W}_l \, d\Gamma = 0$$

If we choose weighting functions $\overline{W}_l$ defined by $\overline{W}_l = N_l|_\Gamma$, then the above becomes

$$\int_0^2 [\hat{\theta}|_{x=3} - \tfrac{1}{2}(9 + y^2)]N_l|_{x=3} \, dy + \int_0^3 [\hat{\theta}|_{y=2} - \tfrac{1}{2}(x^2 + 4)]N_l|_{y=2} \, dx = 0$$

Inserting the expansion for $\hat{\theta}$ and evaluating the resulting integrals produces the standard matrix equation of type (2.55) with

$$\mathbf{K} = \begin{bmatrix} 5 & 12.3333 & -95 \\ 12.3333 & 145 & 98.543 \\ -95 & 98.543 & 18,170.4 \end{bmatrix}$$

$$\mathbf{f} = \begin{bmatrix} 20.8333 \\ 78.1 \\ -539.643 \end{bmatrix}$$

and solution $a_1 = 3.2154$, $a_2 = -0.2749$, and $a_3 = -0.01438$. The three-term approximation $\hat{\theta}$ has thus been determined, and this leads to an equivalent approximation for the original function ϕ by use of the relationship

$$\hat{\phi} = \hat{\theta} - \tfrac{1}{2}(x^2 + y^2)$$

The behavior of $\hat{\phi}$ on the boundaries, where the required solution satisfies $\phi = 0$, is shown in Fig. 2.10. The twisting moment calculated by using this approximation is 75.51, as compared with the exact value of 76.4.

For more general differential equations the choice of the trial functions for a boundary solution method is less obvious. Quite generally, singular functions of the Green's form can be used and the resulting approximation expressed as a set of integral equations. Methods known as boundary integral solutions are of this form and are a useful numerical tool.[5]

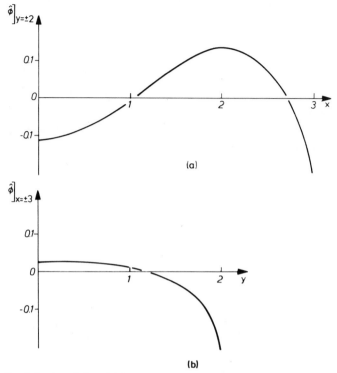

FIGURE 2.10. Behavior of the trial function approximation constructed in Example 2.8 on the lines (*a*) $y = \pm 2$ and (*b*) $x = \pm 3$.

In many texts these boundary solution processes are treated as a completely separate procedure but, as the reader will have noted, they are simply a variant of the general weighted residual trial function approximation method.

2.7. SYSTEMS OF DIFFERENTIAL EQUATIONS

The method of weighted residuals can be applied equally well to the problem of the solution of a set of simultaneous differential equations. Posing this problem in the most general terms, we seek an unknown function ϕ, where

$$\phi^T = (\phi_1, \phi_2 \ldots) \tag{2.67}$$

such that it satisfies, in a region Ω, certain differential equations, which can be expressed in the form

$$A_1(\phi) = 0$$
$$A_2(\phi) = 0 \tag{2.68}$$

$$\vdots \qquad \vdots$$

and which we can write as

$$\mathbf{A}(\phi) = \begin{bmatrix} A_1(\phi) \\ A_2(\phi) \\ \vdots \end{bmatrix} = \mathbf{0} \quad \text{in } \Omega \tag{2.69}$$

The appropriate number of associated conditions, which have to be satisfied on the boundaries Γ of Ω, will be assumed to be given as

$$B_1(\phi) = 0$$

$$B_2(\phi) = 0 \tag{2.70}$$

$$\vdots \qquad \vdots$$

and these we write as

$$\mathbf{B}(\phi) = \begin{bmatrix} B_1(\phi) \\ B_2(\phi) \\ \vdots \end{bmatrix} = \mathbf{0} \quad \text{on } \Gamma \tag{2.71}$$

For each component of the unknown vector ϕ we use a trial function expansion as in Eq. (2.27) and write

$$\phi_1 \simeq \hat{\phi}_1 = \psi_1 + \sum_{m=1}^{M} a_{m,1} N_{m,1}$$

$$\tag{2.72}$$

$$\phi_2 \simeq \hat{\phi}_2 = \psi_2 + \sum_{m=1}^{M} a_{m,2} N_{m,2}$$

and so on. These expressions may be written in the more compact matrix form

$$\phi \simeq \hat{\phi} = \psi + \sum_{m=1}^{M} \mathbf{N}_m \mathbf{a}_m \tag{2.73}$$

in which

$$\psi^T = (\psi_1, \psi_2, \psi_3, \dots)$$

$$\tag{2.74a}$$

$$\mathbf{a}_m^T = (a_{m,1}, a_{m,2}, a_{m,3}, \dots)$$

and

$$\mathbf{N}_m = \begin{bmatrix} N_{m,1} & & & 0 \\ & N_{m,2} & & \\ & & N_{m,3} & \\ 0 & & & \ddots \end{bmatrix} \qquad (2.74\text{b})$$

It is apparent that the standard forms of approximation introduced in the previous sections of this chapter can again be used in this case, on the understanding that $\mathbf{N}_m$ now represents a diagonal matrix assembly of shape functions and the parameter $\mathbf{a}_m$ is a vector with as many components as the number of unknown functions in the definition of ϕ.

To obtain the generalized weighted residual statement for a problem of this type, we can consider each of Eqs. (2.68) and their associated boundary conditions in turn and weight them with appropriate functions. Then, in place of statement (2.41), we now have the equation set

$$\int_\Omega W_{l,1} A_1(\hat{\phi})\, d\Omega + \int_\Gamma \overline{W}_{l,1} B_1(\hat{\phi})\, d\Gamma = 0$$

$$\int_\Omega W_{l,2} A_2(\hat{\phi})\, d\Omega + \int_\Gamma \overline{W}_{l,2} B_2(\hat{\phi})\, d\Gamma = 0 \qquad (2.75)$$

$$\vdots \qquad\qquad \vdots$$

Once again, if diagonal matrices of weighting functions $\mathbf{W}_l$ and $\overline{\mathbf{W}}_l$ are introduced, defined by

$$\mathbf{W}_l = \begin{bmatrix} W_{l,1} & & & 0 \\ & W_{l,2} & & \\ & & W_{l,3} & \\ 0 & & & \ddots \end{bmatrix}$$

$$\qquad\qquad\qquad (2.76)$$

$$\overline{\mathbf{W}}_l = \begin{bmatrix} \overline{W}_{l,1} & & & 0 \\ & \overline{W}_{l,2} & & \\ & & \overline{W}_{l,3} & \\ 0 & & & \ddots \end{bmatrix}$$

the set of weighted residual equations (2.75) can be written in the compact form

$$\int_{\Omega} \mathbf{W}_l \mathbf{A}(\hat{\phi}) \, d\Omega + \int_{\Gamma} \overline{\mathbf{W}}_l \mathbf{B}(\hat{\phi}) \, d\Gamma = 0 \qquad (2.77)$$

and an appropriate approximation obtained.

The use of such an approach to problems involving coupled systems of differential equations will be illustrated in the following example.

Example 2.9

In Chapter 1 we have shown how the general problem of heat conduction can be written in terms of a set of simultaneous first-order equations (1.2) and (1.4). For the case of one-dimensional steady heat conduction, these equations reduce to

$$q + k\frac{d\phi}{dx} = 0$$

$$\frac{dq}{dx} - Q = 0$$

where q is the flux of heat in the x direction, ϕ is the temperature, Q is the heat generation rate within the material, and k is the material thermal conductivity. This pair of equations can be solved for q and ϕ simultaneously by writing

$$\phi = \begin{bmatrix} q \\ \phi \end{bmatrix}$$

and the governing equations as

$$\mathbf{A}(\phi) = \begin{bmatrix} q + k\dfrac{d\phi}{dx} \\ \dfrac{dq}{dx} - Q \end{bmatrix} = \mathbf{0}$$

or

$$\mathcal{L}\phi + \mathbf{p} = \begin{bmatrix} 1 & k\dfrac{d}{dx} \\ \dfrac{d}{dx} & 0 \end{bmatrix} \phi + \begin{bmatrix} 0 \\ -Q \end{bmatrix} = \mathbf{0}$$

For the particular example of a region $0 \leqslant x \leqslant 1$ for which $k = 1$, with Q defined by $Q = 1$ for $0 \leqslant x \leqslant \frac{1}{2}$ and $Q = 0$ otherwise and boundary conditions $\phi = 0$ at $x = 0$ and $q = 0$ at $x = 1$, the problem is properly posed and can be solved. We note in passing that only one condition is specified for each of the unknowns on the boundary and that gradient boundary conditions do not have to be applied.

Using an approximation, as in Eq. (2.72),

$$\hat{q} = \psi_1 + \sum_{m=1}^{M} a_{m,1} N_{m,1}$$

$$\hat{\phi} = \psi_2 + \sum_{m=1}^{M} a_{m,2} N_{m,2}$$

we see that the boundary conditions are automatically satisfied by taking $\psi_1 = \psi_2 = 0$ and choosing the trial function sets to be such that

$$N_{m,1} = 0 \quad \text{at } x = 1; \ N_{m,2} = 0 \quad \text{at } x = 0, \quad m = 1, 2, \ldots, M$$

For this example we shall satisfy these conditions by using the function sets defined by

$$N_{m,1} = x^{m-1}(1 - x); \qquad N_{m,2} = x^m, \qquad m = 1, 2, \ldots, M$$

The approximating equation (2.77) becomes now simply

$$\int_0^1 \mathbf{W}_l \mathcal{L} \hat{\phi} \, dx + \int_0^1 \mathbf{W}_l \mathbf{p} \, dx = 0$$

where $\mathcal{L}$ and $\mathbf{p}$ are as defined above and inserting the approximations leads to the standard equation system $\mathbf{Ka} = \mathbf{f}$, where

$$\mathbf{K} = \begin{bmatrix} \mathbf{K}_{11} & \mathbf{K}_{12} & \mathbf{K}_{13} & \cdots \\ \mathbf{K}_{21} & \mathbf{K}_{22} & \mathbf{K}_{23} & \cdots \\ \vdots & \vdots & \vdots & \\ \mathbf{K}_{M1} & \mathbf{K}_{M2} & \mathbf{K}_{M3} & \cdots \end{bmatrix} \qquad \mathbf{f} = \begin{bmatrix} \mathbf{f}_1 \\ \mathbf{f}_2 \\ \vdots \\ \mathbf{f}_M \end{bmatrix}$$

A typical submatrix $\mathbf{K}_{lm}$ of the matrix $\mathbf{K}$ is the 2×2 matrix defined by

$$\mathbf{K}_{lm} = \int_0^1 \mathbf{W}_l \mathcal{L} \mathbf{N}_m \, dx$$

and the subvector $\mathbf{f}_l$ is the 2×1 column vector given by

$$\mathbf{f}_l = -\int_0^1 \mathbf{W}_l \mathbf{p} \, dx$$

If a Galerkin-type weighting is used, so that $\mathbf{W}_l = \mathbf{N}_l$, we can write

$$\mathbf{K}_{lm} = \int_0^1 \begin{bmatrix} N_{l,1} N_{m,1} & N_{l,1} \dfrac{dN_{m,2}}{dx} \\[2ex] N_{l,2} \dfrac{dN_{m,1}}{dx} & 0 \end{bmatrix} dx$$

Then using a two-term expansion for both q and ϕ, the matrix coefficients can be evaluated to produce the equation

$$\begin{bmatrix} \frac{1}{3} & \frac{1}{2} & \frac{1}{12} & \frac{1}{3} \\[1ex] -\frac{1}{2} & 0 & -\frac{1}{6} & 0 \\[1ex] \frac{1}{12} & \frac{1}{6} & \frac{1}{30} & \frac{1}{6} \\[1ex] -\frac{1}{3} & 0 & -\frac{1}{6} & 0 \end{bmatrix} \begin{bmatrix} a_{1,1} \\[1ex] a_{1,2} \\[1ex] a_{2,1} \\[1ex] a_{2,2} \end{bmatrix} = \begin{bmatrix} 0 \\[1ex] \frac{1}{8} \\[1ex] 0 \\[1ex] \frac{1}{24} \end{bmatrix}$$

which can be solved for the values of the unknown parameters. It should be noted that application of the Galerkin method has not resulted in a symmetrical equation system in this case but such symmetry could be enforced here by using an integration by parts. In Fig. 2.11 the results of using both one- and two-term approximations are compared with the exact solution for this problem. For certain problems it may prove advantageous to use a different number of parameters in the expansion for ϕ than is adopted in the expansion for q. This clearly could be accomplished without loss of the general format of the solution, by setting the relevant parameters to zero and deleting the appropriate rows and columns from the standard equation system. A formulation of the kind discussed in this example is frequently referred to as *mixed* since the equations to be solved can, if required, be reduced in number by suitable manipulation. (Here q may be eliminated from the two-equation system and a single equation produced for ϕ.) At times such formulations can be more accurate than *irreducible* formulations but, as seen above, they lead to more complicated numerical equations.

2.7.1. Two-Dimensional Plane Stress Problems in Elasticity

As a further example of the formulation just described, we consider the problem of two-dimensional plane stress in elasticity.[6] Here the basic unknowns will be considered to be the displacements u and v in the x and y

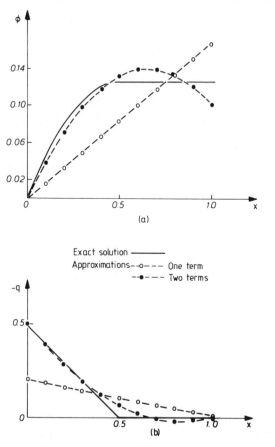

FIGURE 2.11.   Comparison between the exact and the approximate solutions to Example 2.9 for (a) the temperature ϕ and (b) the heat flux q.

directions, that is,

$$\phi^T = (u, v) \tag{2.78}$$

Strains, and hence stresses, can be expressed in terms of these displacements. The strains are written as

$$\varepsilon = \begin{bmatrix} \varepsilon_x \\ \varepsilon_y \\ \varepsilon_{xy} \end{bmatrix} = \begin{bmatrix} \dfrac{\partial u}{\partial x} \\ \dfrac{\partial v}{\partial y} \\ \dfrac{\partial u}{\partial y} + \dfrac{\partial v}{\partial x} \end{bmatrix} = \mathcal{L}\phi \tag{2.79}$$

in which

$$
\mathcal{L} =
\begin{bmatrix}
\dfrac{\partial}{\partial x} & 0 \\[2mm]
0 & \dfrac{\partial}{\partial y} \\[2mm]
\dfrac{\partial}{\partial y} & \dfrac{\partial}{\partial x}
\end{bmatrix}
\tag{2.80}
$$

and, for the particular case of plane stress, the stresses are given by

$$
\boldsymbol{\sigma} =
\begin{bmatrix}
\sigma_x \\
\sigma_y \\
\sigma_{xy}
\end{bmatrix}
= \frac{E}{1 - \nu^2}
\begin{bmatrix}
1 & \nu & 0 \\
\nu & 1 & 0 \\
0 & 0 & \dfrac{1-\nu}{2}
\end{bmatrix}
\boldsymbol{\varepsilon} = \mathbf{D}\boldsymbol{\varepsilon}
\tag{2.81}
$$

Here E is Young's modulus and ν is Poisson's ratio for the material under consideration. It remains to solve the equilibrium equation system which, using Eqs. (2.79)–(2.81), can be written as

$$
\mathbf{A}(\boldsymbol{\phi}) =
\begin{bmatrix}
\dfrac{\partial \sigma_x}{\partial x} + \dfrac{\partial \sigma_{xy}}{\partial y} + X \\[3mm]
\dfrac{\partial \sigma_{xy}}{\partial x} + \dfrac{\partial \sigma_y}{\partial y} + Y
\end{bmatrix}
= \mathcal{L}^T \mathbf{D} \mathcal{L} \boldsymbol{\phi} + \mathbf{X} = \mathbf{0}
\tag{2.82}
$$

in a two-dimensional region Ω. In this equation X and Y are external forces per unit volume, $\mathbf{X}^T = (X, Y)$.

The boundary conditions for a typical two-dimensional elasticity problem may well be defined as specified surface tractions or displacements, and so we could have

$$
\mathbf{B}(\boldsymbol{\phi}) =
\begin{bmatrix}
\sigma_x n_x + \sigma_{xy} n_y - \bar{t}_x \\
\sigma_{xy} n_x + \sigma_y n_y - \bar{t}_y
\end{bmatrix}
= \mathbf{0} \qquad \text{on } \Gamma_\sigma
\tag{2.83}
$$

and

$$
\mathbf{B}(\boldsymbol{\phi}) =
\begin{bmatrix}
u - \bar{u} \\
v - \bar{v}
\end{bmatrix}
= \mathbf{0} \qquad \text{on } \Gamma_\phi
\tag{2.84}
$$

in which n_x and n_y are the direction cosines of the outward normal to Γ and $\bar{t}_x$, $\bar{t}_y$, $\bar{u}$, and $\bar{v}$ are specified boundary tractions and displacements. With $\mathbf{A}$ and $\mathbf{B}$ so defined, the weighted residual statement is then as in Eq. (2.77).

For this set of equations it is of considerable interest to once again identify the relevant natural boundary conditions.

If we can find functions ψ_1 and ψ_2 such that

$$\psi_1 = \bar{u}, \qquad \psi_2 = \bar{v} \qquad \text{on } \Gamma_\phi \tag{2.85}$$

and use trial functions which vanish on Γ_ϕ, then a trial expansion

$$\hat{\phi} = \begin{bmatrix} \hat{u} \\ \hat{v} \end{bmatrix} = \psi + \sum_{m=1}^{M} \mathbf{N}_m \mathbf{a}_m \tag{2.86}$$

where $\psi^{\mathrm{T}} = (\psi_1, \psi_2)$, automatically satisfies the boundary conditions on Γ_ϕ. Defining the weighting functions as

$$\mathbf{W}_l = \begin{bmatrix} W_{l,1} & 0 \\ 0 & W_{l,2} \end{bmatrix}, \qquad \overline{\mathbf{W}}_l = \begin{bmatrix} \overline{W}_{l,1} & 0 \\ 0 & \overline{W}_{l,2} \end{bmatrix} \tag{2.87}$$

the weighted residual statement for the equilibrium equation in terms of the stresses can then be written as

$$\int_\Omega \left(\frac{\partial \hat{\sigma}_x}{\partial x} + \frac{\partial \hat{\sigma}_{xy}}{\partial y} + X \right) W_{l,1} \, d\Omega + \int_{\Gamma_\sigma} \left(n_x \hat{\sigma}_x + n_y \hat{\sigma}_{xy} - \bar{t}_x \right) \overline{W}_{l,1} \, d\Gamma = 0$$

$$\int_\Omega \left(\frac{\partial \hat{\sigma}_{xy}}{\partial x} + \frac{\partial \hat{\sigma}_y}{\partial y} + Y \right) W_{l,2} \, d\Omega + \int_{\Gamma_\sigma} \left(n_x \hat{\sigma}_{xy} + n_y \hat{\sigma}_y - \bar{t}_y \right) \overline{W}_{l,2} \, d\Gamma = 0$$

$$\tag{2.88}$$

where $\hat{\sigma} = \mathbf{D}\mathcal{L}\hat{\phi}$. On using Green's lemma [Eq. (2.50)], this statement becomes

$$-\int_\Omega \left(\hat{\sigma}_x \frac{\partial W_{l,1}}{\partial x} + \hat{\sigma}_{xy} \frac{\partial W_{l,1}}{\partial y} - W_{l,1} X \right) d\Omega + \int_{\Gamma_\phi + \Gamma_\sigma} \left(\hat{\sigma}_x n_x + \hat{\sigma}_{xy} n_y \right) W_{l,1} \, d\Gamma$$

$$+ \int_{\Gamma_\sigma} \left(\hat{\sigma}_x n_x + \hat{\sigma}_{xy} n_y - \bar{t}_x \right) \overline{W}_{l,1} \, d\Gamma = 0$$

$$\tag{2.89}$$

$$-\int_\Omega \left(\hat{\sigma}_{xy} \frac{\partial W_{l,2}}{\partial x} + \hat{\sigma}_y \frac{\partial W_{l,2}}{\partial y} - W_{l,2} Y \right) d\Omega + \int_{\Gamma_\phi + \Gamma_\sigma} \left(\hat{\sigma}_{xy} n_x + \hat{\sigma}_y n_y \right) W_{l,2} \, d\Gamma$$

$$+ \int_{\Gamma_\sigma} \left(\hat{\sigma}_{xy} n_x + \hat{\sigma}_y n_y - \bar{t}_y \right) \overline{W}_{l,2} \, d\Gamma = 0$$

Limiting now the choice of the weighting functions, so that

$$W_{l,1} = W_{l,2} = 0 \qquad \text{on } \Gamma_\phi \tag{2.90a}$$

$$\overline{W}_{l,1} = -W_{l,1}|_{\Gamma_\sigma}$$

$$\overline{W}_{l,2} = -W_{l,2}|_{\Gamma_\sigma} \tag{2.90b}$$

the above equation can be written in the compact form

$$\int_\Omega (\mathcal{L}\mathbf{W}_l)^T \hat{\boldsymbol{\sigma}} \, d\Omega - \int_\Omega \mathbf{W}_l \mathbf{X} \, d\Omega - \int_{\Gamma_\sigma} \mathbf{W}_l \bar{\mathbf{t}} \, d\Gamma = 0 \tag{2.91}$$

where $\bar{\mathbf{t}}^T = (\bar{t}_x, \bar{t}_y)$.

Expressing the stress in terms of the displacement, means that Eq. (2.91) becomes

$$\int_\Omega (\mathcal{L}\mathbf{W}_l)^T \mathbf{D}\mathcal{L}\hat{\boldsymbol{\phi}} \, d\Omega = \int_\Omega \mathbf{W}_l \mathbf{X} \, d\Omega + \int_{\Gamma_\sigma} \mathbf{W}_l \bar{\mathbf{t}} \, d\Gamma \tag{2.92}$$

which is the weighted residual statement for Eq. (2.82). The traction boundary condition, involving as it does differentiation of the displacement field, is thus a natural boundary condition for this problem. Following the insertion of the approximation of Eq. (2.86) for the displacement field, Eq. (2.92) becomes

$$\sum_{m=1}^M \left(\int_\Omega (\mathcal{L}\mathbf{W}_l)^T \mathbf{D}\mathcal{L}\mathbf{N}_m \, d\Omega \right) \mathbf{a}_m$$

$$= \int_\Omega \mathbf{W}_l \mathbf{X} \, d\Omega + \int_{\Gamma_\sigma} \mathbf{W}_l \bar{\mathbf{t}} \, d\Gamma - \int_\Omega (\mathcal{L}\mathbf{W}_l)^T \mathbf{D}\mathcal{L}\boldsymbol{\psi} \, d\Omega \tag{2.93}$$

and immediately an equation of the form

$$\mathbf{Ka} = \mathbf{f} \tag{2.94}$$

can be identified. In this case, the use of the Galerkin approximation results in a symmetric matrix $\mathbf{K}$ in which the submatrices $\mathbf{K}_{lm}$ are given by

$$\mathbf{K}_{lm} = \int_\Omega (\mathcal{L}\mathbf{N}_l)^T \mathbf{D}\mathcal{L}\mathbf{N}_m \, d\Omega \tag{2.95}$$

The above form is completely general and can be used for the solution of many plane elasticity problems. Indeed, with suitable definition of displacements, stresses, and strains it can be extended to all linear elastic situations.

In the following example we shall consider the application of this method to a sample problem. However, before proceeding to this illustrative example, it is appropriate to add another comment which arises from the form of Eq. (2.91). This equation could have been derived from the principle of virtual work[7] in which it is stated that equilibrium is satisfied at all points of a body if the internal and external work performed by internal stresses and external forces during an arbitrary or "virtual" displacement of the body are equal.

If thus

$$\phi^* = \begin{bmatrix} u^* \\ v^* \end{bmatrix}, \qquad \phi^*|_{\Gamma_\phi} = \overline{\phi}|_{\Gamma_\phi} \tag{2.96}$$

is a displacement of an arbitrary type and the corresponding strains are

$$\varepsilon^* = \mathcal{L}\phi^* \tag{2.97}$$

then it follows from the principle of virtual work that

$$\int_\Omega \varepsilon^{*T}\sigma \, d\Omega = \int_\Omega \phi^{*T} \mathbf{X} \, d\Omega + \int_{\Gamma_\sigma} \phi^{*T} \overline{t} \, d\Gamma \tag{2.98}$$

Applying the above statement to the problem at hand, where approximate stresses, $\sigma = \hat{\sigma}$ are used, and limiting the arbitrariness by taking

$$\phi^* = \mathbf{W}_l \delta_l^* \tag{2.99}$$

where only δ_l^* is arbitrary, we have immediately

$$\delta_l^{*T} \int_\Omega (\mathcal{L}\mathbf{W}_l)^T \hat{\sigma} \, d\Omega = \delta_l^{*T} \left(\int_\Omega \mathbf{W}_l \mathbf{X} \, d\Omega + \int_{\Gamma_\sigma} \mathbf{W}_l \overline{t} \, d\Gamma \right) \tag{2.100}$$

As this has to be true for all values of δ_l^*, it follows that the weighted residual equation (2.91) results.

Example 2.10

A square plate, of Young's modulus E and Poisson's ratio ν ($= 0.25$), occupies the region defined by $-1 \leqslant x, y \leqslant 1$ and is held with its edges $y = \pm 1$ fixed and loaded such that $\overline{t}_x = E(1 - y^2)/(1 + \nu)$, $\overline{t}_y = 0$ on the edges $x = \pm 1$. It is required to determine approximations to the resulting displacement and stress fields.

From symmetry considerations we choose trial functions

$$N_{1,1} = x(1 - y^2), \qquad N_{2,1} = x^3(1 - y^2), \qquad N_{3,1} = xy^2(1 - y^2)$$

and so on, to represent the displacement u in the x direction and trial functions

$$N_{1,2} = y(1 - y^2), \qquad N_{2,2} = x^2 y(1 - y^2), \qquad N_{3,2} = y^3(1 - y^2)$$

and so on, to represent the displacement v, in the y direction. Defining three-term approximations by

$$\hat{u} = a_{1,1} N_{1,1} + a_{2,1} N_{2,1} + a_{3,1} N_{3,1}$$

$$\hat{v} = a_{1,2} N_{1,2} + a_{2,2} N_{2,2} + a_{3,2} N_{3,2}$$

it can be seen that the boundary condition of zero displacement on the edges $y = \pm 1$ is automatically satisfied. The weighted residual statement is then of the form given in Eq. (2.91), namely,

$$\int_{-1}^{1} \int_{-1}^{1} (\mathcal{L} \mathbf{W}_l)^T \hat{\boldsymbol{\sigma}} \, dx \, dy = \int_{-1}^{1} \mathbf{W}_l \Big|_{x=1} \bar{\mathbf{t}} \, dy - \int_{1}^{-1} \mathbf{W}_l \Big|_{x=-1} \bar{\mathbf{t}} \, dy$$

assuming zero external force.

If we write

$$\mathbf{a}_m = \begin{bmatrix} a_{m,1} \\ a_{m,2} \end{bmatrix}, \qquad \mathbf{N}_m = \begin{bmatrix} N_{m,1} & 0 \\ 0 & N_{m,2} \end{bmatrix}$$

then

$$\hat{\boldsymbol{\phi}} = \begin{bmatrix} \hat{\phi}_1 \\ \hat{\phi}_2 \end{bmatrix} = \sum_{m=1}^{3} \mathbf{N}_m \mathbf{a}_m$$

and

$$\hat{\boldsymbol{\sigma}} = \mathbf{D} \mathcal{L} \hat{\boldsymbol{\phi}}$$

Further, if we choose our weighting functions $\mathbf{W}_l$ to be such that $\mathbf{W}_l = \mathbf{N}_l$, then the weighted residual statement, written in component form, becomes

$$\sum_{m=1}^{3} \int_0^1 \int_0^1 \left\{ \left(8 \frac{\partial N_{l,1}}{\partial x} \frac{\partial N_{m,1}}{\partial x} + 3 \frac{\partial N_{l,1}}{\partial y} \frac{\partial N_{m,1}}{\partial y} \right) a_{m,1} \right.$$

$$\left. + \left(2 \frac{\partial N_{l,1}}{\partial x} \frac{\partial N_{m,2}}{\partial y} + 3 \frac{\partial N_{l,1}}{\partial y} \frac{\partial N_{m,2}}{\partial x} \right) a_{m,2} \right\} dx \, dy$$

$$= 6 \int_0^1 (1 - y^2) N_{l,1} |_{x=1} \, dy$$

$$\sum_{m=1}^{3} \int_0^1 \int_0^1 \left\{ \left(2 \frac{\partial N_{l,2}}{\partial y} \frac{\partial N_{m,1}}{\partial x} + 3 \frac{\partial N_{l,2}}{\partial x} \frac{\partial N_{m,1}}{\partial y} \right) a_{m,1} \right.$$

$$\left. + \left(8 \frac{\partial N_{l,2}}{\partial y} \frac{\partial N_{m,2}}{\partial y} + 3 \frac{\partial N_{l,2}}{\partial x} \frac{\partial N_{m,2}}{\partial x} \right) a_{m,2} \right\} dx \, dy$$

$$= 0$$

Evaluation of these integrals produces a matrix equation of the standard form [Eq. (2.94)] where now

$$\mathbf{a}^T = \left(a_{1,1}, a_{2,1}, a_{3,1}, a_{1,2}, a_{2,2}, a_{3,2}\right)$$

$$\mathbf{f}^T = (3.2, 3.2, 0.457\,143, 0, 0, 0)$$

$$\mathbf{K} = \begin{bmatrix} 5.6 & 5.066667 & 0.876190 & 0.533333 & -0.355556 & 0.228571 \\ & 8.251429 & 0.769524 & 0.533333 & 0 & 0.228571 \\ & & 1.688889 & -0.076190 & 0.050794 & 0.025397 \\ & \text{symmetric} & & 6.4 & 2.133333 & 2.742857 \\ & & & & 1.584762 & 0.914286 \\ & & & & & 2.336508 \end{bmatrix}$$

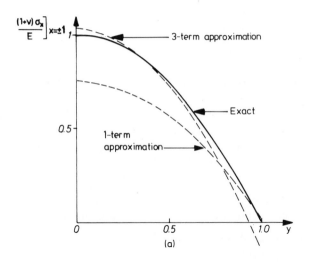

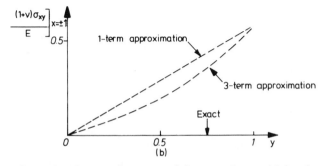

FIGURE 2.12. Comparison between the exact and the approximate trial function solutions to Example 2.10 showing (a) the normal stress and (b) the tangential stress on the lines $x = \pm 1$.

with solution

$$\mathbf{a}^T = (0.574\,778, 0.052\,828, -0.070\,674, -0.177\,255, 0.367\,795, 0.003\,533)$$

The resulting displacement field $\hat{\phi}$ can be used to calculate an approximation to the stress field via the relationship $\hat{\sigma} = \mathbf{D}\mathcal{L}\hat{\phi}$. The behavior of the solution on the lines $x = \pm 1$, where the boundary condition is natural, is shown in Fig. 2.12.

EXERCISES

2.19. The equation representing the deflection of the loaded beam resting on an elastic foundation in Exercise 1.20 can be split into the pair of second-order equations

$$EI\frac{d^2\phi}{dx^2} = -M, \qquad \frac{d^2M}{dx^2} - k\phi = -w$$

where ϕ is the deflection and M the bending moment. By assuming expansions for both M and ϕ, obtain an approximate solution for a beam of unit length when $EI = k = w = 1$ and subject to the boundary conditions that (1) both ends of the beam are clamped and (2) both ends of the beam are simply supported. Compare your answers with those obtained in Exercises 2.6 and 2.15.

2.20. A square plate $0 \leqslant x, y \leqslant 1$ is clamped along its edges and carries unit external forces per unit area in both the x and y directions. Obtain the displacement field by the weighted residual method.

2.21. In two-dimensional inviscid incompressible irrotational fluid flow the velocity components u, v in the x, y directions and the velocity potential ϕ satisfy the equations

$$u = \frac{\partial\phi}{\partial x}, \qquad v = \frac{\partial\phi}{\partial y}, \qquad \frac{\partial u}{\partial x} + \frac{\partial v}{\partial y} = 0$$

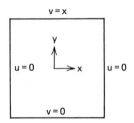

By constructing approximations for u, v, and ϕ, obtain approximations to the velocity field for flow in the square region $-1 \leqslant x, y \leqslant 1$ subject to the boundary conditions shown in the diagram.

2.8. NONLINEAR PROBLEMS

It has been noted in Section 1.5 that the modeling of physical problems frequently produces governing differential equations and/or boundary conditions which are nonlinear in character. So far in this chapter we have described the application of weighted residual methods to problems that are linear, but the method can equally well be applied to nonlinear problems. Then, however, following through the normal weighted residual procedure produces now not the standard linear system of equations (2.94), but a nonlinear equation set, which may be written as

$$\mathbf{K(a)a = f} \qquad (2.101)$$

and an equation of this type can, as we have noted in Chapter 1, be solved by a suitable iterative technique.

The construction of the matrix equation (2.101) can be illustrated by returning again to the problem of two-dimensional steady heat transfer in a medium with temperature-dependent thermal conductivity. The governing equation is

$$\frac{\partial}{\partial x}\left(k(\phi)\frac{\partial \phi}{\partial x}\right) + \frac{\partial}{\partial y}\left(k(\phi)\frac{\partial \phi}{\partial y}\right) + Q = 0 \qquad \text{in } \Omega \qquad (2.102)$$

where $k(\phi)$ is a given function, and general boundary conditions of interest are the Dirichlet condition

$$\phi = \bar{\phi} \qquad \text{on } \Gamma_\phi \qquad (2.103\text{a})$$

and the Neumann condition

$$k(\phi)\frac{\partial \phi}{\partial n} = -\bar{q} \qquad \text{on } \Gamma_q \qquad (2.103\text{b})$$

Seeking an approximate solution in the form

$$\hat{\phi} = \psi + \sum_{m=1}^{M} a_m N_m \qquad (2.104)$$

with ψ and N_m chosen in the usual manner so as to ensure automatic satisfaction of the boundary condition on Γ_ϕ, it is still possible to proceed as in

Eqs. (2.49)–(2.54) and produce the weighted residual statement

$$\int_{\Omega}\left(\frac{\partial W_l}{\partial x} k(\hat{\phi}) \frac{\partial \hat{\phi}}{\partial x} + \frac{\partial W_l}{\partial y} k(\hat{\phi}) \frac{\partial \hat{\phi}}{\partial y} \right) dx\, dy - \int_{\Omega} W_l Q dx\, dy + \int_{\Gamma_q} W_l \bar{q}\, d\Gamma = 0$$

$$l = 1, 2, \ldots, M \quad (2.105)$$

This nonlinear equation set requires iterative solution, and any one of the many standard methods can be applied. Following the simple iteration method used in Section 1.5, we could start from some initial guess

$$\mathbf{a} = \mathbf{a}^0 = \left(a_1^0, a_2^0, \ldots, a_M^0 \right)^T \quad (2.106)$$

with corresponding approximate solution $\hat{\phi}^0$, and obtain an improved solution $\mathbf{a}^1$ by solving the linear equation

$$\mathbf{K}(\mathbf{a}^0)\mathbf{a}^1 = \mathbf{f}^0 \quad (2.107)$$

where

$$K_{lm}(\mathbf{a}^0) = \int_{\Omega}\left(\frac{\partial W_l}{\partial x} k(\hat{\phi}^0) \frac{\partial N_m}{\partial x} + \frac{\partial W_l}{\partial y} k(\hat{\phi}^0) \frac{\partial N_m}{\partial y} \right) dx\, dy \quad (2.108)$$

and

$$f_l^0 = \int_{\Omega} W_l Q\, dx\, dy - \int_{\Gamma_q} W_l \bar{q}\, d\Gamma - \int_{\Omega}\left(\frac{\partial W_l}{\partial x} k(\hat{\phi}^0) \frac{\partial \psi}{\partial x} + \frac{\partial W_l}{\partial y} k(\hat{\phi}^0) \frac{\partial \psi}{\partial y} \right) dx\, dy$$

$$(2.109)$$

The general iteration scheme

$$\mathbf{K}(\mathbf{a}^{n-1})\mathbf{a}^n = \mathbf{f}^{n-1} \quad (2.110)$$

is then repeated until covergence, to within a suitable tolerance, is obtained.

Example 2.11

Consider the problem of one-dimensional heat conduction with generation in a region with temperature-dependent thermal conductivity. Such a problem, governed by the equation

$$\frac{d}{dx}\left(k \frac{d\phi}{dx} \right) = -10x$$

and with $\phi = 0$ at $x = 0$ and at $x = 1$, and $k = 1 + 0.1\phi$ was solved by the finite difference method in Example 1.4. A solution of this problem by the weighted residual method can be obtained by following the steps outlined above. For example, with the trial function set

$$N_m = x^m(1 - x), \qquad m = 1, 2, \ldots$$

the expansion

$$\hat{\phi} = \sum_{m=1}^{M} a_m x^m(1 - x)$$

automatically satisfies the boundary conditions at $x = 0$ and at $x = 1$. The weighted residual statement is

$$\int_0^1 \left\{ \frac{d}{dx}\left(k\frac{d\hat{\phi}}{dx} \right) + 10x \right\} W_l \, dx = 0, \qquad l = 1, 2, \ldots, M$$

and, with a two-term approximation and the point collocation method, this produces an equation of the form of Eq. (2.101), where

$$\mathbf{K}(\mathbf{a}) = \begin{bmatrix} -\dfrac{d}{dx}\left(k(\hat{\phi})\dfrac{dN_1}{dx} \right)\Big|_{x=x_1} & -\dfrac{d}{dx}\left(k(\hat{\phi})\dfrac{dN_2}{dx} \right)\Big|_{x=x_1} \\[4mm] -\dfrac{d}{dx}\left(k(\hat{\phi})\dfrac{dN_1}{dx} \right)\Big|_{x=x_2} & -\dfrac{d}{dx}\left(k(\hat{\phi})\dfrac{dN_2}{dx} \right)\Big|_{x=x_2} \end{bmatrix}$$

$$\mathbf{f} = \begin{bmatrix} 10x_1 \\ 10x_2 \end{bmatrix}$$

The collocation points $x_1 = \frac{1}{3}$ and $x_2 = \frac{2}{3}$ are chosen. Starting from an initial guess $a_1 = a_2 = 0$ (i.e., $\mathbf{a}^0 = \mathbf{0}$) gives

$$\mathbf{K}(\mathbf{a}^0) = \begin{bmatrix} 2 & 0 \\ 2 & 2 \end{bmatrix}$$

and the solution is

$$\mathbf{a}^1 = \begin{bmatrix} 1.66667 \\ 1.66667 \end{bmatrix}$$

These values are used to construct the matrix

$$\mathbf{K}^1 = \begin{bmatrix} 2.06061 & -0.037037 \\ 1.10494 & 2.12346 \end{bmatrix}$$

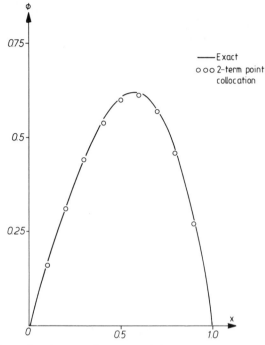

FIGURE 2.13.   Comparison between the exact and the two-term point collocation solutions to Example 2.11.

and the solution now is

$$\mathbf{a}^2 = \begin{bmatrix} 1.64477 \\ 1.50911 \end{bmatrix}$$

Repeating the above process gives, after two further cycles, the converged solution to three decimal places as

$$\mathbf{a} = \begin{bmatrix} 1.644 \\ 1.517 \end{bmatrix}$$

The behavior of the two-term approximation obtained in this way is compared with the exact solution to the problem in Fig. 2.13.

EXERCISES

2.22.   Repeat the calculation of Example 2.11 using the Galerkin method and compare the accuracy achieved with that of the point collocation method and that produced by the finite difference method in Example 1.4.

2.23.   Return to Exercise 1.11 and obtain the solution by the point collocation and Galerkin methods.

2.24.   Produce a solution to Exercise 1.12 by using the Galerkin method.

2.25.   Repeat Exercise 1.13 using the point collocation and Galerkin methods.

2.9.   CONCLUDING REMARKS

We have introduced in this chapter a trial function approximation process for the solution of ordinary and partial differential equations, which is an alternative to the finite difference method discussed in the first chapter. The usefulness of the method is apparent, as for some problems it has been shown to yield a solution that is more accurate than that given by the use of the finite difference method with the same number of unknowns. Unfortunately, major difficulties still remain when attempting to use these weighted residual trial function methods. For two- or three-dimensional regions the finite difference method appears to be the more versatile process, as the trial functions used here limit us to rectangles, parallelepipeds, or other simple shapes if we desire to prescribe the boundary conditions exactly. In addition, the matrix **K** of the algebraic equation system produced by the method [e.g., Eq. (2.55)] can sometimes become numerically ill-conditioned as the number of terms adopted in the approximation increases. This can be seen in Example 2.6, where the matrix **K** is similar in form to the well-known Hilbert matrix. The problem is associated with the choice of the trial function set, and it can normally be removed by using trial functions which are more strongly orthogonal.[8] As a result, in Example 2.6 if polynomial trial functions are required, then a better choice might be the Legendre polynomials (which are discussed in a different context in Chapter 4) or the Chebyshev polynomials.[9]

   In the next chapter we address ourselves to eliminating these deficiencies in the trial function solution method, and then it will become apparent that these procedures, if properly handled, give the largest range of possibilities.

REFERENCES

[1]R. Courant, *Differential and Integral Calculus*, vol. I, 2nd ed., Blackie and Son, London, 1937.

[2]A full discussion of the various possibilities can be found in B. A. Finlayson, *The Method of Weighted Residuals and Variational Principles*, Academic, New York, 1972.

[3]Proof of this lemma is given in O. C. Zienkiewicz, *The Finite Element Method*, 3rd ed., McGraw-Hill, New York, 1977.

[4]Such trial function procedures were originally introduced by Trefftz in 1926 (see, for example, S. A. Mikhlin, *Variational Methods in Mathematical Physics*, Macmillan, New York 1964) and are often associated with his name.

[5]A full discussion of such procedures can be found in O. C. Zienkiewicz, D. W. Kelly, and P. Bettess, Marriage à la mode—The best of both worlds (finite elements and boundary integrals), in

Energy Methods in Finite Element Analysis, edited by R. Glowinski, E. Y. Rodin, and O. C. Zienkiewicz, Wiley-Interscience, New York, 1979.

[6]S. Timoshenko and J. N. Goodier, *Theory of Elasticity*, 2nd ed., McGraw-Hill, New York, 1951.

[7]This principle is discussed in T. H. Richards, *Energy Methods in Stress Analysis*, Ellis-Horwood, Chichester, 1977.

[8]G. Strang and G. J. Fix, *An Analysis of the Finite Element Method*, Prentice-Hall, Englewood Cliffs, N.J., 1973.

[9]L. Fox and I. B. Parker, *Chebyshev Polynomials in Numerical Analysis*, Oxford University Press, New York, 1968.

SUGGESTED FURTHER READING

C. A. Brebbia and S. Walker, *Boundary Element Techniques in Engineering*, Newnes-Butterworths, London, 1980.

L. Collatz, *The Numerical Treatment of Differential Equations*, Springer-Verlag, Berlin, 1960.

S. H. Crandall, *Engineering Analysis*, McGraw-Hill, New York, 1956.

_____

CHAPTER THREE

Piecewise Defined Trial Functions and the Finite Element Method

3.1. INTRODUCTION — THE FINITE ELEMENT CONCEPT

In the approximation methods of the previous chapter we assumed implicitly that the trial functions N_m of the expansion

$$\phi \simeq \hat{\phi} = \psi + \sum_{m=1}^{M} a_m N_m \tag{3.1}$$

were defined by a single expression valid throughout the whole domain Ω and that the integrals of the approximating equations, such as Eqs. (2.6) and (2.41), were evaluated in one operation over this domain.

An alternative approach is to divide the region Ω into a number of nonoverlapping *subdomains* or *elements* Ω^e and then to construct the approximation $\hat{\phi}$ in a *piecewise* manner over each subdomain. The trial functions used in the approximation process can then also be defined in a piecewise manner by using different expressions in the various subdomains Ω^e from which the total domain is developed. In such a case, the definite integrals occurring in the approximating equations can be obtained simply by summing the contributions from each subdomain or element as

$$\int_{\Omega} W_I R_{\Omega} \, d\Omega = \sum_{e=1}^{E} \int_{\Omega^e} W_I R_{\Omega} \, d\Omega \tag{3.2a}$$

$$\int_{\Gamma} \overline{W}_I R_{\Gamma} \, d\Gamma = \sum_{e=1}^{E} \int_{\Gamma^e} \overline{W}_I R_{\Gamma} \, d\Gamma \tag{3.2b}$$

provided that $\sum_{e=1}^{E} \Omega^e = \Omega$, $\sum_{e=1}^{E} \Gamma^e = \Gamma$. Here E denotes the total number of subdivisions of the region and Γ^e denotes that portion of the boundary of Ω^e which lies on Γ. Summations involving Γ^e are therefore taken only over those elements Ω^e which lie immediately adjacent to the boundary.

If the subdomains are of a relatively simple shape and if the definition of the trial functions over these subdomains can be made in a repeatable manner, it is possible to deal in this fashion with assembled regions of complex shapes quite readily. It is here that the essential idea of the *finite element* process lies, and indeed the reader will note that the processes of the previous chapter are then but a special case of the finite element method in which a single element only is used.

The piecewise definition of the trial or shape functions means that discontinuities in the approximating function or in its derivatives will occur. Some degree of such discontinuity is permissible, and it will be shown how this governs the choice of formulation used.

If the trial functions are to be defined in a piecewise manner, it is advantageous to assign to them a narrow "base" and make their value zero everywhere except in the element in question and in the subdomains immediately adjacent to this element. This, as we shall see later, will give banded matrices in the final approximation equations, resulting in yet another advantage of the finite element process.

3.2. SOME TYPICAL LOCALLY DEFINED NARROW-BASE SHAPE FUNCTIONS

To illustrate the use of the finite element method, we consider how an approximation to an arbitrary function $\phi(x)$ might be constructed over the one-dimensional domain $\Omega = [0, L_x]$. The subdivision of Ω into $E(= M_n - 1)$ nonoverlapping subregions is simply accomplished by choosing a suitable set of points $\{x_I; \, I = 1, 2, \ldots, M_n\}$ in Ω, with $x_1 = 0$ and $x_{M_n} = L_x$, and defining the element Ω^e to be the interval $x_e \leqslant x \leqslant x_{e+1}$.

In Fig. 3.1a it is shown how the point collocation method can then be used to approximate a given function $\phi(x)$ by means of a function $\hat{\phi}(x)$ which takes a constant value on each element. The resulting approximation is not continu-

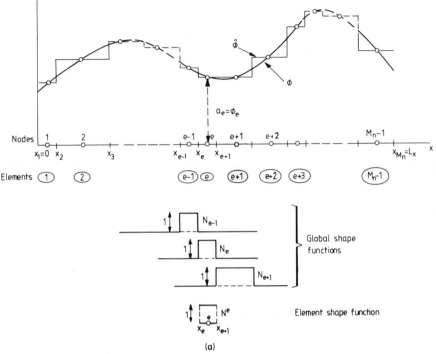

FIGURE 3.1. Approximating a given function in one dimension using point collocation and (a) piecewise constant elements and (b) piecewise linear elements.

ous and changes discontinuously at the element connection points $\{x_l; \; l = 2, 3,\ldots, M_n - 1\}$. The mid points of the elements have been chosen as the collocation points, and these points are termed *nodes*. In finite element processes the nodes and elements are numbered, and here we adopt the obvious numbering system shown in the figure in which node m is the node belonging to element m. The function $\hat{\phi}(x)$ can be written in the standard form of Eq. (3.1) by associating with each node m a piecewise constant, discontinuous, *global shape function* N_m, where N_m is defined to have the value unity on element m and zero on all the other elements. Then we can write

$$\phi \simeq \hat{\phi} = \sum_{m=1}^{M_n - 1} \phi_m N_m \qquad \text{in } \Omega \qquad (3.3a)$$

as $a_m = \phi_m$, where ϕ_m is the value of the function ϕ at node m, and we note that, with the finite element method, the parameters in our approximation now have an easily identifiable meaning. The arbitrary function ψ of Eq. (3.1) has been omitted here, and so this approximation will not, in general, be equal to

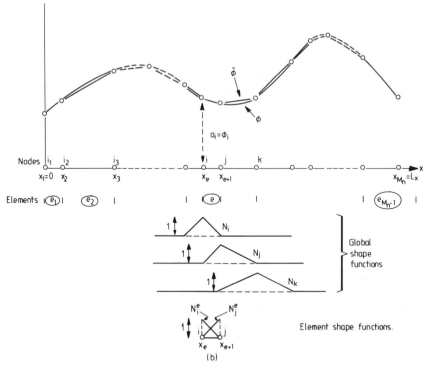

FIGURE 3.1. (*continued*).

the value of the function ϕ at the two ends $x = 0$, $x = L_x$ of the region. However, these end values can be satisfied as closely as required in this representation by suitable reduction in the length of the elements at $x = 0$ and $x = L_x$.

Over any element e the global approximation of Eq. (3.3a) can be expressed in terms of the element nodal value ϕ_e and the *element shape function N^e* as

$$\phi \simeq \hat{\phi} = \phi_e N^e = \phi_e \qquad \text{on element } e \qquad (3.3b)$$

where N^e is just defined for element e and has unit value on this element.

In Fig. 3.1*b* the same element subdivision of the domain $0 \leqslant x \leqslant L_x$ has been adopted, but now an improved approximation has been produced by using an approximating function that varies linearly with x over each element. In this case the points of connection of the elements are now the (numbered) nodes, and the approximation is accomplished by associating a piecewise linear global shape function N_i with each node i.

These global shape functions have the property that N_i is nonzero only on the elements associated with node i, while $N_i = 1$ at node i and is zero at all other nodes. It can be observed that the only global shape functions that are then nonzero on any element are those associated with the nodes of that particular element.

With the nodes taken to be the collocation points, the global approximation can now be written as

$$\phi \simeq \hat{\phi} = \sum_{m=1}^{M_n} \phi_m N_m \qquad \text{in } \Omega \qquad (3.4a)$$

where again ϕ_m denotes the value of ϕ at node m. Inserting the appropriate nodal values at $x = 0$ and $x = L_x$ ensures that this representation automatically assumes the required values at the two ends of the interval, and explicit use of the function ψ is not required. On any element e, with nodes i and j, the approximation can be expressed simply in terms of two linear element shape functions N_i^e, N_j^e and the nodal values ϕ_i, ϕ_j as

$$\phi \simeq \hat{\phi} = \phi_i N_i^e + \phi_j N_j^e \qquad \text{on element } e \qquad (3.4b)$$

The linear variation of the approximation over each element is then obvious as both element shape functions are linear.

It will be apparent from these two examples that the numbering of nodes and elements is a feature of the finite element method. No attention has been paid here to the method of numbering that should be adopted, but we shall see later that the manner in which the nodes and elements are numbered affects the bandwidth of the matrix equation system which is produced in general finite element processes, and this fact can be computationally important.

The two trial function sets used to construct these approximations can be seen to be complete in that, by increasing the number of subdivisions, they are capable of producing any well-behaved function shape as closely as desired. Similar piecewise constant and piecewise linear shape functions may be defined for the purpose of function approximation over two-dimensional domains. This is illustrated in Fig. 3.2 where a triangular subdivision of a two-dimensional region has been constructed. Again the point collocation method is used and, with the piecewise constant approximation, the centroids of the triangles have been taken as the nodes, while the vertices of the triangles are the nodes for the piecewise linear approximation.

The approximation of Eqs. (3.3) and (3.4) have been constructed by using the point collocation method, but it is possible to use similar expressions to construct approximations via the general weighted residual method of the preceding chapter. In Fig. 3.3 the Galerkin method has been used to approximate a given function in one dimension, using both piecewise constant and piecewise linear elements. The given function was approximated by continuous functions in Figs. 2.2–2.4. For this example, the weighting functions are defined by $W_l = N_l$, since the Galerkin procedure is adopted, and the weighted residual approximation equations then become

$$\int_0^1 N_l(\phi - \hat{\phi}) \, dx = 0 \qquad (3.5)$$

With ϕ represented as in Eq. (3.3a) or (3.4a), this leads to a simultaneous

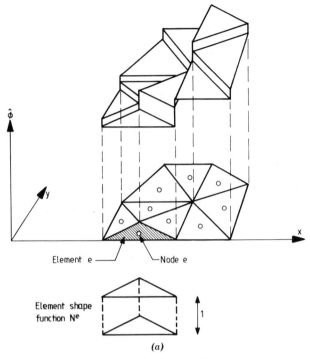

FIGURE 3.2. Approximating a given function in two dimensions using point collocation and (a) piecewise constant triangular elements and (b) piecewise linear triangular elements.

equation system of the usual form

$$\mathbf{K}\boldsymbol{\phi} = \mathbf{f} \tag{3.6}$$

in which

$$K_{lm} = \int_0^1 N_l N_m \, dx$$
$$f_l = \int_0^1 \phi N_l \, dx \tag{3.7a}$$

and

$$\boldsymbol{\phi}^T = (\phi_1, \phi_2, \ldots) \tag{3.7b}$$

The components of the vector $\boldsymbol{\phi}$ will again be the nodal values of the approximation $\hat{\phi}$ and are therefore approximations to the value of the given function ϕ at the nodes. Clearly, as we have already anticipated in Eq. (3.2), the global integrals appearing in Eq. (3.7a) can be evaluated by summing the

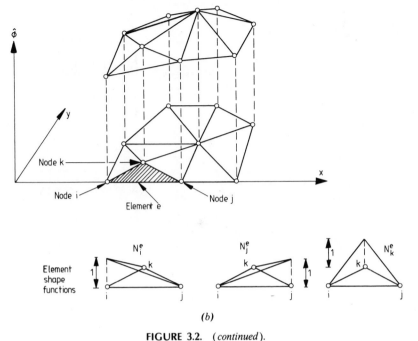

(b)

FIGURE 3.2. (*continued*).

contributions from the individual elements, that is,

$$K_{lm} = \sum_{e=1}^{E} K_{lm}^e$$

$$f_l = \sum_{e=1}^{E} f_l^e$$

(3.8)

where K_{lm}^e and f_l^e are obtained by performing the appropriate integration over the single element e only.

For an approximation by piecewise constant trial functions, the equation system (3.6) is diagonal, since

$$N_l N_m = 0, \qquad l \neq m$$

(3.9)

In Fig. 3.3a we illustrate an approximation constructed in this manner, and with the nodes and elements numbered as shown, the solution is

$$\phi_l = \frac{\int_{x_l}^{x_{l+1}} \phi \, dx}{\int_{x_l}^{x_{l+1}} dx}$$

(3.10)

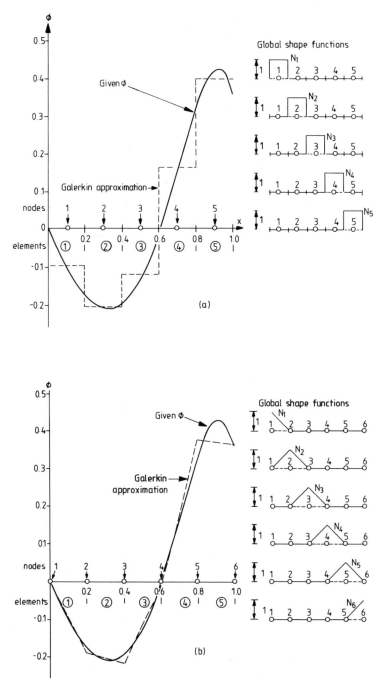

FIGURE 3.3. Approximating a given function in one dimension by the Galerkin method using (*a*) piecewise constant elements and (*b*) piecewise linear elements.

The result of constructing an approximation to the same function by using piecewise linear functions is shown in Fig. 3.3*b*. With the nodes numbered as illustrated, the equation system (3.6) is now tridiagonal and symmetric, just as in many problems considered previously. The precise details of how best to determine the matrix coefficients in this case will not be described here but should be clear when the reader has worked through the examples in Section 3.5.

3.3. APPROXIMATION TO SOLUTIONS OF DIFFERENTIAL EQUATIONS AND CONTINUITY REQUIREMENTS

The process of function approximation defined in Sections 3.1 and 3.2 can be used for the solution of problems governed by differential equations in the manner discussed in Chapter 2.

Writing such a differential equation again as

$$A(\phi) = \mathcal{L}\phi + p = 0 \qquad \text{in } \Omega \tag{3.11}$$

with boundary conditions

$$B(\phi) = \mathfrak{M}\phi + r = 0 \qquad \text{on } \Gamma \tag{3.12}$$

we shall obtain our discrete approximation equations in weighted residual form as

$$\int_\Omega W_l R_\Omega \, d\Omega + \int_\Gamma \overline{W}_l R_\Gamma \, d\Gamma = 0 \tag{3.13}$$

with

$$R_\Omega = \mathcal{L}\hat{\phi} + p$$
$$R_\Gamma = \mathfrak{M}\hat{\phi} + r \tag{3.14}$$

In the process of finite element function approximation of Section 3.2, discontinuous shape functions and continuous shape functions with discontinuous derivatives were successfully used. We shall now pose the question: are such shape functions allowed in the present context, as Eq. (3.13) involves shape function derivatives?

To answer this question we consider the behavior of three types of one-dimensional shape functions N_m near a junction A of two elements, as illustrated in Fig. 3.4. The first function is discontinuous at the point A, while the second shows a discontinuity in the slope dN_m/dx, at the same point, and the third a discontinuity of the second derivative d^2N_m/dx^2. Clearly the

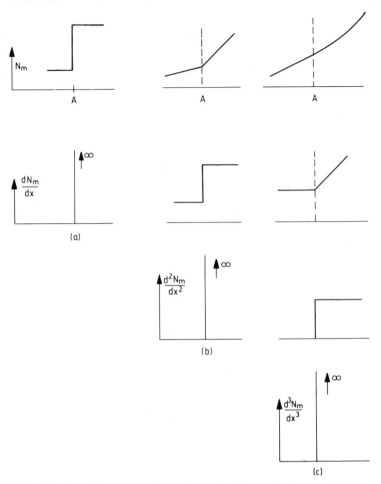

FIGURE 3.4. Behavior of three types of one-dimensional shape functions and their derivatives near the junction A of two elements.

functions illustrated will give infinite values for the first, second, and third derivatives, respectively, at the point where these discontinuities occur.

Now if integrals of the weighted residual type in the form of Eq. (3.13) are to be evaluated, it is desirable to avoid such infinite values. Indeed, at times such infinities may result in indeterminate integrals. To avoid such difficulties it is necessary that, if the integrals in Eq. (3.13) contain derivatives of order s (i.e., the operators $\mathcal{L}$ or $\mathcal{M}$ contain such derivatives), we must ensure that derivatives of the order $s - 1$ are continuous in the trial functions N_m used in the approximation. Introducing here a mathematical notation, we say that we shall require that the shape functions then show C^{s-1} continuity.

For example, if as in Section 3.2 we are simply fitting a prescribed function and no differential operators are involved, we have $s = 0$, and a discontinuity of trial functions of the form given in Fig. 3.4*a* is permissible. Similarly if first derivatives occur in $\mathcal{L}$ or $\mathcal{M}$, that is, $s = 1$, then C^0 continuity, demonstrated by the function type of Fig. 3.4*b*, is necessary. Further, if second derivatives occur, then $s = 2$, and a C^1 continuity, as in Fig. 3.4*c*, is required.

The continuity requirements imposed here on the trial functions are also applicable to the weighting function W_l. Thus in general, for Eq. (3.13) to be valid, infinite values of W_l must be avoided and ordinary discontinuities in this function are allowed. In Section 2.2 we did in fact admit the use of a weighting function W_l which was infinite at a certain point, but whose integral across that point was unity (the Dirac delta function). This form of weighting led to the possibility of point collocation. Clearly we have here violated the rule just introduced, but this exception is permissible if the value of the residual at the point in question is finite. In the general finite element process to be developed here such special weighting forms will seldom be used, and the rules expounded are normally sufficient.

3.4. WEAK FORMULATION AND THE GALERKIN METHOD

In the previous chapter it was seen how a term involving a weighted domain residual of the form

$$\int_{\Omega} W_l \mathcal{L} N_m \, d\Omega \tag{3.15a}$$

could often be replaced by [see Eq. (2.45)]

$$\int_{\Omega} (\mathcal{C} W_l)(\mathcal{D} N_m) \, d\Omega + \text{boundary terms} \tag{3.15b}$$

where the operators $\mathcal{C}$ and $\mathcal{D}$ have a lower order of differentiation than that involved in the original operator $\mathcal{L}$. Such a reformulation is obviously advantageous when locally defined trial functions are used, as lower orders of continuity will now have to be demanded. The reader will immediately, however, observe that now a higher order of continuity is needed in specifying the weighting function W_l. As the operators $\mathcal{C}$ and $\mathcal{D}$ usually involve the same order of differentiation, the identical specification of trial and weighting functions involved in the Galerkin process then has much to commend it. In the majority of practical applications mentioned in the remainder of this book we will find that the Galerkin rule, that is,

$$W_l = N_l \tag{3.16}$$

will be used with piecewise specified trial functions. In a few isolated problems it has been found that better approximations can sometimes be achieved by not applying this rule, although it is usually desirable that W_l and N_l possess the same type of continuity.

3.5.   SOME ONE-DIMENSIONAL PROBLEMS

To make the ideas so far expressed more concrete, we shall consider in this section some one-dimensional examples in which piecewise defined shape functions are used to solve problems governed by ordinary second-order differential equations. In the first two examples the derivatives occurring in the residual mean that the piecewise constant shape functions considered previously are not readily usable, and the piecewise continuous, linear functions of the type illustrated in Fig. 3.1b will be used. It has already been mentioned that a reformulation will be necessary to reduce the order of differentiation of the integrand, and for this reason we shall be using the Galerkin approximation of Eq. (3.16). The third example serves to illustrate how piecewise constant elements can be used in the finite element solution of second-order equations when a mixed formulation of the type considered in Example 2.9 is adopted.

Example 3.1.

In the previous two chapters the equation $d^2\phi/dx^2 - \phi = 0$, $0 \leqslant x \leqslant 1$, has been solved by various methods, subject to the boundary conditions $\phi = 0$ at $x = 0$ and $\phi = 1$ at $x = 1$. We shall now attempt to solve this problem by the finite element method. A set of $M + 1$ nodes is chosen in the region $0 \leqslant x \leqslant 1$, as shown in Fig. 3.5, and with each node m we associate a piecewise linear global shape function N_m. An approximation

$$\phi \simeq \hat{\phi} = \sum_{m=1}^{M+1} \phi_m N_m, \qquad 0 \leqslant x \leqslant 1$$

can then be constructed, where ϕ_m is the value of the approximation at node m. The boundary conditions at $x = 0$ and $x = 1$ can therefore be satisfied immediately by specification of the appropriate nodal values. However, it is more convenient to formulate the weighted residual equations without initially prescribing the boundary values of ϕ and then to insert these known conditions when solving the final equation set. Thus with $\phi_1, \phi_2, \ldots, \phi_M, \phi_{M+1}$ all regarded as being unknown at this stage, the weighted residual form of Eq. (3.13) is then

$$\int_0^1 W_l \left(\frac{d^2\hat{\phi}}{dx^2} - \hat{\phi} \right) dx = 0, \qquad l = 1, 2, \ldots, M + 1$$

The term involving the boundary residual R_Γ is not included, as this will be made identically zero later. In its present form, this statement requires continuity of first derivatives of the trial functions if infinite values are to be avoided. Integration by parts relaxes this requirement on the shape functions and leads to a weak form of the weighted residual statement

$$-\int_0^1 \left(\frac{dW_l}{dx} \frac{d\hat{\phi}}{dx} + W_l\hat{\phi} \right) dx + \left[W_l \frac{d\hat{\phi}}{dx} \right]_0^1 = 0, \qquad l = 1, 2, \ldots, M, M + 1$$

Now it is apparent that only C^0 continuity of $\hat{\phi}$ (and hence of N_m) and W_l is demanded. The piecewise linear shape functions satisfy this requirement, and C^0 continuity of the weighting functions is ensured if the Galerkin method is used. The Galerkin form of this equation set can be expressed immediately as

$$\mathbf{K\phi} = \mathbf{f}$$

where

$$K_{lm} = \int_0^1 \left(\frac{dN_l}{dx} \frac{dN_m}{dx} + N_l N_m \right) dx, \qquad 1 \leqslant l, m \leqslant M + 1$$

$$f_l = \left[N_l \frac{d\hat{\phi}}{dx} \right]_0^1, \qquad 1 \leqslant l \leqslant M + 1$$

We now observe that the contributions to these terms from a typical element e, associated with nodes numbered i and j, as in Fig. 3.5, can be calculated in a general form, and the usefulness of the summation rule of Eq. (3.2) becomes apparent. On such a typical element e we can write, with $\chi = x - x_i$,

$$N_i = N_i^e = \frac{\chi}{h^e}$$

$$N_j = N_j^e = \frac{h^e - \chi}{h^e}$$

where $h^e = x_j - x_i$. The only nonzero global trial functions on element e are N_i

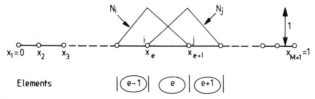

FIGURE 3.5.   General element and node numbering for the problem of Example 3.1.

and N_j, and so $N_l = 0$ on element e if l does not equal i or j, that is, if node l does not belong to element e. Since

$$K_{lm} = \sum_{e=1}^{E} K_{lm}^e$$

we need only evaluate the element contributions to the matrix **K**. These we can write quite generally for the element e as

$$K_{lm}^e = 0, \qquad l, m \neq i, j$$

$$K_{ij}^e = K_{ji}^e = \int_0^{h^e} \left[\frac{dN_i^e}{dx} \frac{dN_j^e}{dx} + N_i^e N_j^e \right] d\chi$$

$$= -\frac{1}{h^e} + \frac{h^e}{6}$$

$$K_{ii}^e = K_{jj}^e = \int_0^{h^e} \left[\left(\frac{dN_i^e}{dx} \right)^2 + (N_i^e)^2 \right] d\chi$$

$$= \frac{1}{h^e} + \frac{h^e}{3}$$

With the components of the element matrix $\mathbf{K}^e$ completely defined in this manner, the matrix **K** then follows by direct summation over all such elements.

This process is normally termed *assembly* and will be illustrated here for the case $E = M = 3$. All the elements will be assumed to be equal in length (i.e., $h^1 = h^2 = h^3 = h = \frac{1}{3}$), and the results can then be compared with the finite difference solution of Example 1.1. If the nodes are numbered sequentially from 1 to 4 and the elements from 1 to 3, as shown in Fig. 3.6, the element

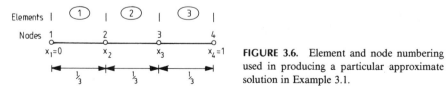

FIGURE 3.6. Element and node numbering used in producing a particular approximate solution in Example 3.1.

matrices become

$$\mathbf{K}^1 = \begin{bmatrix} \frac{1}{h} + \frac{h}{3} & -\frac{1}{h} + \frac{h}{6} & 0 & 0 \\ -\frac{1}{h} + \frac{h}{6} & \frac{1}{h} + \frac{h}{3} & 0 & 0 \\ 0 & 0 & 0 & 0 \\ 0 & 0 & 0 & 0 \end{bmatrix}$$

$$\mathbf{K}^2 = \begin{bmatrix} 0 & 0 & 0 & 0 \\ 0 & \frac{1}{h} + \frac{h}{3} & -\frac{1}{h} + \frac{h}{6} & 0 \\ 0 & -\frac{1}{h} + \frac{h}{6} & \frac{1}{h} + \frac{h}{3} & 0 \\ 0 & 0 & 0 & 0 \end{bmatrix}$$

$$\mathbf{K}^3 = \begin{bmatrix} 0 & 0 & 0 & 0 \\ 0 & 0 & 0 & 0 \\ 0 & 0 & \frac{1}{h} + \frac{h}{3} & -\frac{1}{h} + \frac{h}{6} \\ 0 & 0 & -\frac{1}{h} + \frac{h}{6} & \frac{1}{h} + \frac{h}{3} \end{bmatrix}$$

Making use of the properties of the shape functions, we find that $f_2 = f_3 = 0$, and the assembled form of the equations is

$$\begin{bmatrix} \frac{1}{h} + \frac{h}{3} & -\frac{1}{h} + \frac{h}{6} & 0 & 0 \\ -\frac{1}{h} + \frac{h}{6} & 2\left(\frac{1}{h} + \frac{h}{3}\right) & -\frac{1}{h} + \frac{h}{6} & 0 \\ 0 & -\frac{1}{h} + \frac{h}{6} & 2\left(\frac{1}{h} + \frac{h}{3}\right) & -\frac{1}{h} + \frac{h}{6} \\ 0 & 0 & -\frac{1}{h} + \frac{h}{6} & \frac{1}{h} + \frac{h}{3} \end{bmatrix} \begin{bmatrix} \phi_1 \\ \phi_2 \\ \phi_3 \\ \phi_4 \end{bmatrix} = \begin{bmatrix} -\frac{d\hat{\phi}}{dx}\Big|_{x=0} \\ 0 \\ 0 \\ \frac{d\hat{\phi}}{dx}\Big|_{x=1} \end{bmatrix}$$

Lines 1 and 4 of this equation set, arising as they do from weighting the original equation with trial functions corresponding to nodes where the value of the solution is prescribed, can now be removed from the system and the prescribed values $\phi_1 = 0$ and $\phi_4 = 1$ applied in the remaining equations. We then have

$$2\left(\frac{1}{h} + \frac{h}{3}\right)\phi_2 + \left(-\frac{1}{h} + \frac{h}{6}\right)\phi_3 = 0$$

$$\left(-\frac{1}{h} + \frac{h}{6}\right)\phi_2 + 2\left(\frac{1}{h} + \frac{h}{3}\right)\phi_3 = -\left(-\frac{1}{h} + \frac{h}{6}\right)$$

Solving these equations, with $h = \frac{1}{3}$, gives $\phi_2 = 0.2855$ and $\phi_3 = 0.6098$, whereas the exact solution at the nodes is $\phi_2 = 0.28892$ and $\phi_3 = 0.61024$. This solution is slightly more accurate than that obtained by the finite difference approximation of Example 1.1, although the basic three-point algorithm is very similar.

In this example all the elements could be treated in exactly similar fashion, as the boundary conditions were only applied at the solution stage. In addition, direct substitution of the numerical solution back into the two equations which were deleted from the assembled system will give approximations to the gradient of the solution at $x = 0$ and $x = 1$. This extra information can be of interest in the analysis of physical problems where, as we have already seen, such quantities can sometimes have a particular physical significance. (For example, this would give information on the heat flux across $x = 0$ and $x = 1$ in the problem of one-dimensional heat conduction.) The equations deleted here give

$$\frac{d\hat{\phi}}{dx}\bigg|_{x=0} = -\left(-\frac{1}{h} + \frac{h}{6}\right)\phi_2 = 0.8496$$

$$\frac{d\hat{\phi}}{dx}\bigg|_{x=1} = \left(-\frac{1}{h} + \frac{h}{6}\right)\phi_3 + \left(\frac{1}{h} + \frac{h}{3}\right)\phi_4 = 1.3156$$

and these compare favorably with the exact values of 0.8509 and 1.3130, respectively.

It should be apparent from this example that it is not necessary to calculate each component of the element matrix $\mathbf{K}^e$ as the entrant K^e_{lm} is zero if nodes l and m do not belong to the element. In practice, therefore, only the reduced element matrix $\mathbf{k}^e$, containing the nonzero elements of $\mathbf{K}^e$, is calculated. For this example $\mathbf{k}^e$ will be the 2×2 matrix defined by[*]

$$\mathbf{k}^e = \int_{x_i}^{x_j} \begin{bmatrix} \left(\dfrac{dN_i^e}{dx}\right)^2 + (N_i^e)^2 & \dfrac{dN_i^e}{dx}\dfrac{dN_j^e}{dx} + N_i^e N_j^e \\[3mm] \dfrac{dN_i^e}{dx}\dfrac{dN_j^e}{dx} + N_i^e N_j^e & \left(\dfrac{dN_j^e}{dx}\right)^2 + (N_j^e)^2 \end{bmatrix} dx$$

and inserting the element shape functions and performing the integrations

---

[*]Frequently the entries in the reduced element matrix are simply defined in terms of a compact general expression, for example $\int_{\Omega^e}\left(\dfrac{dN_i^e}{dx}\dfrac{dN_j^e}{dx} + N_i^e N_j^e\right)dx$ in this case. Then it is understood that i and j are only to be assigned the element node numbers in suitable permutation. For clarity we have not adopted this compact form here.

gives

$$
\mathbf{k}^e =
\begin{bmatrix}
\dfrac{1}{h^e} + \dfrac{h^e}{3} & -\dfrac{1}{h^e} + \dfrac{h^e}{6} \\[2ex]
-\dfrac{1}{h^e} + \dfrac{h^e}{6} & \dfrac{1}{h^e} + \dfrac{h^e}{3}
\end{bmatrix}
$$

These matrices can be simply calculated for each element in turn and assembled immediately into the matrix **K**. To ensure that this assembly process is performed correctly, it is helpful to note that the nonzero terms in $\mathbf{K}^e\phi$ are contained in

$$
\mathbf{k}^e\phi^e =
\begin{bmatrix}
\dfrac{1}{h^e} + \dfrac{h^e}{3} & -\dfrac{1}{h^e} + \dfrac{h^e}{6} \\[2ex]
-\dfrac{1}{h^e} + \dfrac{h^e}{6} & \dfrac{1}{h^e} + \dfrac{h^e}{3}
\end{bmatrix}
\begin{bmatrix}
\phi_i \\[1ex] \phi_j
\end{bmatrix}
$$

where $\phi^e = (\phi_i, \phi_j)^T$ is a vector of the nodal values associated with element e. Then, when the contribution from element e is added into **K**, the terms in $\mathbf{k}^e$ must be inserted so as to operate correctly on the nodal variables ϕ_i and ϕ_j as in the above. This process may be illustrated by considering again the solution of this example using the element configuration and numbering system shown in Fig. 3.6. Since the element size $h\ (=\frac{1}{3})$ is equal for all three elements, we have

$$
\mathbf{k}^1 = \mathbf{k}^2 = \mathbf{k}^3 =
\begin{bmatrix}
\frac{28}{9} & -\frac{53}{18} \\[1ex]
-\frac{53}{18} & \frac{28}{9}
\end{bmatrix}
$$

The assembly process then proceeds as follows.

Contribution from Element 1

The nodes associated with this element are nodes 1 and 2, and the nonzero contributions to $K\phi$ from this element are contained in

$$
\mathbf{k}^1\phi^1 =
\begin{bmatrix}
\frac{28}{9} & -\frac{53}{18} \\[1ex]
-\frac{53}{18} & \frac{28}{9}
\end{bmatrix}
\begin{bmatrix}
\phi_1 \\[1ex] \phi_2
\end{bmatrix}
$$

The members of $\mathbf{k}^1$ are then inserted into **K** so as to act correctly on the nodal variables ϕ_1 and ϕ_2, that is,

$$
\mathbf{K}\boldsymbol{\phi} = \begin{bmatrix} \frac{28}{9} & -\frac{53}{18} & 0 & 0 \\ -\frac{53}{18} & \frac{28}{9} & 0 & 0 \\ 0 & 0 & 0 & 0 \\ 0 & 0 & 0 & 0 \end{bmatrix} \begin{bmatrix} \phi_1 \\ \phi_2 \\ \phi_3 \\ \phi_4 \end{bmatrix}
$$

Contribution from Element 2

The nodes associated with this element are nodes 2 and 3, and so the nonzero contributions from this element are contained in

$$
\mathbf{k}^2\boldsymbol{\phi}^2 = \begin{bmatrix} \frac{28}{9} & -\frac{53}{18} \\ -\frac{53}{18} & \frac{28}{9} \end{bmatrix} \begin{bmatrix} \phi_2 \\ \phi_3 \end{bmatrix}
$$

When these contributions are added into $\mathbf{K}$ we find that

$$
\mathbf{K}\boldsymbol{\phi} = \begin{bmatrix} \frac{28}{9} & -\frac{53}{18} & 0 & 0 \\ -\frac{53}{18} & \frac{28}{9} + \frac{28}{9} & -\frac{53}{18} & 0 \\ 0 & -\frac{53}{18} & \frac{28}{9} & 0 \\ 0 & 0 & 0 & 0 \end{bmatrix} \begin{bmatrix} \phi_1 \\ \phi_2 \\ \phi_3 \\ \phi_4 \end{bmatrix}
$$

and it should be observed that K_{22} is formed from the sum of two terms since node 2 belongs to both element 1 and element 2.

Contribution from Element 3

The assembly process is completed by adding in the contribution from this element, noting that the relevant nodes are now node 3 and node 4. Thus

$$
\mathbf{k}^3\boldsymbol{\phi}^3 = \begin{bmatrix} \frac{28}{9} & -\frac{53}{18} \\ -\frac{53}{18} & \frac{28}{9} \end{bmatrix} \begin{bmatrix} \phi_3 \\ \phi_4 \end{bmatrix}
$$

and the addition of these terms into $\mathbf{K}$ gives the final assembled form

$$
\mathbf{K}\boldsymbol{\phi} = \begin{bmatrix} \frac{28}{9} & -\frac{53}{18} & 0 & 0 \\ -\frac{53}{18} & \frac{28}{9} + \frac{28}{9} & -\frac{53}{18} & 0 \\ 0 & -\frac{53}{18} & \frac{28}{9} + \frac{28}{9} & -\frac{53}{18} \\ 0 & 0 & -\frac{53}{18} & \frac{28}{9} \end{bmatrix} \begin{bmatrix} \phi_1 \\ \phi_2 \\ \phi_3 \\ \phi_4 \end{bmatrix}
$$

Again it can be observed that, as node 3 belongs to both element 2 and element 3, K_{33} is obtained by summing a contribution from these two elements.

The matrix $\mathbf{K}$ produced in this manner is identical to the matrix obtained previously, and the solution then proceeds as before.

Example 3.2

In this example we illustrate the application of the finite element method to problems involving derivative boundary conditions. The equation solved will be that just considered in Example 3.1, but the solution will now be sought subject to the boundary conditions $\phi = 0$ at $x = 0$ and $d\phi/dx = 1$ at $x = 1$. This problem was solved by the finite difference method in Examples 1.2 and 1.3. Constructing a finite element approximation $\hat{\phi}$ as in Example 3.1 leads to the weighted residual statement [Eq. (3.13)]

$$\int_0^1 W_I \left[\frac{d^2\hat{\phi}}{dx^2} - \hat{\phi} \right] dx + \left[\overline{W_I} \left(\frac{d\hat{\phi}}{dx} - 1 \right) \right]_{x=1} = 0$$

with the residual at $x = 0$ being omitted, as this will be made identically zero later, as in the previous example. Carrying out integration by parts, and choosing $\overline{W_I}|_{x=1} = -W_I|_{x=1}$ gives

$$\int_0^1 \left[\frac{dW_I}{dx} \frac{d\hat{\phi}}{dx} + W_I\hat{\phi} \right] dx + \left[W_I \frac{d\hat{\phi}}{dx} \right]_{x=0} - [W_I]_{x=1} = 0$$

and the boundary condition to be imposed at $x = 1$ can be seen to be a natural condition for this problem. When the weighting functions are defined by $W_I = N_I$, and with the three equal elements shown in Fig. 3.6, we see that the reduced element matrices $\mathbf{k}^1, \mathbf{k}^2, \mathbf{k}^3$ are just as in Example 3.1, but that the right-hand side vector $\mathbf{f}$ is now changed. Identifying the various terms and performing the assembly process produces the equation system

$$\begin{bmatrix} \frac{1}{h}+\frac{h}{3} & -\frac{1}{h}+\frac{h}{6} & 0 & 0 \\ -\frac{1}{h}+\frac{h}{6} & 2\left(\frac{1}{h}+\frac{h}{3}\right) & -\frac{1}{h}+\frac{h}{6} & 0 \\ 0 & -\frac{1}{h}+\frac{h}{6} & 2\left(\frac{1}{h}+\frac{h}{3}\right) & -\frac{1}{h}+\frac{h}{6} \\ 0 & 0 & -\frac{1}{h}+\frac{h}{6} & \frac{1}{h}+\frac{h}{3} \end{bmatrix} \begin{bmatrix} \phi_1 \\ \phi_2 \\ \phi_3 \\ \phi_4 \end{bmatrix} = \begin{bmatrix} -\frac{d\hat{\phi}}{dx}\Big|_{x=0} \\ 0 \\ 0 \\ 1 \end{bmatrix}$$

The boundary condition at $x = 0$ is now imposed by deleting the first equation from this set and setting $\phi_1 = 0$ in the other equations. The solution of the resulting set, with $h = \frac{1}{3}$, is $\phi_2 = 0.2193$, $\phi_3 = 0.4634$, $\phi_4 = 0.7600$, and these results are considerably closer to the exact solution than those obtained by the central difference algorithm of Example 1.3. Again, the deleted equation can be used to evaluate

$$\left.\frac{d\hat{\phi}}{dx}\right|_{x=0} = -\left(-\frac{1}{h} + \frac{h}{6}\right)\phi_2 = 0.6457$$

which is a good approximation to the exact value of 0.6481.

In this and the previous example the finite element Galerkin method applied to the equation $d^2\phi/dx^2 - \phi = 0$ has been shown to lead to a banded symmetric global matrix $\mathbf{K}$, thus regaining one of the features of the finite difference method which was lost in the general trial function procedure of Chapter 2. In addition, as the number of nodes is increased, the matrix $\mathbf{K}$ now shows none of the ill-conditioning problems associated with the system matrix in Example 2.6.

Example 3.3

It has been shown in Chapter 1 how the second-order heat conduction equation $k(d^2\phi/dx^2) + Q = 0$ can be replaced by an equivalent mixed formulation $k(d\phi/dx) + q = 0;\ dq/dx - Q = 0$. We will illustrate the finite element solution of a mixed formulation of this type by returning to the problem of Example 2.9. If we write

$$q \approx \hat{q} = \sum_{m=1}^{M_q} q_m N_{m,1}$$

$$\phi \approx \hat{\phi} = \sum_{m=1}^{M_\phi} \phi_m N_{m,2}$$

then, as we have seen, a suitable weighted residual statement for this problem is

$$\int_0^1 k\frac{d\hat{\phi}}{dx}N_{l,1}\,dx + \int_0^1 \hat{q}N_{l,1}\,dx = 0, \qquad l = 1,2,\dots,M_q$$

$$\int_0^1 \frac{d\hat{q}}{dx}N_{l,2}\,dx = \int_0^1 QN_{l,2}\,dx, \qquad l = 1,2,\dots,M_\phi$$

Using the linear finite elements shown in Fig. 3.7a to represent both q and ϕ, we have that $M_q = M_\phi = 5$ and $N_{l,1} = N_{l,2} = N_l$. There are then four nodal variables q_i, ϕ_i, q_j, and ϕ_j associated with the general element e of Fig. 3.7b,

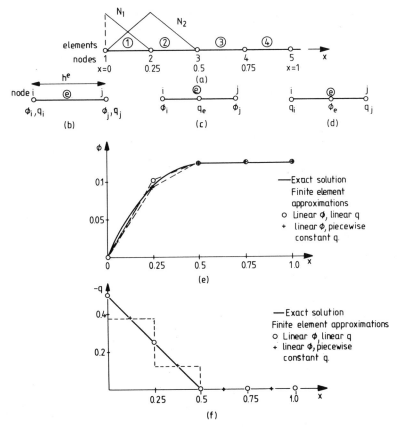

FIGURE 3.7. Solution of Example 3.3 showing (a) the node and element numbering system adopted with (b) piecewise linear ϕ and piecewise linear q. Also shown are the modifications which must be made for (c) piecewise linear ϕ, piecewise constant q, (d) piecewise constant ϕ, piecewise linear q. The graphs show a comparison of the approximations for (e) ϕ and (f)q.

and the nonzero contributions from this element to $\mathbf{K}\boldsymbol{\phi}$ are contained in

$$
\mathbf{k}^e \boldsymbol{\phi}^e = \int_{\Omega^e}
\begin{bmatrix}
(N_i^e)^2 & kN_i^e \dfrac{dN_i^e}{dx} & N_i^e N_j^e & kN_i^e \dfrac{dN_j^e}{dx} \\[3mm]
N_i^e \dfrac{dN_i^e}{dx} & 0 & N_i^e \dfrac{dN_j^e}{dx} & 0 \\[3mm]
N_j^e N_i^e & kN_j^e \dfrac{dN_i^e}{dx} & (N_j^e)^2 & kN_j^e \dfrac{dN_j^e}{dx} \\[3mm]
N_j^e \dfrac{dN_i^e}{dx} & 0 & N_j^e \dfrac{dN_j^e}{dx} & 0
\end{bmatrix}
\begin{bmatrix}
q_i \\[3mm] \phi_i \\[3mm] q_j \\[3mm] \phi_j
\end{bmatrix}
dx
$$

Performing the integration, with $k = 1$ for this problem, results in

$$
\mathbf{k}^e =
\begin{bmatrix}
\dfrac{h^e}{3} & -\dfrac{1}{2} & \dfrac{h^e}{6} & \dfrac{1}{2} \\[2mm]
-\dfrac{1}{2} & 0 & \dfrac{1}{2} & 0 \\[2mm]
\dfrac{h^e}{6} & -\dfrac{1}{2} & \dfrac{h^e}{3} & \dfrac{1}{2} \\[2mm]
-\dfrac{1}{2} & 0 & \dfrac{1}{2} & 0
\end{bmatrix}
$$

and the matrix assembly with $Q = 1$ for $x < \frac{1}{2}$ and $Q = 0$ for $x > \frac{1}{2}$ gives, for the four elements of equal length h,

$$
\begin{bmatrix}
\frac{h}{3} & -\frac{1}{2} & \frac{h}{6} & \frac{1}{2} & 0 & 0 & 0 & 0 & 0 & 0 \\[2mm]
-\frac{1}{2} & 0 & \frac{1}{2} & 0 & 0 & 0 & 0 & 0 & 0 & 0 \\[2mm]
\frac{h}{6} & -\frac{1}{2} & \frac{h}{3}+\frac{h}{3} & \frac{1}{2}-\frac{1}{2} & \frac{h}{6} & \frac{1}{2} & 0 & 0 & 0 & 0 \\[2mm]
-\frac{1}{2} & 0 & \frac{1}{2}-\frac{1}{2} & 0 & \frac{1}{2} & 0 & 0 & 0 & 0 & 0 \\[2mm]
0 & 0 & \frac{h}{6} & -\frac{1}{2} & \frac{h}{3}+\frac{h}{3} & \frac{1}{2}-\frac{1}{2} & \frac{h}{6} & \frac{1}{2} & 0 & 0 \\[2mm]
0 & 0 & -\frac{1}{2} & 0 & \frac{1}{2}-\frac{1}{2} & 0 & \frac{1}{2} & 0 & 0 & 0 \\[2mm]
0 & 0 & 0 & 0 & \frac{h}{6} & -\frac{1}{2} & \frac{h}{3}+\frac{h}{3} & \frac{1}{2}-\frac{1}{2} & \frac{h}{6} & \frac{1}{2} \\[2mm]
0 & 0 & 0 & 0 & -\frac{1}{2} & 0 & \frac{1}{2}-\frac{1}{2} & 0 & \frac{1}{2} & 0 \\[2mm]
0 & 0 & 0 & 0 & 0 & 0 & \frac{h}{6} & -\frac{1}{2} & \frac{h}{3} & \frac{1}{2} \\[2mm]
0 & 0 & 0 & 0 & 0 & 0 & -\frac{1}{2} & 0 & \frac{1}{2} & 0
\end{bmatrix}
\begin{bmatrix}
q_1 \\ \phi_1 \\ q_2 \\ \phi_2 \\ q_3 \\ \phi_3 \\ q_4 \\ \phi_4 \\ q_5 \\ \phi_5
\end{bmatrix}
=
\begin{bmatrix}
0 \\ \frac{h}{2} \\ 0 \\ h \\ 0 \\ \frac{h}{2} \\ 0 \\ 0 \\ 0 \\ 0
\end{bmatrix}
$$

When the boundary conditions $\phi_1 = 0$, $q_5 = 0$ are imposed and the value $h = \frac{1}{4}$ is inserted, this equation set has solution

$$\phi_2 = \tfrac{5}{48}, \qquad \phi_3 = \phi_4 = \phi_5 = \tfrac{1}{8}$$

$$q_1 = -\tfrac{1}{2}, \qquad q_2 = -\tfrac{1}{4}, \qquad q_3 = q_4 = 0$$

which means that q is exactly represented, while $\hat{\phi}$ is a very good approximation to ϕ, as shown in Fig. 3.7e and f.

An alternative approach may be adopted if the second weighted residual statement is integrated by parts (and the sign changed) to give

$$- [\hat{q}N_{l,2}]_0^1 + \int_0^1 \hat{q}\frac{dN_{l,2}}{dx}\,dx = - \int_0^1 QN_{l,2}\,dx, \qquad l = 1,2,\ldots, M_\phi$$

for now it is apparent that a piecewise constant finite element representation for $\hat{q}$ is possible while maintaining a piecewise linear approximation $\hat{\phi}$. Again with the mesh shown in Fig. 3.7a, we now have $M_\phi = 5$ and $M_q = 4$, and there are three variables ϕ_i, ϕ_j, q_e associated with the general element e of Fig. 3.7c. It follows that in this case

$$\mathbf{k}^e \boldsymbol{\phi}^e = \int_{\Omega^e} \begin{bmatrix} 0 & \dfrac{dN_i^e}{dx} & 0 \\ \dfrac{dN_i^e}{dx} & 1 & \dfrac{dN_j^e}{dx} \\ 0 & \dfrac{dN_j^e}{dx} & 0 \end{bmatrix} \begin{bmatrix} \phi_i \\ q_e \\ \phi_j \end{bmatrix} dx$$

and so

$$\mathbf{k}^e = \begin{bmatrix} 0 & -1 & 0 \\ -1 & h^e & 1 \\ 0 & 1 & 0 \end{bmatrix}$$

Assembling the element matrices gives

$$\begin{bmatrix} 0 & -1 & 0 & 0 & 0 & 0 & 0 & 0 & 0 \\ -1 & h & 1 & 0 & 0 & 0 & 0 & 0 & 0 \\ 0 & 1 & 0 & -1 & 0 & 0 & 0 & 0 & 0 \\ 0 & 0 & -1 & h & 1 & 0 & 0 & 0 & 0 \\ 0 & 0 & 0 & 1 & 0 & -1 & 0 & 0 & 0 \\ 0 & 0 & 0 & 0 & -1 & h & 1 & 0 & 0 \\ 0 & 0 & 0 & 0 & 0 & 1 & 0 & -1 & 0 \\ 0 & 0 & 0 & 0 & 0 & 0 & -1 & h & 1 \\ 0 & 0 & 0 & 0 & 0 & 0 & 0 & 1 & 0 \end{bmatrix} \begin{bmatrix} \phi_1 \\ q_1 \\ \phi_2 \\ q_2 \\ \phi_3 \\ q_3 \\ \phi_4 \\ q_4 \\ \phi_5 \end{bmatrix} = \begin{bmatrix} -h/2 - \hat{q}|_{x=0} \\ 0 \\ -h \\ 0 \\ -h/2 \\ 0 \\ 0 \\ 0 \\ 0 \end{bmatrix}$$

with solution

$$\phi_2 = \tfrac{3}{32}, \qquad \phi_3 = \phi_4 = \phi_5 = \tfrac{1}{8}$$

$$q_1 = -\tfrac{3}{8}, \qquad q_2 = -\tfrac{1}{8}, \qquad q_3 = q_4 = 0$$

This method reproduces the exact values of ϕ and q at the nodes, as shown in Fig. 3.7e and f. The first equation can be used to obtain the result $\hat{q}|_{x=0} = -\frac{1}{2}$, which again is exact. In addition, it should be noted that the use of integration by parts and the change of sign has now produced a symmetric equation system.

The final possibility here is to use integration by parts in the first weighted residual statement, and then a piecewise constant representation for $\hat{\phi}$ and a piecewise linear form for $\hat{q}$ can be adopted (Fig. 3.7d). The reader should carry through this analysis and compare the accuracy of the resulting approximations with those just obtained.

EXERCISES

3.1. Use the finite element method to approximate the function $\phi = 1 + \sin(\pi x/2)$ over the range $0 \leqslant x \leqslant 1$. Use both piecewise constant and piecewise linear trial functions and the point collocation and Galerkin methods. Compare your results with those produced in Exercise 2.1.

3.2. Solve the equation $d^2\phi/dx^2 + \phi = 0$, subject to the conditions $\phi = 1$ at $x = 0$, $\phi = 0$ at $x = 1$, by a Galerkin finite element method using four linear elements. Compare the results with those of Exercise 1.1.

3.3. In Example 3.2 use a Taylor series expansion in the equation for node 4 to show that the boundary condition $d\phi/dx = 1$ at $x = 1$ is represented correct to $O(h^3)$. Compare this to the finite difference representation in Example 1.3.

3.4. Return to Exercise 1.2 and obtain the bending moment distribution by a Galerkin finite element method. Compare the answers with the finite difference solution and with the trial function solution of Exercise 2.4.

3.5. Determine the temperature distribution in the problem of Exercise 1.3 by a Galerkin finite element method.

3.6. Repeat Exercise 3.5, but instead of integrating $(U^2\mu/H^4k)(H-y)^2N_i$ exactly over each element, replace $F = (H-y)^2$ by a piecewise linear, finite element approximation $\hat{F}$ satisfying $\hat{F} = F$ at the nodes. Compare the answers with those obtained in Exercise 3.5.

3.7. Repeat the one-dimensional steady heat conduction example of Exercise 2.5 by using a Galerkin finite element method.

3.8. Solve the problem of deflection of a loaded cable on an elastic foundation defined in Exercise 1.4. Use a Galerkin finite element method and compare the results with the exact solution and with those produced by the finite difference method.

3.9. Produce finite element formulations for the nonlinear problems of Exercises 1.11 and 1.12.

3.10. Solve the nonlinear problem of Exercise 1.13 using three linear elements of equal length.

3.6. STANDARD DISCRETE SYSTEM. A PHYSICAL ANALOGUE OF THE EQUATION ASSEMBLY PROCESS

As the weighted residual formulation can be obtained by the summation of appropriate quantities arising from each element [due to the property of integrals given in Eq. (3.2)], the process is analogous to that encountered in the analysis of many discrete physical systems in which the assembled system is formed from the contribution of individual "elements."

In such discrete systems we invariably find the following:

1. A set of quantities a_m can be defined for the system (usually at the nodes where interconnection of elements arises), $m = 1, 2, \ldots, M$.

2. A set of quantities q_l^e can be associated with each node l on an element e, $e = 1, 2, \ldots, E$. Such quantities are often linear functions of a_m and can be written

$$q_l^e = K_{lm}^e a_m - f_l^e \tag{3.17}$$

3. The essential equations for the assembled system are obtained by simple summations of the quantities q_l^e over the elements. Thus the lth equation is, for the linear example, given by

$$\sum_{e=1}^{E} q_l^e = \sum_{e=1}^{E} K_{lm}^e a_m - \sum_{e=1}^{E} f_l^e = 0 \tag{3.18}$$

The equations for the whole system can therefore be written in the usual form

$$\mathbf{Ka} = \mathbf{f} \tag{3.19}$$

with

$$K_{lm} = \sum_{e=1}^{E} K_{lm}^e$$

$$f_l = \sum_{e=1}^{E} f_l^e \tag{3.20}$$

The physical examples of the above process (which we note is identical to that which occurs in the finite element weighted residual procedure where all element matrices are simply added) abound. Probably the best known is the

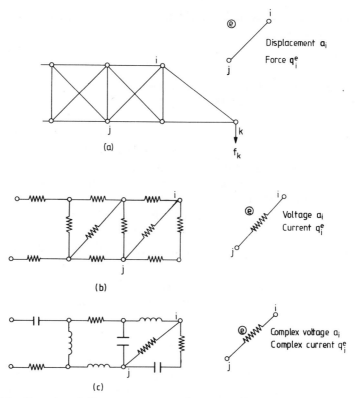

FIGURE 3.8.   Examples of discrete systems. (a) Structure with pin jointed bars. (b) Electric resistance network. (c) Complex electric network.

structural analysis problem where a structure, often of complex shape, is assembled from simple components. The meaning of the various terms is evident (see Fig. 3.8a). The elements of the structure, and the nodes at which the elements are connected, are numbered, and $\mathbf{a}^e = (\mathbf{a}_i, \mathbf{a}_j)^T$ represents the vector of nodal displacements of element e; $\mathbf{q}_i^e$ is the force transmitted from element e to node i. Assuming a linear elastic behavior of the element, it will be possible to write

$$\begin{bmatrix} \mathbf{q}_i^e \\ \mathbf{q}_j^e \end{bmatrix} = \mathbf{k}^e \mathbf{a}^e + \begin{bmatrix} \mathbf{f}_i^e \\ \mathbf{f}_j^e \end{bmatrix} \tag{3.21a}$$

where $\mathbf{f}_i^e$ represents the nodal force at node i required to balance any distributed load acting on the element, and $\mathbf{k}^e$ is an element stiffness matrix. If the structure is also loaded by external forces $\mathbf{r}_l$ applied at node l, then equilibrium at node l requires that

$$\mathbf{r}_l = \sum_{e=1}^{E} \mathbf{q}_l^e \tag{3.21b}$$

where, clearly, only those elements that include node l will contribute nonzero forces to the summation. Combining Eqs. (3.21a) and (3.21b) and assembling all the element equations, we have

$$\mathbf{Ka} = \mathbf{f} + \mathbf{r} \tag{3.22}$$

where $\mathbf{K}$ is the overall system stiffness matrix, $\mathbf{a}$ is the vector of nodal displacements, $\mathbf{f}$ is the vector of nodal forces arising from the distributed loads, and $\mathbf{r}$ is the vector of applied nodal forces.

In Fig. 3.8b we illustrate an analogous case of a resistance network, which may be more familiar to some readers. Here the same procedures apply, and the system variables would be a_l, the voltage applied at node l, and q_l^e, the current leaving element e at node l. Then we have a relationship of the form

$$\begin{bmatrix} q_i^e \\ q_j^e \end{bmatrix} = \mathbf{k}^e \mathbf{a}^e \tag{3.23a}$$

where $\mathbf{k}^e$ is the element conductivity coefficient matrix, and $\mathbf{a}^e$ is the vector of element nodal voltages. The statement

$$r_l = \sum_{e=1}^{E} q_l^e \tag{3.23b}$$

is now simply the Kirchhoff continuity condition of current summation at each node l, where r_l denotes the specified external current input to node l. Combining Eqs. (3.23a) and (3.23b) produces the standard assembled linear equation system.

Finally, in Fig. 3.8c a similar situation is illustrated with more complicated electrical components to which periodic voltages of the form

$$a_l = \tilde{a}_l e^{\sqrt{-1}\,\omega t} \tag{3.24}$$

are applied. Rules and definitions similar to those of the example in Fig. 3.8b again apply, but the quantities a_l, q_l^e are now complex numbers and the element matrices $\mathbf{k}^e$ are the inverse of the impedance matrices.

These three examples illustrate a series of physical problems which have an analogous format to that encountered in the finite element method. Readers should be able to find other such analogies for problems in their particular areas of interest. Such analogies will often be helpful to engineers or physicists pursuing the more abstract formulation of their mathematical problems and may lead to a better understanding of the mathematical operations involved in the solution and even to a better formulation of these. Indeed they may often find that they can use directly a computer program and procedures which have been developed in the context of discrete analysis for the approximate solution of their continuum problems.

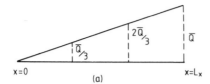

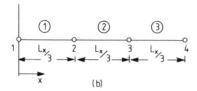

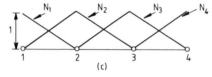

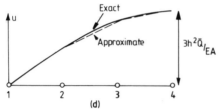

FIGURE 3.9. A finite element solution to the problem of determining the displacement of the axially loaded bar in Example 3.4. (*a*) Distribution of axial body forces. (*b*) Finite element discretization with node and element numbering. (*c*) Piecewise linear shape functions. (*d*) Axial displacement.

We shall conclude this section with a simple one-dimensional example, which we hope illustrates some of the analogy concepts and which compares the solution with that obtained for the same problem by the finite difference method, where the analogy is not so apparent.

Example 3.4

We take the problem of determining the displacement of a bar under the action of axial body forces (Fig. 3.9*a*). Considering the equilibrium of an infinitesimal length dx, we can write the equilibrium condition as

$$\frac{d}{dx}(\sigma A) + Ab = 0$$

where σ is the stress, A the area of cross section, and b the force per unit volume of material in the x direction. (This could for instance be the constant effect of gravity or a variable acceleration due to rotation of the bar.)

The stress is related to the strain ε by $\sigma = E\varepsilon$, where E is Young's modulus for the material in the bar, and thus to the displacement u, which we shall consider as the primary variable, since the strain is defined as $\varepsilon = du/dx$.

Eliminating σ, it is possible to write the equilibrium equation as

$$\frac{d}{dx}\left(k\frac{du}{dx}\right) + Q = 0$$

where $k = AE$ and $Q = Ab$, that is, in the form already encountered for the heat conduction equation of Chapter 1 [Eq. (1.9)].

The boundary conditions of our problem, in which one end of the bar is free and the other end is fixed, are simply, at $x = 0$,

$$u = 0$$

and at $x = L_x$,

$$\sigma = 0 \Rightarrow \frac{du}{dx} = 0$$

For the subdivision and node numbering shown in Fig. 3.9b we take

$$u \simeq \hat{u} = \sum_{m=1}^{4} u_m N_m$$

and the weighted residual statement of Eq. (3.13) becomes

$$\int_0^{L_x}\left[\frac{d}{dx}\left(k\frac{d\hat{u}}{dx}\right) + Q\right]W_l\, dx + \left[\overline{W}_l\frac{d\hat{\phi}}{dx}\right]_{x=L_x} = 0$$

with the condition $u = 0$ at $x = 0$ to be satisfied exactly later.

Integration by parts, with $W_l = N_l$ and $[\overline{W}_l]_{x=L_x} = -[kN_l]_{x=L_x}$, leads to the weak form

$$\int_0^{L_x} k\frac{d\hat{u}}{dx}\frac{dN_l}{dx}\, dx = \int_0^{L_x} N_l Q\, dx - \left[N_l k\frac{d\hat{u}}{dx}\right]_{x=0}$$

and inserting the approximation $\hat{u}$ produces the linear equation system

$$\mathbf{Ku = f}$$

where $\mathbf{u}^T = (u_1, u_2, u_3, u_4)$. The members of the matrices $\mathbf{K}$ and $\mathbf{f}$ can be obtained by summing the contributions from the individual elements.

For a typical element e of length h^e with associated nodes i and j, as in Fig. 3.5,

$$K_{lm}^e = \int_0^{h^e} k\frac{dN_l^e}{dx}\frac{dN_m^e}{dx}\, dx, \qquad k = EA$$

and, as in Example 3.1, we have

$$K^e_{lm} = 0, \qquad l, m \neq i, j$$

$$K^e_{ii} = K^e_{jj} = \frac{EA}{h^e}$$

$$K^e_{ij} = K^e_{ji} = -\frac{EA}{h^e}$$

where we have assumed $k = AE$ to be constant. Similarly,

$$f^e_l = \int_0^{h^e} N^e_l Q \, d\chi$$

and the integral appearing here can be evaluated for any known distribution of $Q = Q(x)$.

Inspection of the above equation in a general form allows us to observe that the analogy here is a physical fact. This follows, as

$$K^e_{ij} u_j = -\frac{EA u_j}{h^e}$$

represents *exactly* the force acting at a node of an element due to a movement u_j at one end, while the other end is held immobile. Indeed, this point could have been our starting point for obtaining the solution. In our particular example (Fig. 3.9c) we shall allow Q to vary in the linear manner shown in Fig. 3.9a. Using the method suggested in Exercise 3.6, we interpolate Q in the same way as u, that is,

$$Q \simeq \hat{Q} = \sum_{m=1}^4 Q_m N_m$$

and this allows us to write

$$f^e_l = \sum_{m=1}^4 S^e_{lm} Q_m$$

$$S^e_{lm} = \int_0^{h^e} N^e_l N^e_m \, d\chi$$

Performing the integration, shows that $S^e_{lm} = 0$ if $l, m \neq i, j$, and

$$S^e_{ii} = S^e_{jj} = \frac{h^e}{3}$$

$$S^e_{ij} = S^e_{ji} = \frac{h^e}{6}$$

The assembled matrix equation becomes, as the reader can verify,

$$\frac{EA}{h}\begin{bmatrix} 1 & -1 & 0 & 0 \\ -1 & 2 & -1 & 0 \\ 0 & -1 & 2 & -1 \\ 0 & 0 & -1 & 1 \end{bmatrix}\begin{bmatrix} u_1 \\ u_2 \\ u_3 \\ u_4 \end{bmatrix} = \begin{bmatrix} -k\,d\hat{u}/dx|_{x=0} + \overline{Q}h/18 \\ \overline{Q}h/3 \\ 2\overline{Q}h/3 \\ 4\overline{Q}h/9 \end{bmatrix}$$

where $h = L_x/3$. Imposition of the boundary condition $u_1 = 0$ in the last three equations produces the solution

$$u_2 = \frac{13h^2\overline{Q}}{9EA}, \qquad u_3 = \frac{23h^2\overline{Q}}{9EA}, \qquad u_4 = \frac{3h^2\overline{Q}}{EA}$$

This can be verified (by direct integration of the original equation) to be the exact answer for the nodal displacements of the problem.* The plot of Fig. 3.9d shows that the results between the nodes are, however, only approximated. The unused equation in the matrix system now gives $AE\,d\hat{u}/dx|_{x=0} = 3h\overline{Q}/2$, which again is exact. Physically this quantity represents the magnitude of the force that must be applied to restrain the displacement at $x = 0$.

If the same problem is to be solved by the finite difference method, the equations can be written, as the reader can readily verify, as

$$\begin{bmatrix} 2 & -1 & 0 \\ -1 & 2 & -1 \\ 0 & -1 & 1 \end{bmatrix}\begin{bmatrix} u_2 \\ u_3 \\ u_4 \end{bmatrix} = \frac{h^2\overline{Q}}{EA}\begin{bmatrix} \frac{1}{3} \\ \frac{2}{3} \\ \frac{1}{2} \end{bmatrix}$$

in which a central difference approximation has been used to represent the boundary condition at $x = 1$. This clearly differs from the finite element equation set and thus does not give the exact solution at the nodes.

The reader will no doubt have observed that the finite difference form could be written down directly without necessity of the apparently tedious summation process. This certainly is true, but on the other hand the finite element computation will allow us, in a very simple manner, to handle problems in which EA varies with x and indeed represents the derivative boundary condition more accurately (Exercise 3.3). The apparent drawback of a more complex derivation presents no difficulty if a computer program is utilised and for simple cases, with constant size elements, the algebra of assembly would not need to be repeated.

---

*It is of interest to observe that in many one-dimensional problems the Galerkin process leads to an exact solution at the nodes, irrespective of the possibly poor approximation elsewhere. This desirable superconvergence is discussed in Ref. 1, and the conditions under which it holds are simply that the trial functions should be capable of identically satisfying the homogeneous part of the differential equation.

EXERCISES

3.11. Obtain, by the finite element method using two equal elements, the steady-state distribution of temperature T in the one-dimensional region $0 \leqslant x \leqslant 1$ subject to the boundary conditions $T = 0$ at $x = 0$ and $T = 1$ at $x = 1$. The thermal conductivity k of the region has the value $k = 1$ for $x < \frac{1}{2}$ and $k = 2$ for $x > \frac{1}{2}$. Compare the solution with the exact temperature distribution. Formulate and solve a finite difference algorithm for this problem.

3.12. Determine the displacement of the axially loaded bar of Example 3.4 when $k = 1 + x/L_x$. Use the finite element method with four linear elements.

3.7. GENERALIZATION OF THE FINITE ELEMENT CONCEPTS FOR TWO- AND THREE-DIMENSIONAL PROBLEMS

3.7.1. General Remarks

One-dimensional examples of the sort we have used in the previous section are of little practical interest as, in many cases, exact solutions are readily obtainable. However, as we have already noted, for two- or three-dimensional problems the situation is altogether different, as an exact solution is then possible only for the simplest domains and boundary conditions, and a numerical solution of practical problems is normally essential.

The choice of finite element shape functions in more than one dimension presents some problems. The use of piecewise linear functions with triangular subdivision of the domain has already been suggested as a possible approach in Fig. 3.2. We now develop this approach and consider also the possibility of using a rectangular element with appropriate shape functions. The extension to three dimensions of these simple elements is also discussed. It is shown that general problems, in which C^0 continuity of the shape functions is demanded, can be analyzed by use of these simple elements. This, of course, limits us to problems in which only first derivatives appear under the integrals in the weak form of approximation given in Eq. (3.15b). Alternative, more advanced, forms for the shape functions and elements are discussed in detail in the next chapter.

3.7.2. The Linear Triangle

The triangle is a particularly useful shape for any two-dimensional analysis, as assemblies of triangles can easily be used to represent accurately regions enclosed by boundaries of quite complex shape. For a typical triangular element e, with nodes numbered anticlockwise as i, j, k, and placed at the vertices of the triangle, as shown in Fig. 3.2b, we shall look for an element shape function $N_i^e(x, y)$ such that N_i^e has the value unity at node i and zero at nodes j and k. In addition the corresponding global shape function N_i must be continuous across element boundaries and only nonzero on elements associated with node i. This continuity requirement can be ensured by assuming a linear form for N_i^e and writing

$$N_i^e = \alpha_i^e + \beta_i^e x + \gamma_i^e y \qquad \text{on element } e \qquad (3.25)$$

The constants in this expression are determined by making $N_i^e(x_i, y_i) = 1$ and $N_i^e(x_j, y_j) = N_i^e(x_k, y_k) = 0$, where the nodal coordinates are specified by the positioning of the nodes in the domain subdivision process. Satisfying these three conditions yields the set of equations

$$\begin{bmatrix} 1 & x_i & y_i \\ 1 & x_j & y_j \\ 1 & x_k & y_k \end{bmatrix} \begin{bmatrix} \alpha_i^e \\ \beta_i^e \\ \gamma_i^e \end{bmatrix} = \begin{bmatrix} 1 \\ 0 \\ 0 \end{bmatrix} \qquad (3.26)$$

with solution

$$\alpha_i^e = \frac{x_j y_k - x_k y_j}{2\Delta^e}$$

$$\beta_i^e = \frac{y_j - y_k}{2\Delta^e} \qquad (3.27a)$$

$$\gamma_i^e = \frac{x_k - x_j}{2\Delta^e}$$

where

$$2\Delta^e = \det \begin{bmatrix} 1 & x_i & y_i \\ 1 & x_j & y_j \\ 1 & x_k & y_k \end{bmatrix} = 2(\text{area of element } e) \qquad (3.27b)$$

With the shape functions specified for a triangle, we can readily compute all the necessary element matrices for any suitably defined problem in which only first derivatives occur in the weak formulation.

In Fig. 3.10 we show a typical domain divided into three-noded triangular elements. The curved left and right side faces have been approximated by straight line segments. This could, for example, represent the profile of a dam which is to be subject to a stress or thermal analysis and a problem of this type leads to the standard equation system (3.6). Without specifying the problem or the boundary conditions, observe that before insertion of the boundary conditions the system matrix $\mathbf{K}$ will have the sparse form given in Fig. 3.11 for the finite element subdivision illustrated. The solid dots represent nonzero terms arising from the connection of elements, recalling that if nodes l and m do not

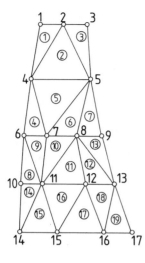

FIGURE 3.10. Triangular finite elements used to represent the profile of a dam.

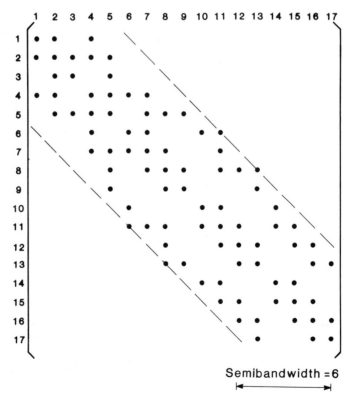

Semibandwidth = 6

FIGURE 3.11. Position of the nonzero entries in the matrix **K** resulting from the assembly of the contributions from the elements shown in Fig. 3.10.

occur in element e then $K_{lm}^e = 0$, as the only trial functions that are nonzero on an element are those associated with the nodes of that element.

We note that, as with the finite difference method, the matrix is now again sparse and banded. This means that the solution of the algebraic equation requires fewer operations than if the matrix were complete, and computer storage can also be reduced.

If a solution procedure is used which utilizes the banded structure[2] of the matrix **K**, then the nodes should be numbered so as to ensure that the bandwidth is a minimum, as this improves the computational efficiency of the solution process. It should be apparent that, in general, if the largest difference in the node numbers in any element is $b - 1$, then the matrix **K** will have a semibandwidth of b. In Fig. 3.10 we see that this largest difference in node numbers is 5 (this occurs in elements $8, 9, 12, 13$), and so the resulting **K** matrix has a semibandwidth of 6.

At this point, assuming that some standard formula exists for the determination of the contributions K_{lm}^e and f_l^e for each element, we want to point out again the following.

1. Such calculations have to be performed once only for each element.
2. The contributions of each element once determined can immediately be entered into the appropriate position in the global matrix and then, if a computer is being employed, deleted from all memory.

Thus the order in which the elements are numbered can also be of importance, and in Fig. 3.12 we show how the contributions from the first four elements enter into the system matrix **K** for the problem of Fig. 3.10.

Examining Fig. 3.12 will show how the completion of a particular row of the assembled matrix occurs as soon as all elements containing the nodal number

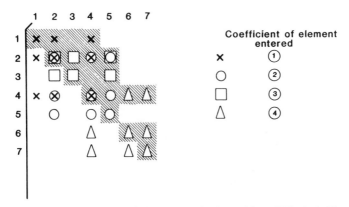

FIGURE 3.12. Process of assembly of elements 1–4 in the problem of Fig. 3.10. The shaded area represents the information that must be stored if the matrix is symmetric.

corresponding to that row have been entered. Thus to complete the equation for node 1 only element 1 is used, and to complete the equations for nodes 2 and 3 all information from elements 2 and 3 has to be entered. This "wave" front of assembly spreads uniformly if elements are entered into the computation in an ordered way—and it permits an elimination of variables to be carried out as soon as assembly of a row is complete. Efficient finite element programs use such *frontal* solution procedures,[3] which minimize storage of information and computer operations, and the order in which the elements (not the nodes) are numbered is then important.

3.7.3.   The Bilinear Rectangle

A simple rectangular element can be produced in a similar fashion. Associating four corner nodes with each element, the nodal shape functions can be derived by taking the product of two linear one-dimensional shape functions, as shown in Fig. 3.13. For instance, with node i of element e we associate the element shape function (relative to coordinate axes with origin at node i)

$$N_i^e = \frac{\left(h_x^e - x\right)}{h_x^e} \frac{\left(h_y^e - y\right)}{h_y^e} \qquad (3.28)$$

The element is termed bilinear because of the presence of the xy term in addition to the linear terms in this expression. The form given for N_i^e in Eq. (3.28) automatically satisfies the usual requirement that N_i^e take the value unity at node i and the value zero at all the other nodes of the element e.

With a similar definition of the global shape function N_i in adjacent elements, it can be seen that C^0 continuity of N_i across element boundaries is assured. The expressions for the element shape functions associated with the other nodes, j, k, l of element e, can be written with equal ease by inspection. More systematic methods for the derivation of rectangular element shape functions are given in Chapter 4.

Rectangular elements of this type are clearly convenient for subdividing regions of square or rectangular shape, but triangular elements are better for representing regions of more general shape in that fewer nodes are required to fit closely complex boundary geometries (Fig. 3.14a and b). Another method

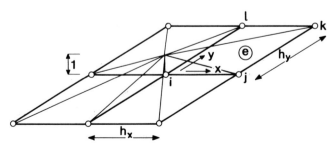

FIGURE 3.13.   Bilinear shape function associated with node i of a typical four-noded rectangular element.

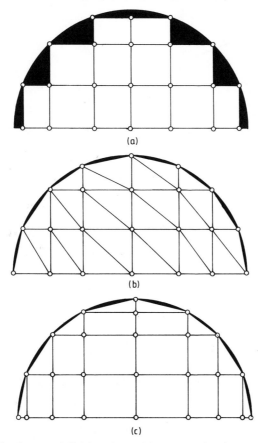

FIGURE 3.14. Finite element subdivision of a semicircular region using (*a*) rectangular elements only, (*b*) triangular elements with the same total number of nodes, (*c*) a combination of triangular and rectangular elements with additional boundary nodes. In each case the dark areas indicate the magnitude of the error made in the representation of the domain.

of dealing with complicated boundary shapes would be to use a mixed finite element mesh consisting of rectangular elements in the interior of the domain and triangular elements near the boundary, as shown in Fig. 3.14c. This problem of accurate representation of boundary curves receives further attention in Chapter 5.

3.7.4. Three-Dimensional Elements of Linear Type

In principle it is simple to extend this discussion to three dimensions and to determine equivalent shape functions which will now be required to exhibit C^0 continuity across interelement faces. For instance, in Fig. 3.15 we illustrate two simple three-dimensional elements for which shape functions can be derived by processes identical to those just described above for two dimensions. Thus for the four-noded tetrahedron of Fig. 3.15*a* we shall associate with node *i* an

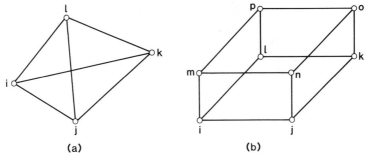

FIGURE 3.15. Three-dimensional elements. (*a*) Four-noded tetrahedron. (*b*) Eight-noded brick.

element shape function

$$N_i^e = \alpha_i^e + \beta_i^e x + \gamma_i^e y + \delta_i^e z \qquad (3.29)$$

where the coefficients are determined, as before, by making

$$N_i^e = 1 \qquad \text{at node } i$$
$$\qquad\qquad\qquad\qquad\qquad\qquad (3.30)$$
$$N_i^e = 0 \qquad \text{at nodes } j, k, l$$

We leave this and the derivation of the shape functions for the eight-noded "brick" type linear elements of Fig. 3.15*b* to the reader, as these problems present no extra difficulty, except for the complications of book keeping. It is evident, however, that in a three-dimensional world the number of nodes required to model any realistic problem will be much larger than that required for two dimensions, and that the costs and effort needed for solution will be much greater. We shall not formulate in detail any three-dimensional situations in this text, since all the principles applicable to two-dimensional analysis can readily be extended by the reader to three dimensions.

3.8. THE FINITE ELEMENT METHOD FOR TWO-DIMENSIONAL HEAT CONDUCTION PROBLEMS

So far in this chapter, the application of the finite element method to two-dimensional problems has only been described in general terms. Now a particular two-dimensional problem is chosen, and the process of determining the coefficients of the corresponding element matrices is described in detail. The problem to be considered is that of solving the now familiar equation for steady-state heat conduction in two dimensions,

$$\frac{\partial}{\partial x}\left(k\frac{\partial \phi}{\partial x}\right) + \frac{\partial}{\partial y}\left(k\frac{\partial \phi}{\partial y}\right) + Q = 0 \qquad \text{in } \Omega \qquad (3.31)$$

with standard boundary conditions

$$\phi = \bar{\phi} \qquad \text{on } \Gamma_\phi$$

$$k\frac{\partial\phi}{\partial n} = -\bar{q} \qquad \text{on } \Gamma_q \qquad (3.32)$$

where $\Gamma_\phi + \Gamma_q = \Gamma$.

Approximating the unknown function ϕ in the usual finite element form as

$$\phi \simeq \hat{\phi} = \sum_{m=1}^{M} \phi_m N_m \qquad (3.33)$$

it can be seen that the boundary condition on Γ_ϕ can be satisfied immediately at the nodes which lie on Γ_ϕ by specification of the appropriate nodal values ϕ_m. As in the one-dimensional examples considered earlier, these nodal values are regarded as unknown at this stage, but are to be exactly specified later. The weak form of the weighted residual statement, using the trial functions themselves as weighting functions, can then be deduced from Eq. (2.54) to be

$$\int_\Omega \left(\frac{\partial N_l}{\partial x} k \frac{\partial\hat{\phi}}{\partial x} + \frac{\partial N_l}{\partial y} k \frac{\partial\hat{\phi}}{\partial y} \right) dx\, dy = \int_\Omega QN_l\, dx\, dy - \int_{\Gamma_q} \bar{q} N_l\, d\Gamma,$$

$$l = 1, 2, \ldots, M \quad (3.34)$$

Inserting the approximation of Eq. (3.33) leads to the standard equation system

$$\mathbf{K}\boldsymbol{\phi} = \mathbf{f} \qquad (3.35)$$

in which the components of the matrices $\mathbf{K}$ and $\mathbf{f}$ are determined by summing the individual element contributions

$$K_{lm}^e = \int_{\Omega^e} \left(\frac{\partial N_l^e}{\partial x} k \frac{\partial N_m^e}{\partial x} + \frac{\partial N_l^e}{\partial y} k \frac{\partial N_m^e}{\partial y} \right) dx\, dy \qquad (3.36a)$$

$$f_l^e = \int_{\Omega^e} QN_l^e\, dx\, dy - \int_{\Gamma_q^e} N_l \bar{q}\, d\Gamma \qquad (3.36b)$$

Here Ω^e is the surface of element e and Γ_q^e is that portion of the element boundary which lies on, or approximates to, a portion of Γ_q. The integral over Γ_q^e in Eq. (3.36b) will thus only appear for elements that are adjacent to Γ_q. This finite element formulation is completely general, and to determine the element matrices for either triangular or rectangular elements we need simply insert the appropriate shape functions.

3.8.1. Triangular Elements

The nodal coordinates are specified by subdividing the domain Ω into a number of three-noded triangular elements. As we have seen in the earlier one-dimensional examples of this chapter, the nonzero contribution to $\mathbf{K}\boldsymbol{\phi}$ in Eq. (3.35) from the general element e, with nodes i, j, k as in Fig. 3.2b, can then be obtained by calculating

$$
\mathbf{k}^e\boldsymbol{\phi}^e = \int_{\Omega^e}
\begin{bmatrix}
k\left[\left(\dfrac{\partial N_i^e}{\partial x}\right)^2 + \left(\dfrac{\partial N_i^e}{\partial y}\right)^2\right] & k\left[\dfrac{\partial N_i^e}{\partial x}\dfrac{\partial N_j^e}{\partial x} + \dfrac{\partial N_i^e}{\partial y}\dfrac{\partial N_j^e}{\partial y}\right] & k\left[\dfrac{\partial N_i^e}{\partial x}\dfrac{\partial N_k^e}{\partial x} + \dfrac{\partial N_i^e}{\partial y}\dfrac{\partial N_k^e}{\partial y}\right] \\[2ex]
k\left[\dfrac{\partial N_i^e}{\partial x}\dfrac{\partial N_j^e}{\partial x} + \dfrac{\partial N_i^e}{\partial y}\dfrac{\partial N_j^e}{\partial y}\right] & k\left[\left(\dfrac{\partial N_j^e}{\partial x}\right)^2 + \left(\dfrac{\partial N_j^e}{\partial y}\right)^2\right] & k\left[\dfrac{\partial N_j^e}{\partial x}\dfrac{\partial N_k^e}{\partial x} + \dfrac{\partial N_j^e}{\partial y}\dfrac{\partial N_k^e}{\partial y}\right] \\[2ex]
k\left[\dfrac{\partial N_i^e}{\partial x}\dfrac{\partial N_k^e}{\partial x} + \dfrac{\partial N_i^e}{\partial y}\dfrac{\partial N_k^e}{\partial y}\right] & k\left[\dfrac{\partial N_j^e}{\partial x}\dfrac{\partial N_k^e}{\partial x} + \dfrac{\partial N_j^e}{\partial y}\dfrac{\partial N_k^e}{\partial y}\right] & k\left[\left(\dfrac{\partial N_k^e}{\partial x}\right)^2 + \left(\dfrac{\partial N_k^e}{\partial y}\right)^2\right]
\end{bmatrix}
dx\,dy
\begin{bmatrix} \phi_i \\[1ex] \phi_j \\[1ex] \phi_k \end{bmatrix}
$$

$$(3.37)$$

where

$$
\boldsymbol{\phi}^e = \begin{bmatrix} \phi_i \\ \phi_j \\ \phi_k \end{bmatrix}
\tag{3.38}
$$

is the vector of element nodal values.

The coefficients in the element shape functions of Eq. (3.25) and the element area Δ^e can be determined by use of Eq. (3.27). The shape function derivatives required in Eq. (3.37) are then evaluated directly as

$$
\frac{\partial N_i^e}{\partial x} = \beta_i^e, \qquad \frac{\partial N_i^e}{\partial y} = \gamma_i^e
\tag{3.39}
$$

with similar expressions for the derivatives of the other element shape functions. We shall assume that the element side joining nodes i and j lies on, or approximates to, the actual boundary Γ_q and that the other sides do not lie on this boundary. If k, Q, and $\bar{q}$ are not constant, it is convenient, though not essential, to take them to be constant within an element. Then k and Q can be replaced by average values k^e and Q^e, respectively, on element e, while $\bar{q}$ can be replaced by an average value $\bar{q}^e$ along that side of the element which represents part of the boundary Γ_q.

Then noting that

$$
\int_{\Omega^e} dx\,dy = \Delta^e
\tag{3.40}
$$

the integration in Eq. (3.37) can be performed directly, and it follows that

$$
\mathbf{k}^e\boldsymbol{\phi}^e = k^e\Delta^e
\begin{bmatrix}
(\beta_i^e)^2 + (\gamma_i^e)^2 & \beta_i^e\beta_j^e + \gamma_i^e\gamma_j^e & \beta_i^e\beta_k^e + \gamma_i^e\gamma_k^e \\[1ex]
\beta_i^e\beta_j^e + \gamma_i^e\gamma_j^e & (\beta_j^e)^2 + (\gamma_j^e)^2 & \beta_j^e\beta_k^e + \gamma_j^e\gamma_k^e \\[1ex]
\beta_i^e\beta_k^e + \gamma_i^e\gamma_k^e & \beta_j^e\beta_k^e + \gamma_j^e\gamma_k^e & (\beta_k^e)^2 + (\gamma_k^e)^2
\end{bmatrix}
\begin{bmatrix} \phi_i \\[1ex] \phi_j \\[1ex] \phi_k \end{bmatrix}
$$

$$(3.41)$$

The element contribution to the load vector **f** is found by carrying out the integration in Eq. (3.36b), noting that

$$\int_{\Omega^e} N_i^e \, dx \, dy = \alpha_i^e + \beta_i^e \overline{X}^e + \gamma_i^e \overline{Y}^e \qquad (3.42a)$$

where

$$\overline{X}^e = \frac{x_i + x_j + x_k}{3}$$

$$(3.42b)$$

$$\overline{Y}^e = \frac{y_i + y_j + y_k}{3}$$

The result is that

$$f_i^e = \tfrac{1}{3} Q^e \Delta^e - \tfrac{1}{2} \overline{q}^e \underline{\sqrt{(x_i - x_j)^2 + (y_i - y_j)^2}}$$

$$f_j^e = \tfrac{1}{3} Q^e \Delta^e - \tfrac{1}{2} \overline{q}^e \underline{\sqrt{(x_i - x_j)^2 + (y_i - y_j)^2}} \qquad (3.43)$$

$$f_k^e = \tfrac{1}{3} Q^e \Delta^e$$

In these expressions the underlined terms only appear as nodes i and j are assumed to lie on Γ_q, and they would therefore be omitted from the load vector arising from interior elements.

In this manner the structure of the element matrices $\mathbf{k}^e$ and $\mathbf{f}^e$ for the problem of steady-state heat conduction in two dimensions has been defined, and a general computer program could now be written for the analysis of such problems using triangular three-noded elements. The correct assembly procedure for these element matrices will be demonstrated shortly in Example 3.5.

3.8.2. Rectangular Elements

A similar formulation can be established when the domain Ω is subdivided into a number of four-noded rectangular elements. The contribution to $\mathbf{K}\phi$ in Eq. (3.35), from the general element e with nodes i, j, k, l as shown in Fig. 3.13, can now be found by calculating

$$
\mathbf{k}^e \boldsymbol{\phi}^e = \int_0^{h_x^e} \int_0^{h_y^e} k
\begin{bmatrix}
\left[\left(\dfrac{\partial N_i^e}{\partial x} \right)^2 + \left(\dfrac{\partial N_i^e}{\partial y} \right)^2 \right] & \left[\dfrac{\partial N_i^e}{\partial x} \dfrac{\partial N_j^e}{\partial x} + \dfrac{\partial N_i^e}{\partial y} \dfrac{\partial N_j^e}{\partial y} \right] \\[3ex]
\left[\dfrac{\partial N_i^e}{\partial x} \dfrac{\partial N_j^e}{\partial x} + \dfrac{\partial N_i^e}{\partial y} \dfrac{\partial N_j^e}{\partial y} \right] & \left[\left(\dfrac{\partial N_j^e}{\partial x} \right)^2 + \left(\dfrac{\partial N_j^e}{\partial y} \right)^2 \right] \\[3ex]
\left[\dfrac{\partial N_i^e}{\partial x} \dfrac{\partial N_k^e}{\partial x} + \dfrac{\partial N_i^e}{\partial y} \dfrac{\partial N_k^e}{\partial y} \right] & \left[\dfrac{\partial N_j^e}{\partial x} \dfrac{\partial N_k^e}{\partial x} + \dfrac{\partial N_j^e}{\partial y} \dfrac{\partial N_k^e}{\partial y} \right] \\[3ex]
\left[\dfrac{\partial N_i^e}{\partial x} \dfrac{\partial N_l^e}{\partial x} + \dfrac{\partial N_i^e}{\partial y} \dfrac{\partial N_l^e}{\partial y} \right] & \left[\dfrac{\partial N_j^e}{\partial x} \dfrac{\partial N_l^e}{\partial x} + \dfrac{\partial N_j^e}{\partial y} \dfrac{\partial N_l^e}{\partial y} \right]
\end{bmatrix}
$$

(3.44)

$$
\begin{bmatrix}
\left[\dfrac{\partial N_i^e}{\partial x} \dfrac{\partial N_k^e}{\partial x} + \dfrac{\partial N_i^e}{\partial y} \dfrac{\partial N_k^e}{\partial y} \right] & \left[\dfrac{\partial N_i^e}{\partial x} \dfrac{\partial N_l^e}{\partial x} + \dfrac{\partial N_i^e}{\partial y} \dfrac{\partial N_l^e}{\partial y} \right] \\[3ex]
\left[\dfrac{\partial N_j^e}{\partial x} \dfrac{\partial N_k^e}{\partial x} + \dfrac{\partial N_j^e}{\partial y} \dfrac{\partial N_k^e}{\partial y} \right] & \left[\dfrac{\partial N_j^e}{\partial x} \dfrac{\partial N_l^e}{\partial x} + \dfrac{\partial N_j^e}{\partial y} \dfrac{\partial N_l^e}{\partial y} \right] \\[3ex]
\left[\left(\dfrac{\partial N_k^e}{\partial x} \right)^2 + \left(\dfrac{\partial N_k^e}{\partial y} \right)^2 \right] & \left[\dfrac{\partial N_k^e}{\partial x} \dfrac{\partial N_l^e}{\partial x} + \dfrac{\partial N_k^e}{\partial y} \dfrac{\partial N_l^e}{\partial y} \right] \\[3ex]
\left[\dfrac{\partial N_k^e}{\partial x} \dfrac{\partial N_l^e}{\partial x} + \dfrac{\partial N_k^e}{\partial y} \dfrac{\partial N_l^e}{\partial y} \right] & \left[\left(\dfrac{\partial N_l^e}{\partial x} \right)^2 + \left(\dfrac{\partial N_l^e}{\partial y} \right)^2 \right]
\end{bmatrix}
\, dx \, dy
\begin{bmatrix}
\phi_i \\[2ex] \phi_j \\[2ex] \phi_k \\[2ex] \phi_l
\end{bmatrix}
$$

The element shape functions can be determined from the results of the previous section. We have that

$$N_i^e = \frac{\left(h_x^e - x\right)\left(h_y^e - y\right)}{h_x^e h_y^e}$$

$$N_j^e = \frac{x\left(h_y^e - y\right)}{h_x^e h_y^e}$$

$$N_k^e = \frac{xy}{h_x^e h_y^e}$$

$$N_l^e = \frac{\left(h_x^e - x\right)y}{h_x^e h_y^e}$$

(3.45)

and the differentiation of these shape functions produces the expressions required in Eq. (3.44), namely,

$$\frac{\partial N_i^e}{\partial x} = -\frac{\left(h_y^e - y\right)}{h_x^e h_y^e}, \qquad \frac{\partial N_i^e}{\partial y} = -\frac{\left(h_x^e - x\right)}{h_x^e h_y^e}$$

$$\frac{\partial N_j^e}{\partial x} = \frac{h_y^e - y}{h_x^e h_y^e}, \qquad \frac{\partial N_j^e}{\partial y} = -\frac{x}{h_x^e h_y^e}$$

(3.46)

$$\frac{\partial N_k^e}{\partial x} = \frac{y}{h_x^e h_y^e}, \qquad \frac{\partial N_k^e}{\partial y} = \frac{x}{h_x^e h_y^e}$$

$$\frac{\partial N_l^e}{\partial x} = -\frac{y}{h_x^e h_y^e}, \qquad \frac{\partial N_l^e}{\partial y} = \frac{h_x^e - x}{h_x^e h_y^e}$$

For illustration it will be assumed that the element side connecting nodes i and j and the element side connecting nodes i and l both form, or approximate to, a portion of Γ_q. Then once again assuming $\bar{q}$ to be constant and equal to $\bar{q}^e$ along these sides, and taking k and Q to be constant over the element, we have from Eq. (3.44)

$$
\mathbf{k}^e \boldsymbol{\phi}^e = \frac{k^e}{3h_x^e h_y^e}
\begin{bmatrix}
\left[(h_x^e)^2 + (h_y^e)^2\right] & -\left[-\frac{1}{2}(h_x^e)^2 + (h_y^e)^2\right] \\[2ex]
-\left[-\frac{1}{2}(h_x^e)^2 + (h_y^e)^2\right] & \left[(h_x^e)^2 + (h_y^e)^2\right] \\[2ex]
-\frac{1}{2}\left[(h_x^e)^2 + (h_y^e)^2\right] & -\left[(h_x^e)^2 - \frac{1}{2}(h_y^e)^2\right] \\[2ex]
-\left[(h_x^e)^2 - \frac{1}{2}(h_y^e)^2\right] & -\frac{1}{2}\left[(h_x^e)^2 + (h_y^e)^2\right]
\end{bmatrix}
$$

(3.47)

$$
\begin{bmatrix}
-\frac{1}{2}\left[(h_x^e)^2 + (h_y^e)^2\right] & -\left[(h_x^e)^2 - \frac{1}{2}(h_y^e)^2\right] \\[2ex]
-\left[(h_x^e)^2 - \frac{1}{2}(h_y^e)^2\right] & -\frac{1}{2}\left[(h_x^e)^2 + (h_y^e)^2\right] \\[2ex]
\left[(h_x^e)^2 + (h_y^e)^2\right] & -\left[-\frac{1}{2}(h_x^e)^2 + (h_y^e)^2\right] \\[2ex]
-\left[-\frac{1}{2}(h_x^e)^2 + (h_y^e)^2\right] & \left[(h_x^e)^2 + (h_y^e)^2\right]
\end{bmatrix}
\begin{bmatrix}
\phi_i \\[2ex]
\phi_j \\[2ex]
\phi_k \\[2ex]
\phi_l
\end{bmatrix}
$$

It should be noted from Eq. (3.45) that, on each element, the sum of the local element shape functions is unity, that is,

$$
N_i^e + N_j^e + N_k^e + N_l^e = 1 \qquad \text{on element } e \tag{3.48}
$$

It therefore follows that

$$
\frac{\partial N_i^e}{\partial x} + \frac{\partial N_j^e}{\partial x} + \frac{\partial N_k^e}{\partial x} + \frac{\partial N_l^e}{\partial x} = \frac{\partial N_i^e}{\partial y} + \frac{\partial N_j^e}{\partial y} + \frac{\partial N_k^e}{\partial y} + \frac{\partial N_l^e}{\partial y} = 0 \qquad \text{on element } e
$$

(3.49)

and so, by Eq. (3.44), we see that the sum of the entries in each row of the matrix $\mathbf{k}^e$ must be zero. This condition is satisfied by the matrix $\mathbf{k}^e$ just derived in Eq. (3.47), and this serves as an easy check on the correctness of the integration that has been performed.

The components of the element load vector are calculated using Eq. (3.36b). For example, for node i we have

$$f_i^e = \int_0^{h_x^e} \int_0^{h_y^e} \frac{Q^e}{h_x^e h_y^e} \left(h_x^e - x \right) \left(h_y^e - y \right) dx\, dy$$

$$- \int_0^{h_x^e} \frac{\bar{q}^e \left(h_x^e - x \right)}{h_x^e}\, dx - \int_0^{h_y^e} \frac{\bar{q}^e \left(h_y^e - y \right)}{h_y^e}\, dy$$

$$= \tfrac{1}{4} Q^e h_x^e h_y^e - \tfrac{1}{2} \bar{q}^e h_x^e - \tfrac{1}{2} \bar{q}^e h_y^e \tag{3.50}$$

where the underlined terms only arise because the sides joining nodes i and j and nodes i and l have been assumed to form part of Γ_q. The other nonzero components of the element load vector $\mathbf{f}^e$ then follow as

$$f_j^e = \tfrac{1}{4} Q^e h_x^e h_y^e - \tfrac{1}{2} \bar{q}^e h_x^e$$

$$f_k^e = \tfrac{1}{4} Q^e h_x^e h_y^e \tag{3.51}$$

$$f_l^e = \tfrac{1}{4} Q^e h_x^e h_y^e - \tfrac{1}{2} \bar{q}^e h_y^e$$

Although the computation of these element matrices is somewhat tedious if carried out by hand, their evaluation within a computer program is obviously easily accomplished. The relationships derived above can be used for arbitrary size and shape of triangles and for any rectangles. The next stage in the solution process is the assembly of the contributions from these element matrices into the global system matrices $\mathbf{K}$ and $\mathbf{f}$ of Eq. (3.35). This procedure is illustrated in the next example, where we assemble, for both triangular and rectangular subdivisions of the domain, a typical equation for an interior node in a homogeneous region.

Example 3.5

For the problem of two-dimensional heat conduction just considered it is required to demonstrate the process of assembly of the element matrices for a mesh in which the nodes are spaced at a regular interval h in the x and y directions. The initial assembly will be for the mesh of triangles shown in Fig 3.16a, and the process will then be repeated for the rectangular mesh of Fig. 3.16b. The thermal conductivity k of the region will be assumed constant. For the triangular elements, Eq. (3.27) can be used to evaluate the coefficients of

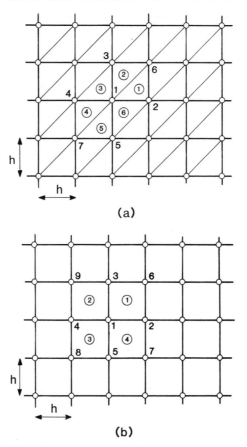

(a)

(b)

FIGURE 3.16. Regular subdivision of the interior of a two-dimensional region using (a) three-noded triangular elements, (b) four-noded rectangular elements.

the element shape functions for element 1 with nodes numbered 1, 2, and 6. With the origin of coordinates taken at node 1, the result is that

$$\alpha_1^1 = 1, \qquad \alpha_2^1 = 0, \qquad \alpha_6^1 = 0$$

$$\beta_1^1 = -1/h, \qquad \beta_2^1 = 1/h, \qquad \beta_6^1 = 0$$

$$\gamma_1^1 = 0, \qquad \gamma_2^1 = -1/h, \qquad \gamma_6^1 = 1/h$$

and $2\Delta^1 = h^2$. Direct substitution of these values into Eq. (3.41) gives

$$\mathbf{k}^1\boldsymbol{\phi}^1 = \frac{k}{2}\begin{bmatrix} 1 & -1 & 0 \\ -1 & 2 & -1 \\ 0 & -1 & 1 \end{bmatrix}\begin{bmatrix} \phi_1 \\ \phi_2 \\ \phi_6 \end{bmatrix}$$

and the nonzero terms in the element load vector are

$$f_1^1 = \frac{Q^1 h^2}{6}$$

$$f_2^1 = \frac{Q^1 h^2}{6}$$

$$f_6^1 = \frac{Q^1 h^2}{6}$$

The contribution from element 1 to the global matrices **K** and **f** has thus been evaluated, and the assembly process begins by inserting these contributions into the equation system, as shown in Fig. 3.17a.

The coefficients in the reduced element matrices and element load vectors for the other elements can be produced in the same manner. In fact, making

$$
\begin{pmatrix}
1 & -1 & & & & & & \\
-1 & 2 & & & -1 & & & \\
& & & & & & & \\
& & & & & & & \\
& -1 & & & 1 & & & \\
& & & & & & & \\
& & & & & & & \\
& & & & & & &
\end{pmatrix}
\begin{pmatrix}
\phi_1 \\ \phi_2 \\ \phi_3 \\ \phi_4 \\ \phi_5 \\ \phi_6 \\ \phi_7 \\ \phi_8
\end{pmatrix}
= \frac{2h^2}{6k}
\begin{pmatrix}
Q^1 \\ Q^1 \\ \\ \\ \\ Q^1 \\ \\
\end{pmatrix}
\qquad (a)
$$

$$
\begin{pmatrix}
2 & -1 & -1 & & & & & \\
-1 & 2 & & & -1 & & & \\
-1 & & 2 & & -1 & & & \\
& & & & & & & \\
& -1 & -1 & & 2 & & & \\
& & & & & & & \\
& & & & & & & \\
& & & & & & &
\end{pmatrix}
\begin{pmatrix}
\phi_1 \\ \phi_2 \\ \phi_3 \\ \phi_4 \\ \phi_5 \\ \phi_6 \\ \phi_7 \\ \phi_8
\end{pmatrix}
= \frac{2h^2}{6k}
\begin{pmatrix}
Q^1 + Q^2 \\ Q^1 \\ Q^2 \\ \\ \\ Q^1 + Q^2 \\ \\
\end{pmatrix}
\qquad (b)
$$

$$
\begin{pmatrix}
4 & -1 & -2 & -1 & & & & \\
-1 & 2 & & & -1 & & & \\
-2 & & 3 & & -1 & & & \\
-1 & & & 1 & & & & \\
& -1 & -1 & & 2 & & & \\
& & & & & & & \\
& & & & & & & \\
& & & & & & &
\end{pmatrix}
\begin{pmatrix}
\phi_1 \\ \phi_2 \\ \phi_3 \\ \phi_4 \\ \phi_5 \\ \phi_6 \\ \phi_7 \\ \phi_8
\end{pmatrix}
= \frac{2h^2}{6k}
\begin{pmatrix}
Q^1 + Q^2 + Q^3 \\ Q^1 \\ Q^2 + Q^3 \\ Q^3 \\ \\ Q^1 + Q^2 \\ \\
\end{pmatrix}
\qquad (c)
$$

FIGURE 3.17. Stages in the assembly process for the problem of two-dimensional heat conduction using the triangular mesh of Fig. 3.16a, with (a)–(f) showing the assembly of the non-zero entries for elements 1–6.

$$\begin{bmatrix} 5 & -1 & -2 & 2 & & & \\ -1 & 2 & & & -1 & & \\ -2 & & 3 & & -1 & & \\ -2 & & & 3 & & -1 & \\ & & & & & & \\ & -1 & -1 & & 2 & & \\ & & & -1 & & 1 & \end{bmatrix} \begin{Bmatrix} \phi_1 \\ \phi_2 \\ \phi_3 \\ \phi_4 \\ \phi_5 \\ \phi_6 \\ \phi_7 \\ \phi_8 \end{Bmatrix} = \frac{2h^2}{6k} \begin{Bmatrix} Q^1+Q^2+Q^3+Q^4 \\ Q^1 \\ Q^2+Q^3 \\ Q^3+Q^4 \\ \\ Q^1+Q^2 \\ Q^4 \end{Bmatrix} \qquad (d)$$

$$\begin{bmatrix} 6 & -1 & -2 & -2 & -1 & & & \\ -1 & 2 & & & & -1 & & \\ -2 & & 3 & & & -1 & & \\ -2 & & & 3 & & & -1 & \\ -1 & & & & 2 & & -1 & \\ & -1 & -1 & & & 2 & & \\ & & & -1 & -1 & & 2 & \end{bmatrix} \begin{Bmatrix} \phi_1 \\ \phi_2 \\ \phi_3 \\ \phi_4 \\ \phi_5 \\ \phi_6 \\ \phi_7 \\ \phi_8 \end{Bmatrix} = \frac{2h^2}{6k} \begin{Bmatrix} Q^1+Q^2+Q^3+Q^4+Q^5 \\ Q^1 \\ Q^2+Q^3 \\ Q^3+Q^4 \\ Q^5 \\ Q^1+Q^2 \\ Q^4+Q^5 \end{Bmatrix} \qquad (e)$$

$$\begin{bmatrix} 8 & -2 & -2 & -2 & -2 & & & \\ -2 & 3 & & & & -1 & & \\ -2 & & 3 & & & -1 & & \\ -2 & & & 3 & & & -1 & \\ -2 & & & & 3 & & -1 & \\ & -1 & -1 & & & 2 & & \\ & & & -1 & -1 & & 2 & \end{bmatrix} \begin{Bmatrix} \phi_1 \\ \phi_2 \\ \phi_3 \\ \phi_4 \\ \phi_5 \\ \phi_6 \\ \phi_7 \\ \phi_8 \end{Bmatrix} = \frac{2h^2}{6k} \begin{Bmatrix} Q^1+Q^2+Q^3+Q^4+Q^5+Q^6 \\ Q^1+Q^6 \\ Q^2+Q^3 \\ Q^3+Q^4 \\ Q^5+Q^6 \\ Q^1+Q^2 \\ Q^4+Q^5 \end{Bmatrix} \qquad (f)$$

FIGURE 3.17. (*continued*).

use of the similar properties of the triangles, it is possible to write immediately for this example

$$k^2\phi^2 = \frac{k}{2}\begin{bmatrix} 1 & -1 & 0 \\ -1 & 2 & -1 \\ 0 & -1 & 1 \end{bmatrix}\begin{bmatrix} \phi_6 \\ \phi_3 \\ \phi_1 \end{bmatrix}$$

$$k^3\phi^3 = \frac{k}{2}\begin{bmatrix} 1 & -1 & 0 \\ -1 & 2 & -1 \\ 0 & -1 & 1 \end{bmatrix}\begin{bmatrix} \phi_4 \\ \phi_1 \\ \phi_3 \end{bmatrix}$$

$$k^4\phi^4 = \frac{k}{2}\begin{bmatrix} 1 & -1 & 0 \\ -1 & 2 & -1 \\ 0 & -1 & 1 \end{bmatrix}\begin{bmatrix} \phi_1 \\ \phi_4 \\ \phi_7 \end{bmatrix}$$

$$\mathbf{k}^5\boldsymbol{\phi}^5 = \frac{k}{2}\begin{bmatrix} 1 & -1 & 0 \\ -1 & 2 & -1 \\ 0 & -1 & 1 \end{bmatrix}\begin{bmatrix} \phi_7 \\ \phi_5 \\ \phi_1 \end{bmatrix}$$

$$\mathbf{k}^6\boldsymbol{\phi}^6 = \frac{k}{2}\begin{bmatrix} 1 & -1 & 0 \\ -1 & 2 & -1 \\ 0 & -1 & 1 \end{bmatrix}\begin{bmatrix} \phi_2 \\ \phi_1 \\ \phi_5 \end{bmatrix}$$

and so on.

The assembly of the contribution from elements 2 to 6 is shown in Fig. 3.17b–f. It may be observed from Fig. 3.16a that node 1 belongs to elements 1–6 only so that the assembly of further elements will not alter the equation for this node. Thus the completed equation for node 1 can be obtained from Fig. 3.17f as

$$4\phi_1 - \phi_2 - \phi_3 - \phi_4 - \phi_5 - \frac{h^2}{6}\left(Q^1 + Q^2 + Q^3 + Q^4 + Q^5 + Q^6\right) = 0$$

For the square elements illustrated in Fig. 3.16b the assembly of the complete equation for node 1 is simpler. Using Eq. (3.47) for element 1, which has nodes numbered 1, 2, 6, and 3, gives

$$\mathbf{k}^1\boldsymbol{\phi}^1 = \frac{k}{3}\begin{bmatrix} 2 & -\frac{1}{2} & -1 & -\frac{1}{2} \\ -\frac{1}{2} & 2 & -\frac{1}{2} & -1 \\ -1 & -\frac{1}{2} & 2 & -\frac{1}{2} \\ -\frac{1}{2} & -1 & -\frac{1}{2} & 2 \end{bmatrix}\begin{bmatrix} \phi_1 \\ \phi_2 \\ \phi_6 \\ \phi_3 \end{bmatrix}$$

and the nonzero components of the element load vector follow from Eq. (3.51) and are given by

$$f_1^1 = \frac{h^2 Q^1}{4} \qquad f_2^1 = \frac{h^2 Q^1}{4}$$

$$f_6^1 = \frac{h^2 Q^1}{4} \qquad f_3^1 = \frac{h^2 Q^1}{4}$$

The reduced element matrices for the remaining elements can be obtained directly from $\mathbf{k}^1$, and we can write

$$\mathbf{k}^2\boldsymbol{\phi}^2 = \frac{k}{3}\begin{bmatrix} 2 & -\frac{1}{2} & -1 & -\frac{1}{2} \\ -\frac{1}{2} & 2 & -\frac{1}{2} & -1 \\ -1 & -\frac{1}{2} & 2 & -\frac{1}{2} \\ -\frac{1}{2} & -1 & -\frac{1}{2} & 2 \end{bmatrix}\begin{bmatrix} \phi_4 \\ \phi_1 \\ \phi_3 \\ \phi_9 \end{bmatrix}$$

$$\mathbf{k}^3\boldsymbol{\phi}^3 = \frac{k}{3}\begin{bmatrix} 2 & -\frac{1}{2} & -1 & -\frac{1}{2} \\ -\frac{1}{2} & 2 & -\frac{1}{2} & -1 \\ -1 & -\frac{1}{2} & 2 & -\frac{1}{2} \\ -\frac{1}{2} & -1 & -\frac{1}{2} & 2 \end{bmatrix}\begin{bmatrix} \phi_8 \\ \phi_5 \\ \phi_1 \\ \phi_4 \end{bmatrix}$$

$$\mathbf{k}^4\boldsymbol{\phi}^4 = \frac{k}{3}\begin{bmatrix} 2 & -\frac{1}{2} & -1 & -\frac{1}{2} \\ -\frac{1}{2} & 2 & -\frac{1}{2} & -1 \\ -1 & -\frac{1}{2} & 2 & -\frac{1}{2} \\ -\frac{1}{2} & -1 & -\frac{1}{2} & 2 \end{bmatrix}\begin{bmatrix} \phi_5 \\ \phi_7 \\ \phi_2 \\ \phi_1 \end{bmatrix}$$

and so on.

FIGURE 3.18. Stages in the assembly process for the problem of two-dimensional heat conduction using the rectangular mesh of Fig. 3.16b with (a)–(d) showing the assembly of the non-zero entries for elements 1–4.

$$\begin{bmatrix}
6 & -\tfrac{1}{2} & -1 & -1 & -\tfrac{1}{2} & -1 & & -1 & -1 \\
-\tfrac{1}{2} & 2 & -1 & & & -\tfrac{1}{2} & & & \\
-1 & -1 & 4 & -1 & & -\tfrac{1}{2} & & & -\tfrac{1}{2} \\
-1 & & -1 & 4 & -1 & & & -\tfrac{1}{2} & -\tfrac{1}{2} \\
-\tfrac{1}{2} & & -1 & & 2 & & & -\tfrac{1}{2} & \\
-1 & -\tfrac{1}{2} & -\tfrac{1}{2} & & & 2 & & & \\
& & & & & & & & \\
-1 & & & -\tfrac{1}{2} & -\tfrac{1}{2} & & & 2 & \\
-1 & & -\tfrac{1}{2} & -\tfrac{1}{2} & & & & & 2
\end{bmatrix}
\begin{Bmatrix}\phi_1\\\phi_2\\\phi_3\\\phi_4\\\phi_5\\\phi_6\\\phi_7\\\phi_8\\\phi_9\end{Bmatrix}
= \frac{3h^2}{4k}
\begin{Bmatrix}Q^1+Q^2+Q^3\\Q^1\\Q^1+Q^2\\Q^2+Q^3\\Q^3\\Q^1\\ \\Q^3\\Q^2\end{Bmatrix} \qquad (c)$$

$$\begin{bmatrix}
8 & -1 & -1 & -1 & -1 & -1 & -1 & -1 & -1 \\
-1 & 4 & -1 & & -1 & -\tfrac{1}{2} & -\tfrac{1}{2} & & \\
-1 & -1 & 4 & -1 & & -\tfrac{1}{2} & & & -\tfrac{1}{2} \\
-1 & & -1 & 4 & -1 & & & -\tfrac{1}{2} & -\tfrac{1}{2} \\
-1 & -1 & & -1 & 4 & & & -\tfrac{1}{2} & -\tfrac{1}{2} \\
-1 & -\tfrac{1}{2} & -\tfrac{1}{2} & & & 2 & & & \\
1 & -\tfrac{1}{2} & & & & & -\tfrac{1}{2} & & 2 \\
-1 & & & -\tfrac{1}{2} & -\tfrac{1}{2} & & & 2 & \\
-1 & & -\tfrac{1}{2} & -\tfrac{1}{2} & & & & & 2
\end{bmatrix}
\begin{Bmatrix}\phi_1\\\phi_2\\\phi_3\\\phi_4\\\phi_5\\\phi_6\\\phi_7\\\phi_8\\\phi_9\end{Bmatrix}
= \frac{3h^2}{4k}
\begin{Bmatrix}Q^1+Q^2+Q^3+Q^4\\Q^1+Q^4\\Q^1+Q^2\\Q^2+Q^3\\Q^3+Q^4\\Q^1\\Q^4\\Q^3\\Q^2\end{Bmatrix} \qquad (d)$$

FIGURE 3.18. (*continued*).

The assembly of the contributions from elements 1–4 into the global matrices **K** and **f** is shown in Fig. 3.18. Again it may be observed that the equation corresponding to node 1 will not be affected by the assembly of the contributions from the other elements as node 1 is only associated with elements 1–4. Thus the complete equation for node 1 can be deduced from Fig. 3.18*d* and is

$$\tfrac{8}{3}\phi_1 - \tfrac{1}{3}\left(\phi_2 + \phi_3 + \phi_4 + \phi_5 + \phi_6 + \phi_7 + \phi_8 + \phi_9\right)$$

$$-\frac{h^2}{4k}\left(Q^1 + Q^2 + Q^3 + Q^4\right) = 0$$

It is of interest to compare these assembled equations for the regular finite element meshes with the finite difference equivalent, which may be deduced from the results of Chapter 1 to be at node 1

$$4\phi_1 - \phi_2 - \phi_3 - \phi_4 - \phi_5 - h^2 Q_1 = 0$$

The reader will observe that the finite difference expression here is similar to the assembled equation at this node for triangular elements. However, we note that in the finite difference expression the value of Q at node 1 is taken in the loading term, while in the finite element form a suitable weighted average of loads contributed by the adjacent elements is used, and if Q is not constant, a different expression results.

The reader can pursue the assembly further and will note that near a boundary where a nonzero flux $\bar{q}$ is specified the assembled equations become of a very different nature from those used in the finite difference approximation. Indeed the finite element formulation here is easier and its accuracy superior. If the two types of assembled equations are used in the problem given in Example 1.5 we find that, with a triangular mesh, results identical to those given previously by the finite difference approximations are obtained (as only zero-flux boundaries exist and Q is constant throughout). However, with the rectangular elements the solution is somewhat different, and the reader should compare the results produced with those obtained by the use of triangular elements (see Exercise 3.16).

EXERCISES

3.13. The square region shown is divided into 18 equal triangular three-noded elements. Obtain, by the finite element method, the steady-state distribution of temperature ϕ when the sides are maintained at the temperatures shown.

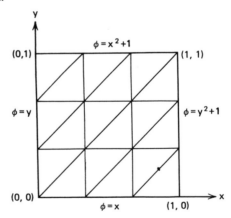

3.14. Repeat Exercise 3.13 using the finite element method and a mesh of nine equal square elements.

3.15. In Exercise 3.13 replace the boundary condition on the line $x = 1$ by the flux condition $\partial\phi/\partial x = 1 - 0.1\phi$. Solve the problem by the finite element method on the triangular mesh, using the weighted residual statement produced in Exercise 2.16. Repeat the calculation using the mesh of nine square elements. As in Exercise 3.3, use Taylor series to investigate the accuracy of representation of the flux condition in both cases.

3.16. Solve the torsion problem of Example 1.5 using the mesh of 12 triangular elements illustrated.

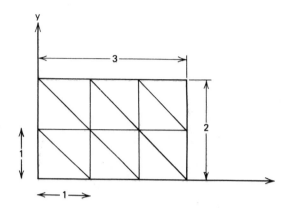

Repeat the calculation using a mesh of six square elements and compare the values obtained for the twisting moment with those produced by other methods in the preceding chapters.

3.17. Solve the problem of torsion of a cylinder of circular cross section by the finite element method using the mesh of four triangular elements illustrated.

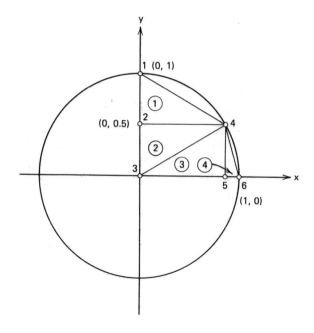

3.18.   Obtain, by the finite element method, the steady-state distribution of temperature ϕ in the sector of the circle shown. Use the triangular mesh and boundary conditions illustrated in the diagram. Repeat the problem by replacing the triangular elements 2 and 3 by a single rectangular element.

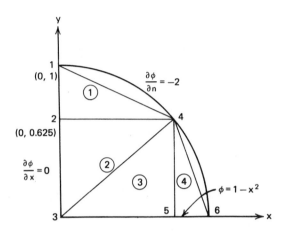

3.19.   Return to Section 3.8.2 and determine the element load vector when Q is not taken as constant over the rectangular element but is interpolated in terms of its nodal values, that is, $Q \approx Q_i N_i^e + Q_j N_j^e + Q_k N_k^e + Q_l N_l^e$ on element e. Use a similar interpolation for $\bar{q}$ along any element side which joins nodes placed on the boundary curve Γ_q.

3.9.   TWO-DIMENSIONAL ELASTIC STRESS ANALYSIS USING TRIANGULAR ELEMENTS

The problem of elastic stress analysis in two dimensions has already been discussed in Section 2.7 as an example of a problem involving two unknowns and requiring the solution of a coupled pair of differential equations. Here we consider the same problem, but we replace the trial functions used previously by piecewise linear trial functions defined locally over three-noded triangular elements. This is a classical problem as it represented the first application of the finite element method in 1956.[4] The formulation to be presented here is still widely used, although in many applications the simple triangular element has been superseded by some of the more complex elements to be dealt with in the next chapter.

Following the notation of Section 2.7.1, the displacement field is defined in terms of its two components u and v in the x and y directions, respectively. These components will be approximated as in Eq. (2.86), but using here the same finite element shape functions to represent both u and v. Omitting the arbitrary function ψ which can, as we have shown, be incorporated into the general finite element formulation, we write

$$\phi \simeq \hat{\phi} = \begin{bmatrix} \hat{u} \\ \hat{v} \end{bmatrix} = \sum_{m=1}^{M} N_m \phi_m \tag{3.52}$$

where

$$\phi_m = \begin{bmatrix} u_m \\ v_m \end{bmatrix} \tag{3.53}$$

is the vector whose components are the approximations to the displacements at node m.

The strains at any point then follow as

$$\varepsilon = \mathcal{L} \sum_{m=1}^{M} N_m \phi_m = \sum_{m=1}^{M} \mathbf{B}_m \phi_m \tag{3.54}$$

where the operator $\mathcal{L}$ has been defined in Eq. (2.80). The components of the matrix $\mathbf{B}_m$ can be identified as

$$\mathbf{B}_m = \mathcal{L} \mathbf{N}_m = \begin{bmatrix} \dfrac{\partial N_m}{\partial x} & 0 \\ 0 & \dfrac{\partial N_m}{\partial y} \\ \dfrac{\partial N_m}{\partial y} & \dfrac{\partial N_m}{\partial x} \end{bmatrix} \tag{3.55}$$

With the weighting functions of Eq. (2.87) defined by

$$W_{l,1} = W_{l,2} = N_l$$
$$\overline{W}_{l,1} = \overline{W}_{l,2} = -N_l|_{\Gamma_o} \tag{3.56}$$

thereby satisfying the conditions of Eq. (2.90), the weighted residual statement of Eq. (2.92) becomes

$$\sum_{m=1}^{M} \left[\int_{\Omega} \mathbf{B}_l^T \mathbf{D} \mathbf{B}_m \, d\Omega \right] \phi_m = \int_{\Omega} N_l \mathbf{X} \, d\Omega + \int_{\Gamma_\sigma} N_l \bar{\mathbf{t}} \, d\Gamma \qquad (3.57)$$

Again this equation set may be expressed in terms of our standard algebraic equation system

$$\mathbf{K}\phi = \mathbf{f} \qquad (3.58)$$

with the matrices $\mathbf{K}$ and $\mathbf{f}$ obtained by summing the contribution from the individual element matrices whose components are defined by

$$\mathbf{K}_{lm}^e = \int_{\Omega^e} \mathbf{B}_l^{eT} \mathbf{D} \mathbf{B}_m^e \, d\Omega$$

$$\mathbf{f}_l^e = \int_{\Omega^e} N_l^e \mathbf{X} \, d\Omega + \int_{\Gamma_\sigma^e} N_l^e \bar{\mathbf{t}} \, d\Gamma \qquad (3.59)$$

In this equation the term involving the integral over Γ_σ^e will only appear if a side of the element forms part of the boundary curve Γ_σ.

This presentation is completely general and applicable to any element form, but now we shall restrict consideration to the special case of the three-noded triangular element with shape functions as defined in Section 3.7.2. For a general element e, with nodes numbered i, j, and k, the nonzero contributions to the vector $\mathbf{K}^e\phi$ can then be obtained by evaluating

$$\mathbf{k}^e\phi^e = \int_{\Omega^e} \begin{bmatrix} \mathbf{B}_i^{eT} \mathbf{D} \mathbf{B}_i^e & \mathbf{B}_i^{eT} \mathbf{D} \mathbf{B}_j^e & \mathbf{B}_i^{eT} \mathbf{D} \mathbf{B}_k^e \\ \mathbf{B}_j^{eT} \mathbf{D} \mathbf{B}_i^e & \mathbf{B}_j^{eT} \mathbf{D} \mathbf{B}_j^e & \mathbf{B}_j^{eT} \mathbf{D} \mathbf{B}_k^e \\ \mathbf{B}_k^{eT} \mathbf{D} \mathbf{B}_i^e & \mathbf{B}_k^{eT} \mathbf{D} \mathbf{B}_j^e & \mathbf{B}_k^{eT} \mathbf{D} \mathbf{B}_k^e \end{bmatrix} \begin{bmatrix} \phi_i \\ \phi_j \\ \phi_k \end{bmatrix} d\Omega \qquad (3.60)$$

Using Eq. (3.55) and recalling that for these elements $\partial N_i^e/\partial x = \beta_i^e$ and $\partial N_i^e/\partial y = \gamma_i^e$, we observe that

$$\mathbf{B}_i^e = \begin{bmatrix} \beta_i^e & 0 \\ 0 & \gamma_i^e \\ \gamma_i^e & \beta_i^e \end{bmatrix} \qquad (3.61)$$

with similar expressions for $\mathbf{B}_j^e$ and $\mathbf{B}_k^e$, and so these matrices are constant over the element. If we now make the assumption that the elastic properties are constant within the element (which appears to be consistent with the approximation generally), then we can perform the integration in Eq. (3.60) and obtain

$$\mathbf{k}^e \boldsymbol{\phi}^e = \Delta^e \begin{bmatrix} \mathbf{B}_i^{eT} \mathbf{D}^e \mathbf{B}_i^e & \mathbf{B}_i^{eT} \mathbf{D}^e \mathbf{B}_j^e & \mathbf{B}_i^{eT} \mathbf{D}^e \mathbf{B}_k^e \\ \mathbf{B}_j^{eT} \mathbf{D}^e \mathbf{B}_i^e & \mathbf{B}_j^{eT} \mathbf{D}^e \mathbf{B}_j^e & \mathbf{B}_j^{eT} \mathbf{D}^e \mathbf{B}_k^e \\ \mathbf{B}_k^{eT} \mathbf{D}^e \mathbf{B}_i^e & \mathbf{B}_k^{eT} \mathbf{D}^e \mathbf{B}_j^e & \mathbf{B}_k^{eT} \mathbf{D}^e \mathbf{B}_k^e \end{bmatrix} \begin{bmatrix} \boldsymbol{\phi}_i \\ \boldsymbol{\phi}_j \\ \boldsymbol{\phi}_k \end{bmatrix} \qquad (3.62)$$

where Δ^e is the area of element e. In a similar manner, by taking the body force $\mathbf{X}$ and the boundary traction $\bar{\mathbf{t}}$ as constants over the element, we can obtain the nonzero components of the element load vector as

$$\mathbf{f}_i^e = \mathbf{X}^e \int_{\Omega^e} N_i^e \, d\Omega + \bar{\mathbf{t}}^e \int_{\Gamma_\sigma^e} N_i^e \, d\Gamma$$

$$\mathbf{f}_j^e = \mathbf{X}^e \int_{\Omega^e} N_j^e \, d\Omega + \bar{\mathbf{t}}^e \int_{\Gamma_\sigma^e} N_j^e \, d\Gamma \qquad (3.63)$$

$$\mathbf{f}_k^e = \mathbf{X}^e \int_{\Omega^e} N_k^e \, d\Omega + \bar{\mathbf{t}}^e \int_{\Gamma_\sigma^e} N_k^e \, d\Gamma$$

Integrals of this form have already been evaluated in the previous section and so, if we make the assumption that only the nodes i and j lie on the boundary Γ_σ, we can write immediately

$$\mathbf{f}_i^e = \tfrac{1}{3}\Delta^e \mathbf{X}^e + \tfrac{1}{2}\bar{\mathbf{t}}^e \sqrt{(x_i - x_j)^2 + (y_i - y_j)^2}$$

$$\mathbf{f}_j^e = \tfrac{1}{3}\Delta^e \mathbf{X}^e + \tfrac{1}{2}\bar{\mathbf{t}}^e \sqrt{(x_i - x_j)^2 + (y_i - y_j)^2} \qquad (3.64)$$

$$\mathbf{f}_k^e = \tfrac{1}{3}\Delta^e \mathbf{X}^e$$

There is little more to be added here in the context of the formulation of two-dimensional elastic problems. The reader can usefully now proceed to derive the element matrices for the four-noded rectangular elements using the results of the preceding section. Before leaving this topic, however, we now include one example in which the notion of using nodal variables of a vector kind in a two-dimensional context is illustrated.

Example 3.6

We shall consider a two-dimensional elastic stress analysis problem in which the region of interest is covered by a regular mesh of triangular elements as shown in Fig. 3.16a. The elastic properties and the body force will be assumed to be constant throughout the region, and it is required to evaluate the reduced element matrix for element 1 with nodes 1, 2, and 6.

We proceed as in Example 3.5 and note that for element 1 we can write

$$\alpha_1^1 = 1, \qquad\qquad \alpha_2^1 = 0, \qquad\qquad \alpha_6^1 = 0$$

$$\beta_1^1 = -1/h, \qquad \beta_2^1 = 1/h, \qquad \beta_6^1 = 0$$

$$\gamma_1^1 = 0, \qquad\qquad \gamma_2^1 = -1/h, \qquad \gamma_6^1 = 1/h$$

$$2\Delta^1 = h^2$$

and it follows immediately that

$$\mathbf{B}_1^1 = \frac{1}{h}\begin{bmatrix} -1 & 0 \\ 0 & 0 \\ 0 & -1 \end{bmatrix}, \qquad \mathbf{B}_2^1 = \frac{1}{h}\begin{bmatrix} 1 & 0 \\ 0 & -1 \\ -1 & 1 \end{bmatrix}, \qquad \mathbf{B}_6^1 = \frac{1}{h}\begin{bmatrix} 0 & 0 \\ 0 & 1 \\ 1 & 0 \end{bmatrix}$$

With constant material properties the matrix $\mathbf{D}$ will be constant throughout the region. Defining this matrix in a completely general way as

$$\mathbf{D} = \begin{bmatrix} D_{11} & D_{12} & D_{13} \\ & D_{22} & D_{23} \\ \text{symmetric} & & D_{33} \end{bmatrix}$$

the reduced element matrix $\mathbf{k}^1$ may then be obtained from Eq. (3.62) as

$$\mathbf{k}^1\boldsymbol{\phi}^1 = \frac{h^2}{2}\begin{bmatrix} \mathbf{B}_1^{1T}\mathbf{D}\mathbf{B}_1^1 & \mathbf{B}_1^{1T}\mathbf{D}\mathbf{B}_2^1 & \mathbf{B}_1^{1T}\mathbf{D}\mathbf{B}_6^1 \\ \mathbf{B}_2^{1T}\mathbf{D}\mathbf{B}_1^1 & \mathbf{B}_2^{1T}\mathbf{D}\mathbf{B}_2^1 & \mathbf{B}_2^{1T}\mathbf{D}\mathbf{B}_6^1 \\ \mathbf{B}_6^{1T}\mathbf{D}\mathbf{B}_1^1 & \mathbf{B}_6^{1T}\mathbf{D}\mathbf{B}_2^1 & \mathbf{B}_6^{1T}\mathbf{D}\mathbf{B}_6^1 \end{bmatrix}\begin{bmatrix} \boldsymbol{\phi}_1 \\ \boldsymbol{\phi}_2 \\ \boldsymbol{\phi}_6 \end{bmatrix}$$

Performing the matrix multiplications it is found that

$$\mathbf{k}^1 = \tfrac{1}{2}\begin{bmatrix}
D_{11} & D_{13} & -D_{11}+D_{13} & D_{12}-D_{13} & -D_{13} & -D_{12} \\
D_{13} & D_{33} & -D_{13}+D_{33} & D_{23}-D_{33} & -D_{33} & -D_{23} \\
-D_{11}+D_{13} & -D_{13}+D_{33} & D_{11}-2D_{13}+D_{33} & D_{13}+D_{23}-D_{12}-D_{33} & D_{13}-D_{33} & D_{12}-D_{23} \\
D_{12}-D_{13} & D_{23}-D_{33} & D_{13}+D_{23}-D_{12}-D_{33} & D_{22}-2D_{23}+D_{33} & -D_{23}+D_{33} & -D_{22}+D_{23} \\
-D_{13} & -D_{33} & D_{13}-D_{33} & -D_{23}+D_{33} & D_{33} & D_{23} \\
-D_{12} & -D_{23} & D_{12}-D_{23} & -D_{22}+D_{23} & D_{23} & D_{22}
\end{bmatrix}$$

The reader can verify that, because of the properties of the shape functions, the sum of the entries in each row of $\mathbf{k}^1$ is identically zero, and this can be used as a simple check on the correctness of the above expression.

EXERCISES

3.20. Solve the fourth-order beam deflection problem of Exercise 2.19 by splitting the governing equation into two second-order equations and using a Galerkin finite element method.

3.21. Formulate the matrix equation which results from applying the Galerkin finite element solution procedure to the problem of determining the deflection of a loaded plate described in Example 2.10.

3.10. ARE FINITE DIFFERENCES A SPECIAL CASE OF THE FINITE ELEMENT METHOD?

By now the reader will have observed that, in several one- and two-dimensional applications, the use of simple finite elements has led on assembly to equations that were either identical or extremely similar to those obtainable by the finite difference approximations of Chapter 1. It is therefore natural to inquire whether one of the methods is a special case of the other. Indeed, having shown in Chapter 2 that the method of weighted residuals allows a fairly large generality, it is perhaps fair to pose the question more specifically and to ask whether the finite difference equations can always be expressed as a special case of the weighted residual process with locally defined shape functions.

In Chapter 1 we noted that the process of finite difference approximation arose from a differential equation of the form

$$A(\phi) = \mathcal{L}\phi + p = 0 \tag{3.65}$$

in which

1. The approximation of each of the derivatives occurring in the operator $\mathcal{L}$ was made by using a local Taylor series expansion in terms of adjacent nodal values.

2. The equation was formulated independently at each of the "nodal" points of the domain.

Clearly, this could be achieved by the use of a collocation process in which the weighting function is the Dirac delta function, that is,

$$W_l = \delta(\mathbf{x} - \mathbf{x}_l) \tag{3.66}$$

and the use of shape functions which represent the derivatives in the same way as the finite difference expressions.

With the above definition it is necessary for the trial functions N_m to be capable of defining all the derivatives present in the operator $\mathfrak{L}$ at the point $\mathbf{x}_l$ where the weighting is taken. With these trial functions, however, discontinuities are allowed at other points of the domain, provided that

$$\int_-^+ W_l \mathfrak{L} N_m \, d\Omega \to 0 \tag{3.67}$$

as $W_l \to 0$, where the limits in this integral simply span the particular discontinuity. This condition is readily achieved even with discontinuous functions.

As an example consider the one-dimensional problem of the equation

$$\frac{d^2\phi}{dx^2} - \phi = 0, \qquad 0 \leqslant x \leqslant 1 \tag{3.68}$$

leading to the weighted form

$$\int_0^1 W_l \left(\frac{d^2\hat{\phi}}{dx^2} - \hat{\phi} \right) dx = 0, \qquad \hat{\phi} = \sum_{m=1}^{M} N_m \phi_m \tag{3.69}$$

Approximating to the function ϕ in the manner shown in Fig. 3.19a by a set of interlacing parabolic arcs is equivalent to using the set of discontinuous shape functions illustrated in Fig. 3.19b.

To prove that the condition of Eq. (3.67) is satisfied, consider W_l to be constant in the vicinity of the discontinuity. Thus putting N_m into the operator we have

$$\int_-^+ W_l \left(\frac{d^2N_m}{dx^2} - N_m \right) dx$$

$$= \lim_{\varepsilon \to 0} \left[W_l \int_-^{x_l - \varepsilon} \left(\frac{d^2N_m}{dx^2} - N_m \right) dx + W_l \int_{x_l + \varepsilon}^+ \left(\frac{d^2N_m}{dx^2} - N_m \right) dx \right]$$

$$= W_l \lim_{\varepsilon \to 0} \left[\frac{dN_m}{dx} \Big|_-^{x_l - \varepsilon} - \int_-^{x_l - \varepsilon} N_m \, dx + \frac{dN_m}{dx} \Big|_{x_l + \varepsilon}^+ - \int_{x_l + \varepsilon}^+ N_m \, dx \right] \tag{3.70}$$

Clearly, as $W_l \to 0$, the condition of Eq. (3.67) is satisfied because the form of the chosen function N_m ensures that the limit of the terms in square brackets is finite.

With W_l given by Eq. (3.66), the weighted form of Eq. (3.69) therefore reduces to

$$\left[\frac{d^2\hat{\phi}}{dx^2} - \hat{\phi} \right]_{x=x_l} = 0 \tag{3.71}$$

where

$$\hat{\phi} = \phi_{l-1} N_{l-1}^e + \phi_l N_l^e + \phi_{l+1} N_{l+1}^e \tag{3.72}$$

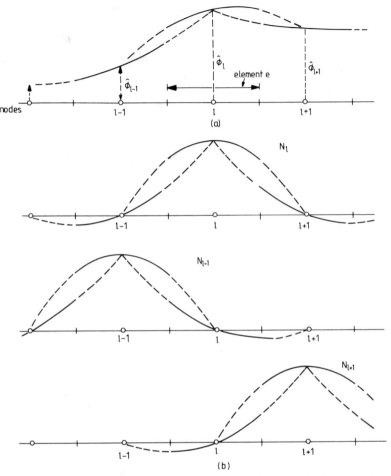

FIGURE 3.19. (*a*) Approximation of a function in one dimension using (*b*) shape functions in the form of discontinuous parabolic arcs.

over the element *e* surrounding the node *l* of Fig. 3.19*a*. With respect to an origin of coordinate x at node *l*, these parabolic shape functions can be expressed as

$$N^e_{l-1} = -\frac{x(h^e - x)}{2(h^e)^2}$$

$$N^e_l = \frac{(h^e - x)(h^e + x)}{(h^e)^2}$$

$$N^e_{l+1} = \frac{x(h^e + x)}{2(h^e)^2}$$

(3.73)

and then

$$\frac{d^2 N_l^e}{dx^2} = -\frac{2}{(h^e)^2}$$

$$\frac{d^2 N_{l+1}^e}{dx^2} = \frac{1}{(h^e)^2} = \frac{d^2 N_{l-1}^e}{dx^2}$$

$$(3.74)$$

The standard finite difference representation of Eq. (3.68)

$$\frac{1}{(h^e)^2}\phi_{l-1} - \frac{2}{(h^e)^2}\phi_l + \frac{1}{(h^e)^2}\phi_{l+1} - \phi_l = 0 \qquad (3.75)$$

then follows by direct substitution of Eq. (3.72) into Eq. (3.71).

The above derivation of finite difference expressions can be extended to two- and three-dimensional problems. In each case we shall observe that standard finite difference expressions can be interpreted as a particular case of the weighted residual method. The derivation of such finite difference expressions by the above process is not however recommended due to its complexity, but the point is now proven—that the finite difference process is but a particular case of the general finite element weighted residual methodology.

We shall not pursue the matter further here except to point out that, in all the examples so far mentioned, the finite element Galerkin formulation always gave results which were at least as accurate as those produced by the very similar finite difference equations. This, combined with the ease of imposing gradient boundary conditions, is undoubtedly one of the reasons for the widespread development and use of the Galerkin finite element approximation. On the other hand it must be mentioned, that the ease of writing down finite difference expressions for homogeneous situations and the avoidance of the assembly process makes the direct application of the latter very popular with many users.

3.11.   CONCLUDING REMARKS

This chapter, together with the previous two chapters, has introduced the students to the ingredients necessary for the understanding of the present-day finite element processes. The formulation of new problems using simple element forms is now in the readers' reach and they will note that a very wide range of practical problems can readily be dealt with. The process can be extended to deal with nonlinear problems following the method outlined in Section 2.8, but the readers are referred elsewhere[5] for full details of the elaborate techniques which have been developed to deal with such problems.

To conclude this section we quote some examples of practical finite element application from which the reader will observe the way in which simple finite elements can be used to represent general two-dimensional regions.

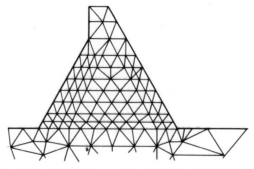

FIGURE 3.20. Triangular elements used in the stress analysis of a gravity dam.

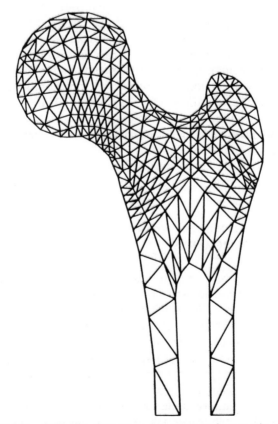

FIGURE 3.21. Triangular elements used to represent the upper human femur.

158

In Fig. 3.20 a gravity dam is presented in which triangular elements are used. Here element sizes are made to vary so as to place small, closely spaced elements near points of expected rapid stress variation. The loading in this example might include body forces due to gravity and external boundary loads due to water pressure.

In Fig. 3.21 we see how triangular elements can be used to give a two-dimensional representation of an upper human femur which is to be subjected to an analysis aimed at gaining information about the development of stresses in the cartilage.

Finally in Fig. 3.22 the subdivision for an electromagnetic problem is illustrated in which Eq. (3.31) of heat conduction applies, but with k indicating now magnetic permeability and Q the intensity of current in the conductors.

The wide variety of problems which can be solved by a unified finite element procedure—and indeed a single computer program—is something the reader should note.

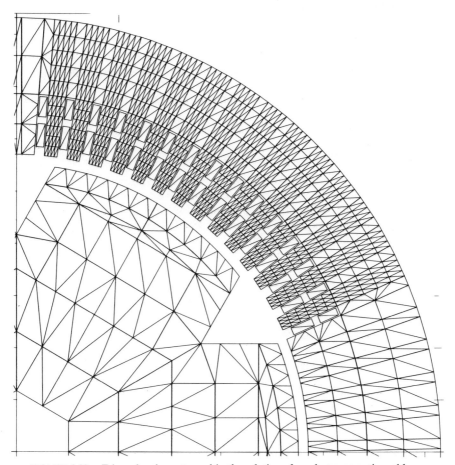

FIGURE 3.22.   Triangular elements used in the solution of an electromagnetic problem.

REFERENCES

[1]P. Tong, Exact solutions of certain problems by the finite element method, *AIAA J.* **7**, 179–180 (1969).

[2]A description of how the banded structure of a matrix may be exploited in computer calculations is given by E. B. Becker, G. F. Carey, and J. T. Oden, *Finite Elements. An Introduction*, Vol. 1, Prentice-Hall, Englewood Cliffs, N.J., 1981.

[3]B. M. Irons, A frontal solution program, *Int. J. Num. Meth. Eng.* **2**, 5–32 (1970).

[4]M. J. Turner, R. W. Clough, H. C. Martin, and L. J. Topp, Stiffness and deflection analysis of complex structures, *J. Aeronaut. Sci.* **23**, 805–823 (1956).

[5]See, for example, pp. 450–458 in O. C. Zienkiewicz, *The Finite Element Method*, 3rd ed., McGraw-Hill, New York, 1977.

SUGGESTED FURTHER READING

A. J. Davies, *The Finite Element Method: A First Approach*, Clarendon, Oxford, 1980.

I. Fried, *Numerical Solution of Differential Equations*, Academic, New York, 1979.

E. Hinton and D. R. J. Owen, *An Introduction to Finite Element Computations*, Pineridge Press, Swansea, 1979.

K. H. Huebner and E. A. Thornton, *The Finite Element Method for Engineers*, 2nd ed., Wiley, New York, 1982.

A. R. Mitchell and R. Wait, *The Finite Element Method in Partial Differential Equations*, Wiley, London, 1977.

O. C. Zienkiewicz, *The Finite Element Method*, 3rd ed., McGraw-Hill, New York, 1977.

CHAPTER FOUR ⸻⸻⸻⸻⸻⸻⸻⸻

Higher Order Finite Element Approximation

4.1. INTRODUCTION

In the previous chapters we introduced the reader to trial function approximation methods in which the trial functions were defined either as in Chapter 2, by smooth expressions over the whole domain, or as in Chapter 3, by simple, local functions defined element by element. Indeed in the latter approach only linear and bilinear polynomials were used, leading to the simplest finite element forms. In the present chapter we explore the possibility of using more complex trial functions, defined however still on the finite element (i.e., local) basis.

The motivation for the introduction of such higher order shape functions is obviously one of achieving a better approximation to the solution of the problem at hand. With the use of the simple finite elements so far described, refinement of the solution can only be achieved by successive creation of finer meshes of elements. On the other hand with the introduction of successively higher order functions, a refinement can be achieved with a constant mesh subdivision.

From the practical point of view, obviously, the best choice will be the one achieving highest accuracy at least computational expense. Anticipating the discussion that follows, we shall find that, in general, higher order approximations appear to be more cost effective (although the actual optimum is very problem dependent).

In the first part of this chapter we attempt to develop a series of higher order elements in one dimension, and polynomial approximations will be used.

We shall find that many alternative methods exist for deriving shape functions of a given polynomial degree and which achieve, therefore, the same order of approximation. In this context we discuss the so-called *hierarchical* forms in which successive higher order functions are additive. Such formulations are usually found to be computationally most efficient.

In the second part of the chapter these arguments are extended to deal with the derivation of higher order two- and three-dimensional polynomial shape functions for geometrically simple domains.

4.2. DEGREE OF POLYNOMIAL IN TRIAL FUNCTIONS AND CONVERGENCE RATES

The question of "order of error" was touched upon in the first chapter dealing with finite difference approximation, but it was not subsequently mentioned in Chapters 2 and 3, in which trial function methods were first introduced. Indeed, in the context of such approximations, we have merely mentioned that convergence will occur if the trial functions are sufficiently complete to be capable in the limit of approximating, to any degree of accuracy, the unknown function and the derivatives with which we are concerned in the formulation. The reason for this omission was simply that we have, in these chapters, been concerned with a variety of trial function sets, and a general estimate of convergence rates was difficult. If our concern is limited only to polynomial approximations—even if these are locally based—the matter of convergence rates and indeed of completeness becomes much easier to discuss.

Let us consider a domain Ω which is subdivided into finite elements Ω^e, each of a characteristic dimension h, and let the trial functions defined in each element subdomain contain a complete polynomial of degree p.

It is immediately evident that if the solution for an unknown function (or functions) ϕ is itself a polynomial of a degree less than or equal to p, then the approximation must yield the exact answer, irrespective of the type of weighting we have used to obtain it.

Normally the solution for ϕ will not be an exact polynomial, but if the solution does not contain singularities making some or all derivatives infinite, we can always expand it locally as a Taylor series. For example, with two independent dimensions we could write in the vicinity of some point O

$$\phi(\Delta x, \Delta y) = \phi|_{O} + \Delta x \left.\frac{\partial \phi}{\partial x}\right|_{O} + \Delta y \left.\frac{\partial \phi}{\partial y}\right|_{O} + \cdots \qquad (4.1)$$

where Δx and Δy are the differences between the coordinates of the point in question and the origin at point O. If now the approximating polynomial of degree p is used, and since this can exactly reproduce the Taylor series up to degree p, then within an element of size h the maximum error E in the approximation will satisfy

$$E = O(h^{p+1}) \tag{4.2}$$

as h is the maximum value which can be attained by Δx or Δy on the element.

In a similar manner we note that the approximation to the first derivative of ϕ will be accurate to $O(h^p)$ and the approximation to the dth derivative will be accurate to $O(h^{p+1-d})$.

The above result is of paramount importance not only in explaining why high-degree polynomials show improved error characteristics, but also in establishing the necessary conditions of completeness required in all weighted residual processes. Thus returning to our general problem of Chapter 2, in which we attempt to solve a differential equation written as

$$A(\phi) = \mathcal{L}\phi + p = 0 \quad \text{in } \Omega \tag{4.3}$$

with boundary conditions

$$B(\phi) = \mathcal{M}\phi + r = 0 \quad \text{on } \Gamma \tag{4.4}$$

by the use of a trial function expansion

$$\hat{\phi} = \sum_{m=1}^{M} a_m N_m \tag{4.5}$$

and a weighted form [namely, Eq. (2.41)]

$$\int_\Omega W_l \left[\mathcal{L}\hat{\phi} + p \right] d\Omega + \int_\Gamma \overline{W}_l \left[\mathcal{M}\hat{\phi} + r \right] d\Gamma = 0 \tag{4.6}$$

we observe that for convergence not only ϕ, but all the derivatives occurring in the operators $\mathcal{L}$ and $\mathcal{M}$, have to be represented correctly in the limit as $h \to 0$. If the order of such derivatives is d, clearly the minimum-degree polynomial expansion which can be used for $\hat{\phi}$ over any element must be such that the error incurred in the representations is at least of $O(h)$. Thus for completeness, we require that

$$p + 1 - d \geqslant 1 \tag{4.7a}$$

or

$$p - d \geqslant 0 \tag{4.7b}$$

The completeness requirement just stated underlines the usefulness of the weak formulation which we have discussed in Section 3.4 [namely, Eq. (3.15)]. Now the reduction in the order d of the operators involved not only reduces the necessary continuity requirements, but it also reduces the minimum degree

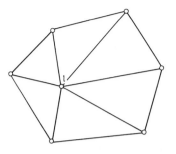

FIGURE 4.1. As assembled field of elements which could be used for the application of a patch test at node *l*.

p of the trial polynomials needed to ensure completeness, as lower order derivatives of $\hat{\phi}$ now occur.

4.3.   THE PATCH TEST

The discussion of the previous section leads directly to a very useful test which finite element practitioners often apply to check the correctness of their formulations (and their arithmetic) when deriving finite element equations. As we have stated already at the beginning of the last section, if the exact solution *φ* is a polynomial of degree *p* or less and the polynomial used to construct the approximation $\hat{\phi}$ is of degree *p*, then an exact solution must be obtained. It follows therefore that if we impose, on the field of finite elements, nodal values corresponding to such an exact solution, the weighted residual equation should be exactly satisfied.

To perform such a test it is simply necessary to assemble a set of elements over which the weighting function W_l associated with node *l* is nonzero and to check that the resulting weighted residual equation is satisfied identically. In Fig. 4.1 we show a field of two-dimensional elements that might be used for the application of the patch test.

Commonly the patch test is used just to ascertain whether lowest order convergence is satisfied. This requires, to ensure an error of $O(h^2)$, that a linear function be imposed at the nodes and an exact result reproduced. The use of the patch test to ascertain the convergence order predicted for an element is also possible.

Now we simply impose on the assembled elements a polynomial solution of degree *p* and check whether, using an appropriate load term arising from Eq. (4.3), an exact satisfaction of the discretised equations is obtained. The highest value of *p* for which this is true determines the order of error as $O(h^{p+1})$.

4.4.   STANDARD HIGHER ORDER SHAPE FUNCTIONS FOR ONE-DIMENSIONAL ELEMENTS WITH C^0 CONTINUITY

Consider a set of one-dimensional elements as illustrated in Fig. 4.2. In Fig. 4.2*a* we show a typical two-noded linear element *e* of the type introduced in the last chapter, and the two nodes are assumed here to be numbered as node 0

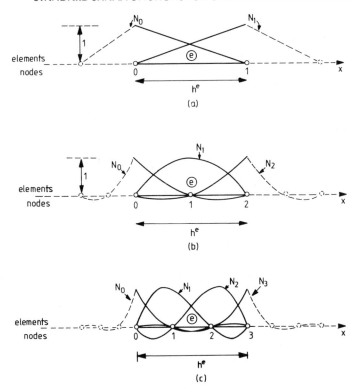

FIGURE 4.2.   One-dimensional elements and associated standard shape functions of (a) linear, (b) quadratic, and (c) cubic form.

and node 1 (to avoid proliferation of symbolic letters). On this element we associate with each node a simple linear shape function as illustrated. The form of these functions on the adjacent elements is also indicated in the figure. The use of such shape functions ensures that:

1. The approximation $\hat{\phi}$ is a linear function of x over each element.
2. The parameter a_l, associated with the shape function N_l in the expansion for $\hat{\phi}$ [see Eq. (4.5)], is simply the approximation to the value of ϕ at node l.

Interelement C^0 continuity of the expansion for $\hat{\phi}$ is also automatically ensured as the shape functions themselves are C^0 continuous.

This process can be extended to obtain higher order shape functions for a typical one-dimensional element. Over the three-noded element of Fig. 4.2b a quadratic representation for $\hat{\phi}$ is possible, and this can be achieved by associating a quadratic shape function with each node of the element. If the element shape function associated with a particular node is required to take the value unity at that node and zero at the two other nodes of the element, then condition 2 above remains satisfied. For such an element, the exact position of the extra node is immaterial, and we note immediately that the shape function associated with this extra node is internal to the element and does not extend to other adjacent elements. The extension is carried one step further in Fig. 4.2c, where we show a four-noded element and the corresponding cubic element shape functions.

In general, the approximation $\hat{\phi}$ will reduce to a polynomial of degree p over an element that has $(p + 1)$ nodes. If we consider such an element, with nodes numbered 0 to p placed at the points $x_0, x_1, x_2, \ldots, x_{p-1}, x_p$, then the element shape function N_l^e associated with node l will be the polynomial of degree p which takes the value unity at node l and the value zero at all the other nodes of the element. We could then derive an expression for such a shape function in the form

$$N_l^e = \alpha_0 + \alpha_1 x + \alpha_2 x^2 + \cdots + \alpha_p x^p \tag{4.8}$$

where the unknown coefficients are given by solution of the equation set

$$x = x_0, \quad N_l^e = \alpha_0 + \alpha_1 x_0 + \alpha_2 x_0^2 + \cdots + \alpha_p x_0^p = 0$$

$$x = x_1, \quad N_l^e = \alpha_0 + \alpha_1 x_1 + \alpha_2 x_1^2 + \cdots + \alpha_p x_1^p = 0$$

$$\vdots \qquad \qquad \vdots$$

$$x = x_l, \quad N_l^e = \alpha_0 + \alpha_1 x_l + \alpha_2 x_l^2 + \cdots + \alpha_p x_l^p = 1 \tag{4.9}$$

$$\vdots \qquad \qquad \vdots$$

$$x = x_p, \quad N_l^e = \alpha_0 + \alpha_1 x_p + \alpha_2 x_p^2 + \cdots + \alpha_p x_p^p = 0$$

However, this process would be tedious and, in fact, unnecessary, for it is obvious by inspection that

$$N_l^e = \Lambda_l^p = \frac{(x - x_0)(x - x_1) \cdots (x - x_{l-1})(x - x_{l+1}) \cdots (x - x_p)}{(x_l - x_0)(x_l - x_1) \cdots (x_l - x_{l-1})(x_l - x_{l+1}) \cdots (x_l - x_p)}$$

$$\tag{4.10}$$

is the polynomial with the desired characteristics. This polynomial, Λ_l^p, is well known as a Lagrange interpolation polynomial[1] of degree p.

Using this definition, it is now possible to write down the element shape functions for the linear, quadratic, and cubic elements of Fig. 4.2. With the nodes equally spaced within the element and numbered as shown in the figure, we shall express the functions in terms of a normalized local element coordinate ξ, defined by

$$\xi = \frac{2(x - x_c^e)}{h^e} \tag{4.11}$$

Here x_c^e is the coordinate of the center of the element, h^e is the element length, and the element is then defined by the range $-1 \leqslant \xi \leqslant 1$. The element shape functions are easily found to be:

Linear (Fig. 4.2a)

$$N_0^e = -\frac{\xi - 1}{2}, \qquad N_1^e = \frac{\xi + 1}{2} \tag{4.12a}$$

Quadratic (Fig. 4.2b)

$$N_0^e = -\frac{\xi(\xi - 1)}{2}, \qquad N_1^e = -(\xi - 1)(\xi + 1), \qquad N_2^e = \frac{\xi(\xi + 1)}{2} \tag{4.12b}$$

Cubic (Fig. 4.2c)

$$N_0^e = -\tfrac{9}{16}(\xi + \tfrac{1}{3})(\xi - \tfrac{1}{3})(\xi - 1), \qquad N_1^e = \tfrac{27}{16}(\xi + 1)(\xi - \tfrac{1}{3})(\xi - 1);$$

$$N_2^e = -\tfrac{27}{16}(\xi + 1)(\xi + \tfrac{1}{3})(\xi - 1), \qquad N_3^e = \tfrac{9}{16}(\xi + 1)(\xi + \tfrac{1}{3})(\xi - \tfrac{1}{3}) \tag{4.12c}$$

To illustrate the use of such higher order shape functions we return again to the solution of the problem considered in Example 3.1.

Example 4.1

The problem is to solve the equation $d^2\phi/dx^2 - \phi = 0$ with $\phi = 0$ at $x = 0$ and $\phi = 1$ at $x = 1$.

If the Galerkin method is employed, then the expression for a typical component of the element matrix $\mathbf{K}^e$ is, as in Example 3.1, given by

$$K_{lm}^e = \int_{\Omega^e} \left(\frac{dN_l^e}{dx} \frac{dN_m^e}{dx} + N_l^e N_m^e \right) dx$$

With the element shape functions expressed in terms of the normalized local coordinate ξ of Eq. (4.11), we note that

$$\frac{dN_l^e}{dx} = \frac{dN_l^e}{d\xi} \frac{d\xi}{dx} = \frac{2}{h^e} \frac{dN_l^e}{d\xi}, \qquad dx = \frac{h^e}{2} d\xi$$

and so we can write

$$K_{lm}^e = \int_{-1}^{1} \left(\frac{2}{h^e} \frac{dN_l^e}{d\xi} \frac{dN_m^e}{d\xi} + \frac{h^e}{2} N_l^e N_m^e \right) d\xi$$

Now typical reduced element matrices for the element e of Fig. 4.2 can be evaluated for linear, quadratic, and cubic shape functions in turn.

Leaving details of the arithmetic to the reader, we find that the element matrices can be written as follows.

Linear

$$\mathbf{k}^e \boldsymbol{\phi}^e = \begin{bmatrix} \dfrac{1}{h^e} + \dfrac{h^e}{3} & -\dfrac{1}{h^e} + \dfrac{h^e}{6} \\[3mm] -\dfrac{1}{h^e} + \dfrac{h^e}{6} & \dfrac{1}{h^e} + \dfrac{h^e}{3} \end{bmatrix} \begin{bmatrix} \phi_0 \\[3mm] \phi_1 \end{bmatrix}$$

which is identical to the expression found in Example 3.1.

Quadratic

$$\mathbf{k}^e \boldsymbol{\phi}^e = \begin{bmatrix} \dfrac{7}{3h^e} + \dfrac{2h^e}{15} & -\dfrac{8}{3h^e} + \dfrac{h^e}{15} & \dfrac{1}{3h^e} - \dfrac{h^e}{30} \\[3mm] -\dfrac{8}{3h^e} + \dfrac{h^e}{15} & \dfrac{16}{3h^e} + \dfrac{8h^e}{15} & -\dfrac{8}{3h^e} + \dfrac{h^e}{15} \\[3mm] \dfrac{1}{3h^e} - \dfrac{h^e}{30} & -\dfrac{8}{3h^e} + \dfrac{h^e}{15} & \dfrac{7}{3h^e} + \dfrac{2h^e}{15} \end{bmatrix} \begin{bmatrix} \phi_0 \\[3mm] \phi_1 \\[3mm] \phi_2 \end{bmatrix}$$

Cubic

$$\mathbf{k}^e \boldsymbol{\phi}^e = \begin{bmatrix} \dfrac{37}{10h^e} + \dfrac{8h^e}{105} & -\dfrac{189}{40h^e} + \dfrac{33h^e}{560} & \dfrac{27}{20h^e} - \dfrac{3h^e}{140} & -\dfrac{13}{40h^e} + \dfrac{19h^e}{1680} \\[3mm] -\dfrac{189}{40h^e} + \dfrac{33h^e}{560} & \dfrac{54}{5h^e} + \dfrac{27h^e}{70} & -\dfrac{297}{40h^e} - \dfrac{27h^e}{560} & \dfrac{27}{20h^e} - \dfrac{3h^e}{140} \\[3mm] \dfrac{27}{20h^e} - \dfrac{3h^e}{140} & -\dfrac{297}{40h^e} - \dfrac{27h^e}{560} & \dfrac{54}{5h^e} + \dfrac{27h^e}{70} & -\dfrac{189}{40h^e} + \dfrac{33h^e}{560} \\[3mm] -\dfrac{13}{40h^e} + \dfrac{19h^e}{1680} & \dfrac{27}{20h^e} - \dfrac{3h^e}{140} & -\dfrac{189}{40h^e} + \dfrac{33h^e}{560} & \dfrac{37}{10h^e} + \dfrac{8h^e}{105} \end{bmatrix} \begin{bmatrix} \phi_0 \\[3mm] \phi_1 \\[3mm] \phi_2 \\[3mm] \phi_3 \end{bmatrix}$$

In Example 3.1 we solved the problem by using three linear finite elements.

Now we shall attempt the solution using only one element, but with polynomial shape functions of different degrees.

With $h^e = 1$ and prescribed values of $\phi_0 = 0$ and $\phi_1 = 1$ the linear element merely gives a linear interpolation and no free unknowns exist for the weighted equations. With a quadratic element and prescribed end values $\phi_0 = 0$, $\phi_2 = 1$ we have one unknown, ϕ_1. Inserting the boundary values gives

$$\left(\tfrac{16}{3} + \tfrac{8}{15}\right)\phi_1 = \left(\tfrac{8}{3} - \tfrac{1}{15}\right)\phi_2$$

and so

$$\phi_1 = 0.4432$$

For a cubic approximation two unknowns, ϕ_1 and ϕ_2, arise with the two resulting simultaneous equations yielding a solution

$$\phi_1 = 0.2889$$

$$\phi_2 = 0.6101$$

With the quadratic element we can compute $\hat\phi$ at $x = \tfrac{1}{3}$ and $x = \tfrac{2}{3}$, and this leads to the following comparison.

	Three Linear Elements	One Quadratic Element	One Cubic Element	Exact
$x = \tfrac{1}{3}$	$\hat\phi = 0.2855$	0.2828	0.2889	0.28892
$x = \tfrac{2}{3}$	$\hat\phi = 0.6098$	0.6162	0.6101	0.61020

We note that even a single quadratic approximation gives smooth results which are superior overall to those produced by the use of three linear elements—now at a cost of solving only one equation.

The example just quoted demonstrates that the process of derivation of the element matrices for higher order elements proceeds in an identical manner to that used for linear elements. It also shows that improved results are available here with a smaller total number of unknowns, a fact generally observed.

We have not assembled a typical matrix of several higher order elements in the above example. Such an assembly is left to the reader as an exercise—and again it follows a pattern identical to that described in the previous chapter. We can, however, immediately observe that the internal nodes in an element are only coupled to the other nodes of that element, and hence can be eliminated at the *element level*. This is useful in computation as the reduced element matrix $\mathbf{k}^e$ for the higher order elements of this example can then, by additional computation, be presented simply as a 2×2 matrix. The reader is advised to work through Exercise 4.3 to gain an understanding of this procedure.

EXERCISES

4.1. The equation

$$\frac{d}{dx}\left(k\frac{d\phi}{dx}\right) + \alpha\phi + Q = 0$$

is to be solved by a Galerkin weighted residual method. Determine the element matrices $\mathbf{k}^e$ and $\mathbf{f}^e$ for typical quadratic and cubic elements.

4.2. By performing the assembly process, determine, for both quadratic and cubic elements, typical equations arising in Exercise 4.1 at (a) an element interior node and (b) an element end node.

4.3. It is required to solve the problem of Example 4.1 using two quadratic elements. (a) Assemble the element matrices given in Example 4.1 and produce a solution. (b) At the element level, express the internal nodal value ϕ_1 in terms of the end node values ϕ_0 and ϕ_2. Hence produce a new 2×2 element matrix operating upon the end nodal values only. Assemble these new element matrices and solve the resulting system. Show that the solution produced is identical to that obtained in part (a).

4.4. Repeat Exercise 4.3 using two cubic elements. In part (b) eliminate the two interior nodes from the reduced element matrices.

4.5. Determine an appropriate function Q such that $\phi = \beta_0 + \beta_1 x + \beta_2 x^2$ is a solution of the equation

$$k\frac{d^2\phi}{dx^2} + \alpha\phi + Q = 0$$

where k and α are constants. Assemble the Galerkin weighted residual equations for a field of quadratic elements and show that the patch test will be exactly satisfied for all values of β_0, β_1, and β_2.

4.6. Return to Exercise 1.2 and obtain the bending moment distribution by using first two quadratic and then two cubic elements. Compare the answers with those obtained by other methods in the previous chapters.

4.7. Repeat the one-dimensional steady-state heat conduction example of Exercise 2.5 by using (a) two quadratic elements, (b) two cubic elements, and (c) a mixed finite element mesh consisting of one quadratic plus one cubic element. Compare the answers produced in each case with the exact solution.

4.5.   HIERARCHICAL FORMS OF HIGHER ORDER ONE-DIMENSIONAL ELEMENTS WITH C^0 CONTINUITY

4.5.1.   General Remarks

When finite element shape functions were introduced in Chapter 3, we identified the parameters a_m in the expansion of Eq. (4.5) with the nodal values of the approximation $\hat{\phi}$, that is, $a_m = \phi_m$. Indeed this specification was followed in the last section, resulting necessarily in the requirement of Eq. (4.9).

This identification has been widely followed in the finite element literature and has the merit of assigning a "physical" meaning to the parameters a_m. There is, however, a disadvantage in this, "standard," definition of finite element shape functions.

This is apparent if we examine the shape functions for linear, quadratic, and cubic polynomials given in Eq. (4.12) and the resulting element matrices of Example 4.1. As the shape functions for each order of approximation are completely different, each level of approximation results in a completely new element matrix, and thus the equation set has to be entirely reevaluated if it is decided to resolve a problem using shape functions of a higher degree.

This type of approximation contrasts sharply with the continuous function approximation of Chapter 2 where we considered an expansion of the standard form

$$\hat{\phi} = \psi + \sum_{m=1}^{M} a_m N_m \tag{4.13}$$

in which, when we refined the solution by increasing the total number M of trial functions used, the form of the lower order trial functions remained unaltered. This resulted in the following sequence of approximations for a linear problem:

$$M = 1 \quad K_{11}a_1 = f_1 \tag{4.14a}$$

$$M = 2 \quad \begin{bmatrix} K_{11} & K_{12} \\ K_{21} & K_{22} \end{bmatrix} \begin{bmatrix} a_1 \\ a_2 \end{bmatrix} = \begin{bmatrix} f_1 \\ f_2 \end{bmatrix} \tag{4.14b}$$

$$M = 3 \quad \begin{bmatrix} K_{11} & K_{12} & K_{13} \\ K_{21} & K_{22} & K_{23} \\ K_{31} & K_{32} & K_{33} \end{bmatrix} \begin{bmatrix} a_1 \\ a_2 \\ a_3 \end{bmatrix} = \begin{bmatrix} f_1 \\ f_2 \\ f_3 \end{bmatrix} \tag{4.14c}$$

In each step above it can be seen that, as the approximation is refined, the matrices produced by the previous stage of the approximation reoccur and need not be recomputed.

Further, if the chosen trial functions are strongly orthogonal, the matrices obtained at each stage of refinement now ensure better conditioning of the equation system to be solved and show a heavy diagonal structure. Indeed, when completely orthogonal trigonometric trial functions and the Galerkin method were used in Example 2.2, it was found that the coupling between the equations disappeared completely and the sequence of approximations became simply

$$M = 1 \quad K_{11}a_1 = f_1 \qquad\qquad (4.15a)$$

$$M = 2 \quad \begin{bmatrix} K_{11} & 0 \\ 0 & K_{22} \end{bmatrix} \begin{bmatrix} a_1 \\ a_2 \end{bmatrix} = \begin{bmatrix} f_1 \\ f_2 \end{bmatrix} \qquad\qquad (4.15b)$$

$$M = 3 \quad \begin{bmatrix} K_{11} & 0 & 0 \\ 0 & K_{22} & 0 \\ 0 & 0 & K_{33} \end{bmatrix} \begin{bmatrix} a_1 \\ a_2 \\ a_3 \end{bmatrix} = \begin{bmatrix} f_1 \\ f_2 \\ f_3 \end{bmatrix} \qquad\qquad (4.15c)$$

Now each level of refinement simply results in the previous solution being augmented by $a_m N_m$, where

$$a_m = K_{mm}^{-1} f_m \qquad\qquad (4.16)$$

In this section we shall suggest a manner in which higher order finite elements can be generated, incorporating some of the advantages of the continuous function approximation method of Chapter 2. Now the shape functions will no longer be defined by requirements such as Eq. (4.9), and they will represent simply additive refinements of higher order. We shall call such shape functions *hierarchical*, as their contributions to the approximation will be of diminishing importance.

Although in general complete diagonality of the approximating equation will be difficult or impossible to achieve, we shall strive to achieve as weak a coupling as possible between the various levels of approximation. This has the numerical advantage that, because a certain loss of accuracy always occurs in the computation of large problems due to numerical roundoff, the successive coefficients a_m, decreasing in importance, can then tolerate a larger amount of numerical inaccuracy without seriously affecting the quality of the total solution.

4.5.2. Hierarchical Polynomials

Let us consider once again the generation of shape functions for a general element as illustrated in Fig. 4.3. Clearly, the linear expansion over this element can only be given by the shape functions illustrated in Fig. 4.2a

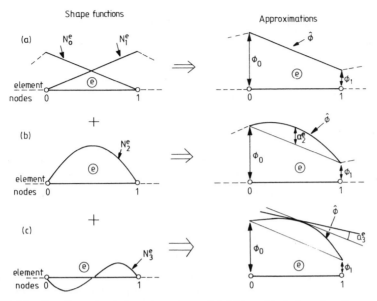

FIGURE 4.3. One-dimensional elements and associated hierarchical shape functions and variables of (a) linear, (b) quadratic, and (c) cubic form.

[namely, Eq. (4.12a)] and cannot be improved upon—as the identification of the connecting nodes between elements guarantees C^0 continuity. However, we can achieve the hierarchical format over this element if we use a quadratic to modify this linear expansion.

If $a_2^e N_2^e$ represents this modification and N_2^e is defined, in terms of the local element coordinate ξ, as a quadratic function of the form

$$N_2^e = \alpha_0 + \alpha_1 \xi + \alpha_2 \xi^2 \qquad (4.17)$$

with coefficients chosen so as to give $N_2^e = 0$ at $\xi = \pm 1$, we shall preserve the required C^0 continuity of the approximation $\hat{\phi}$ between elements. This condition can be satisfied by a symmetric parabolic function with any suitable scaling parameter.

Thus for a quadratic approximation, we can write over the typical element e (Fig. 4.3b)

$$\hat{\phi}^e = \phi_0 N_0^e + \phi_1 N_1^e + a_2^e N_2^e \qquad (4.18)$$

where [see Eq. (4.12a/b)]

$$N_0^e = -\frac{\xi - 1}{2}, \qquad N_1^e = \frac{\xi + 1}{2}, \qquad N_2^e = -(\xi - 1)(\xi + 1) \qquad (4.19)$$

We note immediately that the parameter a_2^e does in fact have a meaning in this case as it is the magnitude of the departure from linearity of the approximation $\hat{\phi}$ at the element center, since N_2^e has been chosen here to have the value unity at that point.

In a similar manner, for a cubic element we simply have to add $a_3^e N_3^e$ to the quadratic expansion of Eq. (4.18), where N_3^e is any cubic of the form

$$N_3^e = \alpha_0 + \alpha_1 \xi + \alpha_2 \xi^2 + \alpha_3 \xi^3 \qquad (4.20)$$

and which has zero values at $\xi = \pm 1$ (i.e., at nodes 0 and 1). Again an infinity of choices exists, and we could select a cubic of the form shown in Fig. 4.3c, which has a zero value at the center of the element and for which $dN_3^e/d\xi = 1$ at the same point. Immediately we can write

$$N_3^e = \xi(1 - \xi^2) \qquad (4.21)$$

as the cubic function with the desired properties. Now the parameter a_3^e denotes the departure of the slope at the center of the element from that of the first approximation.

We note that we could proceed in a similar manner and define the fourth-order hierarchical element shape function as

$$N_4^e = \xi^2(1 - \xi^2) \qquad (4.22)$$

but a physical identification of the parameters now becomes more difficult (even though it is not strictly necessary).

As we have already noted, the above set is not unique and many other possibilities exist. An alternative convenient form for the hierarchical functions is defined by

$$N_p^e(\xi) = \begin{cases} \dfrac{1}{p!}(\xi^p - 1), & p \text{ even} \\[2mm] \dfrac{1}{p!}(\xi^p - \xi), & p \text{ odd} \end{cases} \qquad (4.23)$$

where $p(\geqslant 2)$ is the degree of the introduced polynomial. This yields the set of element shape functions:

$$N_2^e = \tfrac{1}{2}(\xi^2 - 1), \qquad N_3^e = \tfrac{1}{6}(\xi^3 - \xi)$$

$$N_4^e = \tfrac{1}{24}(\xi^4 - 1), \qquad N_5^e = \tfrac{1}{120}(\xi^5 - \xi) \qquad (4.24)$$

We observe that all derivatives of N_p^e of second or higher order have the value zero at $\xi = 0$, apart from $d^p N_p^e/d\xi^p$, which equals unity at that point, and hence, when shape functions of the form given by Eq. (4.23) are used, we can identify the parameters in the approximation as

$$a_p^e = \frac{d^p \hat{\phi}}{d\xi^p}\bigg|_{\xi=0}, \qquad p \geq 2 \tag{4.25}$$

Such identification gives a general physical significance but is by no means necessary.

4.5.3. Hierarchical Polynomials of Nearly Orthogonal Form

As mentioned previously, an optimal form of hierarchical functions is one that results in a diagonal equation system. This can on occasion be achieved, or at least approximated, quite closely.

In many of the problems which we have discussed in the preceding chapters the element matrix $\mathbf{K}^e$ possesses terms of the form

$$K_{lm}^e = \int_{\Omega^e} k \frac{dN_l^e}{dx} \frac{dN_m^e}{dx} dx = \frac{2}{h^e} \int_{-1}^{1} k \frac{dN_l^e}{d\xi} \frac{dN_m^e}{d\xi} d\xi \tag{4.26}$$

If shape function sets containing the appropriate polynomials can be found for which such integrals are zero for $l \neq m$, then orthogonality is achieved and the coupling between successive solutions disappears.

One set of polynomial functions which is known to possess this orthogonality property over the range $-1 \leq \xi \leq 1$ is the set of Legendre polynomials $P_p(\xi)$, and the shape functions could be defined in terms of integrals of these polynomials. Here we define the Legendre polynomial of degree p by

$$P_p(\xi) = \frac{1}{(p-1)!} \frac{1}{2^{p-1}} \frac{d^p}{d\xi^p} \left[(\xi^2 - 1)^p \right] \tag{4.27}$$

and integrating these polynomials for $p = 1, 2, 3, 4$ in turn gives

$$N_2^e = \xi^2 - 1, \qquad\qquad N_3^e = 2(\xi^3 - \xi)$$
$$N_4^e = \tfrac{1}{4}(15\xi^4 - 18\xi^2 + 3), \qquad N_5^e = 7\xi^5 - 10\xi^3 + 3\xi \tag{4.28}$$

These differ from the element shape functions given by Eq. (4.24) only by a multiplying constant up to N_3^e, but for $p \geq 4$ the differences become significant. The reader can easily verify the orthogonality of the derivatives of these functions, which is useful in computation. A plot of these functions and their derivatives is given in Fig. 4.4.

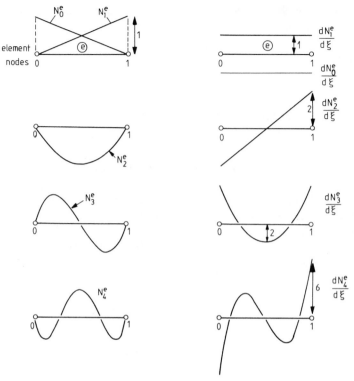

FIGURE 4.4. Hierarchical element shape functions of nearly orthogonal form and their derivatives.

Example 4.2

We shall consider here precisely the same problem as that discussed in Example 4.1, but now we formulate the element matrices using the hierarchical approach. The shape functions given in Eq. (4.28) will be used, and once again the element matrices are given by the expressions

$$K_{lm}^e = \int_{-1}^{1} \left(\frac{2}{h^e} \frac{dN_l^e}{d\xi} \frac{dN_m^e}{d\xi} + \frac{h^e}{2} N_l^e N_m^e \right) d\xi$$

The matrices for the various approximations are as follows.

Linear

$$\mathbf{k}^e \boldsymbol{\phi}^e = \begin{bmatrix} \dfrac{1}{h^e} + \dfrac{h^e}{3} & -\dfrac{1}{h^e} + \dfrac{h^e}{6} \\[3mm] -\dfrac{1}{h^e} + \dfrac{h^e}{6} & \dfrac{1}{h^e} + \dfrac{h^e}{3'} \end{bmatrix} \begin{bmatrix} \phi_0 \\[3mm] \phi_1 \end{bmatrix}$$

Quadratic

$$\mathbf{k}^e \boldsymbol{\phi}^e = \begin{bmatrix} \dfrac{1}{h^e} + \dfrac{h^e}{3} & -\dfrac{1}{h^e} + \dfrac{h^e}{6} & -\dfrac{h^e}{3} \\[2ex] -\dfrac{1}{h^e} + \dfrac{h^e}{6} & \dfrac{1}{h^e} + \dfrac{h^e}{3} & -\dfrac{h^e}{3} \\[2ex] -\dfrac{h^e}{3} & -\dfrac{h^e}{3} & \dfrac{16}{3h^e} + \dfrac{8h^e}{15} \end{bmatrix} \begin{bmatrix} \phi_0 \\[2ex] \phi_1 \\[2ex] a_2^e \end{bmatrix}$$

Cubic

$$\mathbf{k}^e \boldsymbol{\phi}^e = \begin{bmatrix} \dfrac{1}{h^e} + \dfrac{h^e}{3} & -\dfrac{1}{h^e} + \dfrac{h^e}{6} & -\dfrac{h^e}{3} & \dfrac{2h^e}{15} \\[2ex] -\dfrac{1}{h^e} + \dfrac{h^e}{6} & \dfrac{1}{h^e} + \dfrac{h^e}{3} & -\dfrac{h^e}{3} & -\dfrac{2h^e}{15} \\[2ex] -\dfrac{h^e}{3} & -\dfrac{h^e}{3} & \dfrac{16}{3h^e} + \dfrac{8h^e}{15} & 0 \\[2ex] \dfrac{2h^e}{15} & -\dfrac{2h^e}{15} & 0 & \dfrac{164}{5h^e} + \dfrac{32h^e}{105} \end{bmatrix} \begin{bmatrix} \phi_0 \\[2ex] \phi_1 \\[2ex] a_2^e \\[2ex] a_3^e \end{bmatrix}$$

We note now that each successive approximation contains the previous approximation matrices and that the coupling matrices are weak. (Orthogonality of the quadratic and cubic approximations is apparent for this example, but complete orthogonality is not achieved here as the Legendre polynomials generally ensure only the orthogonality of derivatives, and this is manifest in the absence of terms involving $1/h^e$ in the off-diagonal positions.)

Once again applying the general expression to a one-element solution of the problem (i.e., $h^e = 1$), we obtain for the quadratic approximation

$$\left(\tfrac{16}{3} + \tfrac{8}{15} \right) a_2^e = \tfrac{1}{3}$$

giving

$$a_2^e = 0.05682$$

and for the cubic approximation,

$$a_2^e = 0.05682$$

$$a_3^e = 0.01017$$

Evaluating the approximation at $x = \tfrac{1}{3}$ and $x = \tfrac{2}{3}$, we find that the accuracy achieved here is identical to that obtained in Example 4.1. The points made previously, however, about the ease of calculating the matrices and the diminishing effect on the accuracy of successive parameters should be noted.

EXERCISES

4.8. Repeat Exercise 4.1 using hierarchical quadratic and cubic elements.

4.9. Return to Exercise 1.2 and obtain the bending moment distribution by using first two quadratic and then two cubic hierarchical elements. Compare the answers with those obtained in Exercise 4.6.

4.10. Repeat Exercise 4.7 using hierarchical elements.

4.11. Produce the reduced element matrix $\mathbf{k}^e$ of Example 4.2 for a quartic hierarchical element.

4.12. Obtain the weak form of the weighted residual statement for the problem of the deflection of a loaded beam on an elastic foundation, described in Exercise 1.20. Show that the standard theory requires C^1 continuous elements if a solution is attempted by the finite element method. Develop the cubic shape functions for such an element when the approximation over the element is expressed as

$$\hat{\phi}^e = \phi_0 N_0^e + \phi_1 N_1^e + \left.\frac{d\phi}{dx}\right|_0 M_0^e + \left.\frac{d\phi}{dx}\right|_1 M_1^e$$

and where the two nodes 0 and 1 are placed at the ends of the element. Hence obtain a solution to Exercise 1.20 using three equal elements of this type. (Note that N_0^e must satisfy the conditions $N_0^e = 1$, $dN_0^e/dx = 0$ at $\xi = -1$ and $N_0^e = dN_0^e/dx = 0$ at $\xi = 1$ while M_0^e must satisfy $M_0^e = 0$, $dM_0^e/dx = 1$ at $\xi = -1$ and $M_0^e = dM_0^e/dx = 0$ at $\xi = 1$. The functions N_1^e and M_1^e are similarly defined. These requirements mean that the shape functions can be expressed in terms of Hermite polynomials.)

4.6. TWO-DIMENSIONAL RECTANGULAR FINITE ELEMENT SHAPE FUNCTIONS OF HIGHER ORDER

4.6.1. Standard-Type Shape Functions-Lagrangian and Serendipity Elements

In Section 3.7 we have already introduced the reader to a simple rectangular element with four corner nodes. The standard shape functions for this element were derived as a simple product of appropriate linear terms in the x and y directions (Fig. 4.5a).

It is clear that for such elements C^0 continuity of all global shape functions is maintained between elements as the connection of the two nodal values at each corner ensures a unique straight line variation along any element side.

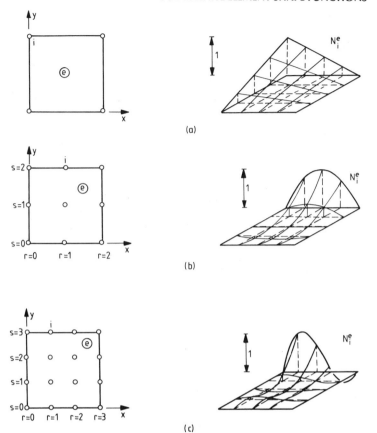

FIGURE 4.5. Some rectangular elements with Lagrangian element shape functions.

If we wish to generate higher order shape functions, we can proceed in a very simple manner if the element nodes are placed on a rectangular grid. For instance, for a quadratic element we can obtain the shape functions corresponding to any node shown on the grid of Fig. 4.5b by taking the standard one-dimensional shape functions in x and y corresponding to that node and establishing their product.

Indeed if we label each point by a double suffix (r, s) $(r = 0, 1, 2; s = 0, 1, 2$ for the quadratic element), we can write the element shape function corresponding to node (r, s) as

$$N_{rs}^e = \Lambda_r^p(x)\Lambda_s^p(y) \tag{4.29}$$

where for the quadratic element $p = 2$. In the above equation the notation Λ_r^p refers to the Lagrangian interpolation polynomial of degree p, as defined in Eq. (4.10).

The above relationship is quite general, and element shape functions of any degree can be obtained in this manner. In Fig. 4.5c we illustrate the extension to a cubic element. Note that N_{rs}^e will be equal to unity only at the node (r, s) and will be identically zero at all other nodes on the element.

It is self-evident that, for polynomial element shape functions of this type and of any degree, continuity of the global shape functions between adjacent elements is assured as the number of parameters associated with any side is sufficient to define the polynomial uniquely there. This in turn ensures C^0 continuity of the global approximation $\hat{\phi}$. Further we note that only the nodes along the sides of each element are connected to other elements, and so the nodes in the element interior can be eliminated from the element matrices at the element level.

It is of interest to examine the terms which occur in the element shape functions of a pth-degree Lagrangian-type element. From Eq. (4.29) the shape

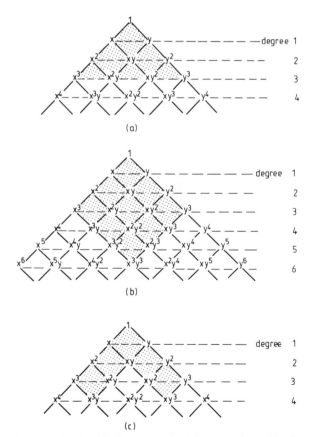

FIGURE 4.6. (*a*) Pascal triangle with the terms present in a complete cubic shown shaded. (*b*) Terms generated by a cubic Lagrangian element. (*c*) Terms generated by a cubic serendipity element.

functions are formed by the product of complete polynomials of degree p in both x and y separately. Clearly, the number of terms in this product is in excess of the number of terms necessary to reproduce a complete polynomial of degree p in x and y. Figure 4.6a shows the well-known Pascal triangle from which the number of terms present in a complete polynomial of any degree can be determined (e.g., 10 terms for a cubic polynomial). In Fig. 4.6b we identify the additional six terms which arise in a cubic Lagrangian shape function of the type defined by Eq. (4.29). This suggests that the number of nodes associated with higher order elements could therefore be reduced by attempting to ensure that the element shape functions are capable of producing more closely only the terms present in a complete polynomial of appropriate degree, and to this end an alternative series of so-called serendipity elements has been developed. Here the nodes are arranged (as far as possible) on the element boundaries, and the shape functions are formed by multiplying terms of degree p in one variable by linear terms in the other. On the boundaries the form of the approximation $\hat{\phi}$ is then identical with that produced by the Lagrangian elements, and so C^0 continuity of the approximation is maintained.

Now we define two local normalized element coordinates (see Fig. 4.7)

$$\xi = \frac{2(x - x_c^e)}{h_x^e}, \qquad d\xi = \frac{2\,dx}{h_x^e}$$

$$\eta = \frac{2(y - y_c^e)}{h_y^e}, \qquad d\eta = \frac{2\,dy}{h_y^e}$$

$$(4.30)$$

where (x_c^e, y_c^e) denotes the position of the element center in the (x, y) coordinate system.

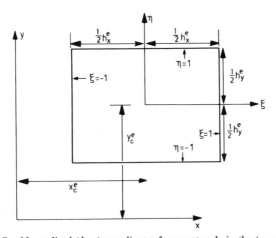

FIGURE 4.7.   Normalised (ξ, η) coordinates for a rectangle in the (x, y) plane.

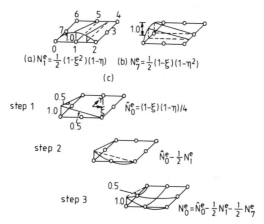

$$(a)\, N_1^e = \tfrac{1}{2}(1-\xi^2)(1-\eta) \qquad (b)\, N_7^e = \tfrac{1}{2}(1-\xi)(1-\eta^2)$$

(c)

step 1 $\hat{N}_0^e = (1-\xi)(1-\eta)/4$

step 2 $\hat{N}_0^e - \tfrac{1}{2}N_1^e$

step 3 $N_0^e = \hat{N}_0^e - \tfrac{1}{2}N_1^e - \tfrac{1}{2}N_7^e$

FIGURE 4.8.   Systematic generation of serendipity shape functions for rectangular elements.

The derivation of a typical serendipity-type function is illustrated in Fig. 4.8 for a quadratic element and the process is almost self-explanatory. The midside shape functions are first obtained *directly* by taking the appropriate Lagrangian expansion of the second order in one direction and multiplying it by a linear function in the other. This achieves the desired order of polynomial on the sides, and the condition that the shape function associated with a particular node should take the value unity at that node and zero at all other nodes is automatically satisfied.

For a corner node this simple process is inadequate as along one side a nonzero value for the shape function will be obtained at a midside node. However, a linear combination of the bilinear shape function of Eq. (3.45) and the quadratic shape functions just constructed for the midside nodes gives readily the necessary function.

The serendipity family of elements is illustrated in Fig. 4.9, and the shape functions for the first three members are given below. To avoid having to write a separate expression for all nodes, we define

$$\tilde{\xi}_l = \xi\xi_l, \qquad \tilde{\eta}_l = \eta\eta_l \tag{4.31}$$

where ξ_l and η_l are the coordinates of node l.

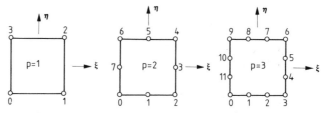

FIGURE 4.9.   Serendipity element family. For $p \geq 4$ internal parameters are required.

The element shape functions are now as follows.

Linear Elements

$$N_l^e = \tfrac{1}{4}\left(1 + \tilde{\xi}_l\right)\left(1 + \tilde{\eta}_l\right), \qquad l = 0, 1, 2, 3 \tag{4.32}$$

which are identical to the first-order Lagrangian element shape functions.

Quadratic Elements

Corner node $\qquad N_l^e = \tfrac{1}{4}(1 + \tilde{\xi}_l)(1 + \tilde{\eta}_l)(\tilde{\xi}_l + \tilde{\eta}_l - 1), \quad l = 0, 2, 4, 6 \qquad (4.33a)$

Midside node $\qquad N_l^e = \tfrac{1}{2}(1 - \xi^2)(1 + \tilde{\eta}_l), \qquad l = 1, 5$

$$N_l^e = \tfrac{1}{2}(1 + \tilde{\xi}_l)(1 - \eta^2), \qquad l = 3, 7 \tag{4.33b}$$

Cubic Elements

Corner node $\qquad N_l^e = \tfrac{1}{32}(1 + \tilde{\xi}_l)(1 + \tilde{\eta}_l)[-10 + 9(\xi^2 + \eta^2)], \qquad l = 0, 3, 6, 9$

$$\tag{4.34a}$$

Midside node $\qquad N_l^e = \tfrac{9}{32}(1 + \tilde{\xi}_l)(1 - \eta^2)(1 + 9\tilde{\eta}_l), \qquad l = 4, 5, 10, 11 \quad (4.34b)$

with the shape functions corresponding to the remaining midside nodes obtained simply by interchanging the ξ and η variables in this expression. The terms which are then present in the approximation $\hat{\phi}^e$ on a typical cubic serendipity element are shown on the Pascal triangle in Fig. 4.6c.

It is of interest to note here that the number of polynomial terms that can be obtained with the use of boundary nodes only is inadequate to give a complete polynomial representation for $\hat{\phi}^e$ when $p \geqslant 4$. For such higher order elements it is therefore again necessary to introduce internal nodes or simply an appropriate hierarchical degree of freedom of the type to be described in the following section.

4.6.2. Hierarchical-Type Shape Functions

With the one-dimensional hierarchical shape functions already established, the generation of hierarchical shape functions for rectangular elements is almost trivial as:

1. The corner node functions are simply the standard bilinear functions.
2. The products of hierarchical functions of the type defined in Sections 4.5.2 and 4.5.3 are always zero at the corner nodes.

The identification (or connection) of hierarchical variables associated with any element side with the same value on the adjacent element will automatically guarantee the uniqueness of the approximation $\hat{\phi}$ along that side, and C^0 continuity will be ensured. Any product of a one-dimensional hierarchical shape function in the local element variable along a side, say ξ, with a linear (or hierarchical) function in the other local element direction, say η, proves to be acceptable, and polynomial terms (of any degree) can be obtained by forming such simple products. Using for instance the one-dimensional functions of Eq. (4.28), the shape functions for the element shown in Fig. 4.10 can be written as follows.

Typical corner node, $\quad N_3^e(\xi, \eta) = \frac{1}{4}(1 + \eta)(1 - \xi)$ $\qquad$ (4.35a)

Typical side 2–3, $\qquad N_{2(2-3)}^e(\xi, \eta) = \frac{1}{2}(1 + \eta)(\xi^2 - 1)$

$$N_{3(2-3)}^e(\xi, \eta) = \frac{1}{6}(1 + \eta)(\xi^3 - \xi)$$

(4.35b)

and this process can be extended up to any desired order.

As we have already indicated in the context of the serendipity elements, to obtain an expansion for $\hat{\phi}^e$, which is complete to order $p \geqslant 4$, it will be necessary to add product shape functions which will be associated with a

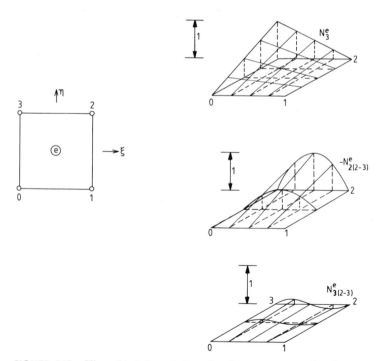

FIGURE 4.10. Hierarchical element shape functions on a rectangular element.

parameter that is not connected between elements. For instance,

$$N_{2(e)}^e = \tfrac{1}{4}(\xi^2 - 1)(\eta^2 - 1) \qquad (4.36)$$

provides a suitable shape function which can be used to add $\xi^2\eta^2$ terms to the expression for $\hat{\phi}^e$.

In passing it should be observed that it is very easy with hierarchical elements to add polynomial expansions locally to achieve a refinement in the region where the unknown function varies most rapidly and where the approximation is therefore prone to the largest errors.

The introduction of such refinements in a finite element mesh is touched upon in the last chapter where we discuss so-called adaptive refinement processes.

The numbering of nodal and hierarchical parameters in a logical sequence presents interesting logistic problems as the label corresponding to a hierarchical degree of freedom can be associated either with an element side or with a single element.

4.7. TWO-DIMENSIONAL SHAPE FUNCTIONS FOR TRIANGLES

4.7.1. Standard-Type Shape Functions — Area Coordinates

If we consider a family of triangles of arbitrary shape as shown in Fig. 4.11, we note that if the nodes follow the intersection pattern of the lines shown in the Pascal triangle (Fig. 4.6), we have at all times just a sufficient number of such nodes to generate a family of complete polynomials. This characteristic, coupled with the ease with which arbitrarily shaped domains may be closely approximated by an assembly of triangles, means that triangular elements are a popular choice for the analysis of problems involving two space dimensions.

Before discussing the details of the possible shape functions, it is desirable to define a "natural" set of coordinates for these elements. Such a convenient set of coordinates L_0, L_1, and L_2 can be defined for the element simply by

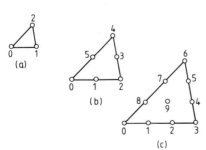

FIGURE 4.11. Standard triangular element family. (*a*) Linear, (*b*) quadratic, and (*c*) cubic.

writing a linear relationship between these coordinates and the Cartesian coordinates (x, y) in the form

$$x = L_0 x_0 + L_1 x_1 + L_2 x_2$$

$$y = L_0 y_0 + L_1 y_1 + L_2 y_2 \qquad (4.37)$$

$$1 = L_0 + L_1 + L_2$$

Here nodes numbered $0, 1, 2$ have been placed at the vertices of the triangle and (x_i, y_i) denotes the Cartesian coordinates of node i.

We note immediately that L_0 must be a function which is zero at vertices 1 and 2 and takes the value unity at vertex 0—just like the shape function for a linear element. Contours of this function are shown in Fig. 4.12. The value of L_0 at any point P can in fact be defined as the ratio of two triangle areas, that is,

$$L_0|_P = \frac{\text{area}(P12)}{\text{area}(012)} \qquad (4.38)$$

and hence the name area coordinates for the system (L_0, L_1, L_2).

The relationship of Eq. (4.37) can be used to relate directly the area coordinates (L_0, L_1, L_2) and the Cartesian coordinates (x, y) as

$$L_0 = \frac{\alpha_0^e + \beta_0^e x + \gamma_0^e y}{2\Delta^e}$$

$$L_1 = \frac{\alpha_1^e + \beta_1^e x + \gamma_1^e y}{2\Delta^e} \qquad (4.39)$$

$$L_2 = \frac{\alpha_2^e + \beta_2^e x + \gamma_2^e y}{2\Delta^e}$$

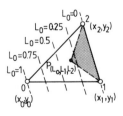

FIGURE 4.12.   Area coordinates for a triangular element.

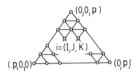

FIGURE 4.13.   General standard triangular element.

where

$$\Delta^e = \tfrac{1}{2}\det\begin{vmatrix} 1 & x_0 & y_0 \\ 1 & x_1 & y_1 \\ 1 & x_2 & y_2 \end{vmatrix} = \text{area}(012) \qquad (4.40a)$$

and

$$\alpha_0^e = x_1 y_2 - x_2 y_1$$

$$\beta_0^e = y_1 - y_2 \qquad\qquad (4.40b)$$

$$\gamma_0^e = x_2 - x_1$$

and the correspondence with the expressions of Eq. (3.27) should be noted.

Thus, for the linear three-noded triangle, the element shape functions can be simply defined by

$$N_0^e = L_0, \qquad N_1^e = L_1, \qquad N_2^e = L_2 \qquad (4.41)$$

For a general triangle with nodes spaced as in Fig. 4.13, we can write an expression for the element shape function associated with a node i, labeled by numbers I, J, and K, in the coordinate directions (L_0, L_1, L_2), as

$$N_i^e = \Lambda_I^I(L_0)\Lambda_J^J(L_1)\Lambda_K^K(L_2) \qquad (4.42)$$

in which Λ_I^I, Λ_J^J, Λ_K^K are the Lagrangian interpolation polynomials defined in Eq. (4.10).

This expression is by no means obvious, but the reader can easily verify its truth by noting that:

1. From the definition of these Lagrangian functions it follows that $N_i^e = 1$ at node i and $N_i^e = 0$ at all other nodes of the element.
2. Since $I + J + K = p$, a constant for a given triangulation, the highest order term in N_i^e will be of the form $L_0^I L_1^J L_2^K$ which, by the linear relation of Eq. (4.39), is then a polynomial in x and y of degree p.

In Fig. 4.14 typical shape functions of this form are shown for linear and quadratic elements.

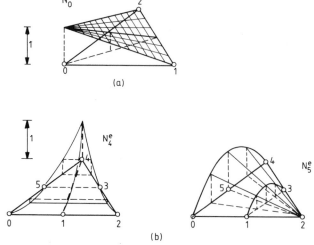

FIGURE 4.14. Triangular elements and associated standard shape functions of (*a*) linear and (*b*) quadratic form.

4.7.2. Hierarchical-Type Shape Functions

The derivation of hierarchical C^0 continuous shape functions for triangular elements is surprisingly easy.[2]

Returning to the triangle of Fig. 4.12 we note that, along the side 1–2, L_0 is identically zero, and therefore we have from Eq. (4.37),

$$[L_1 + L_2]_{1-2} = 1 \qquad (4.43)$$

If ξ, measured along side 1–2, is the usual nondimensional local element coordinate of the type we have used in deriving hierarchical functions for one-dimensional elements, we can write

$$L_1|_{1-2} = \tfrac{1}{2}(1 - \xi), \qquad L_2|_{1-2} = \tfrac{1}{2}(1 + \xi) \qquad (4.44)$$

from which it follows that we have

$$\xi = (L_2 - L_1)_{1-2} \qquad (4.45)$$

This suggests that we could generate hierarchical shape functions over the triangle by generalizing the one-dimensional shape function forms produced earlier. For example, using the expressions of Eq. (4.23), we associate with the side 1–2 a polynomial of degree $p(\geqslant 2)$ defined by

$$N_{p(1-2)}^{e} = \begin{cases} \dfrac{1}{p!}\left[(L_2 - L_1)^p - (L_1 + L_2)^p\right], & p \text{ even} \\[3mm] \dfrac{1}{p!}\left[(L_2 - L_1)^p - (L_2 - L_1)(L_1 + L_2)^{p-1}\right], & p \text{ odd} \end{cases}$$

(4.46)

It follows from Eq. (4.44) that these shape functions are zero at nodes 1 and 2. In addition, it can easily be shown that $N_{p(1-2)}^{e}$ will be zero all along the sides 0–1 and 0–2 of the triangle, and so C^0 continuity of the approximation $\hat{\phi}$ is assured.

It should be noted that in this case for $p \geqslant 3$ the number of hierarchical functions arising from the element sides in this manner is insufficient to define a complete polynomial of degree p and internal hierarchical functions, which are identically zero on the boundaries, need to be introduced; for example, for $p = 3$ the function $L_0 L_1 L_2$ could be used, while for $p = 4$ the three additional functions $L_0^2 L_1 L_2$, $L_0 L_1^2 L_2$, $L_0 L_1 L_2^2$ could be adopted.

In Fig. 4.15 typical hierarchical linear, quadratic, and cubic trial functions for a triangular element are shown. Similar hierarchical shape functions could be generated from the alternative set of one-dimensional shape functions defined in Eq. (4.28).

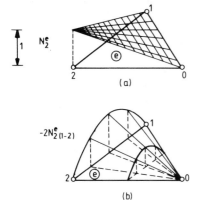

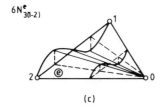

FIGURE 4.15. Triangular elements and associated hierarchial shape functions of (a) linear, (b) quadratic, and (c) cubic form.

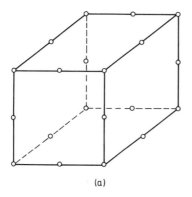

(a)

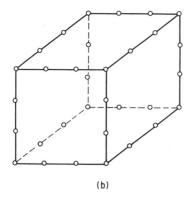

(b)

FIGURE 4.16.   Standard three-dimensional serendipity elements of (*a*) quadratic and (*b*) cubic form.

4.8.   THREE-DIMENSIONAL SHAPE FUNCTIONS

The general procedures developed in this chapter for one- and two-dimensional elements are easily adapted to three-dimensional elements of regular hexa-hedral or tetrahedral types. Once again standard or hierarchical shape functions can be developed. The only detail in which the derivation of such functions differs slightly from the previous analysis is the need for the introduction of face variables in addition to edge and element variables. In Fig. 4.16 we show typical three-dimensional elements of the serendipity family, and the reader can gain experience in the derivation of the relevant element functions by working through the exercises at the end of this chapter.

4.9.   CONCLUDING REMARKS

We have shown in this chapter how higher order element shape functions can be generated in a straightforward manner for geometrically simple elements. In particular we drew the readers' attention to hierarchical forms which possess many merits of simplicity and computational ease. More will be heard of such forms in future finite element programs for the reasons outlined.

While it is "obvious" that convergence will occur in elements of fixed form in which the degree of the polynomial p is successively increased (p convergence), it is not at all clear that this convergence is faster than that obtained with elements of fixed order in which the size h is successively reduced (h convergence). Recent studies show that if the basis of comparison is taken to be the total number of unknown parameters M, the p convergence rate is always higher. This fact can be noted from some examples quoted—but its formal proof is difficult.

In practice a compromise is always necessary as small elements may be needed to model the boundaries or inhomogeneities of the problem. Thus second- or third-degree polynomials often represent the optimal choice.

EXERCISES

4.13. Repeat Exercise 3.13 using (a) four square serendipity quadratic elements and (b) four square hierarchical quadratic elements.

4.14. For Exercise 3.13 evaluate the element matrices for (a) a Lagrangian cubic element and (b) a hierarchical cubic element.

4.15. Solve the torsion problem of Example 1.5, analyzing a quarter section with two equal hierarchical quadratic elements.

4.16. Obtain the shape functions for a standard quadratic triangular element. Repeat the torsion problem of Example 1.5 using a mesh consisting of two such elements. (The formula

$$\int_{\Delta^e} L_0^I L_1^J L_2^J \, d\Omega = \frac{I!J!K!}{(I + J + K + 2)!} 2\Delta^e$$

may be assumed.)

4.17. In Exercise 3.18 elements 1 and 4 are now to be regarded as quadratic triangular elements, while elements 2 and 3 are to be replaced by a single quadratic serendipity element. Formulate the resulting matrix equation governing the approximation to the steady-state temperature distribution.

4.18. Obtain the weak form of the weighted residual statement for the problem of the deflection of a uniform thin elastic plate, described in Exercise 1.21. Show that the standard theory requires C^1 continuous elements if a solution is to be attempted by the finite element method. Produce the element shape functions for a rectangular four-noded cubic element which ensures continuity of $\hat{\phi}$, $\partial\hat{\phi}/\partial x$, $\partial\hat{\phi}/\partial y$, $\partial^2\hat{\phi}/\partial x\partial y$ across element boundaries and evaluate the resulting element matrices for the problem of Exercise 1.21. (Hint: Use element shape functions which are products of Hermite polynomials of the type introduced in Exercise 4.12).

4.19. Repeat Exercise 3.19 for the case of a three-noded linear triangular element and compare the results with those produced in Section 3.8.1. (The integration formula of Exercise 4.16 may be assumed.)

4.20. Obtain the element shape functions for (a) the quadratic and cubic three-dimensional serendipity elements shown in Fig. 4.16 and (b) standard quadratic and cubic tetrahedral elements. For part (a) calculate the appropriate element matrices for use in a problem of steady heat conduction in three dimensions.

4.21. Produce hierarchical element shape functions for (a) quadratic and cubic hexahedral elements and (b) quadratic and cubic tetrahedral elements. For part (a) calculate the appropriate element matrices for use in a problem of steady heat conduction in three dimensions.

REFERENCES

[1] S. D. Conte and C. de Boor, *Elementary Numerical Analysis*, 2nd ed., McGraw-Hill, New York, 1972.

[2] A. Peano, R. Riccioni, A. Pasini, and L. Sardella, *Adaptive Approximations in Finite Element Structural Analysis*, ISMES, Bergamo, Italy, 1978.

SUGGESTED FURTHER READING

E. B. Becker, G. F. Carey, and J. T. Oden, *Finite Elements: An Introduction*, Vol. 1, Prentice-Hall, Englewood Cliffs, N.J., 1981.

A. J. Davies, *The Finite Element Method*, Clarendon, Oxford, 1980.

I. Fried, *Numerical Solution of Differential Equations*, Academic, New York, 1979.

E. Hinton and D. R. J. Owen, *An Introduction to Finite Element Computations*, Pineridge Press, Swansea, 1979.

O. C. Zienkiewicz, *The Finite Element Method*, 3rd ed., McGraw-Hill, London, 1977.

CHAPTER FIVE

Mapping and Numerical Integration

5.1. THE CONCEPT OF MAPPING

5.1.1. General Remarks

The high degree of accuracy which can be achieved with the higher order elements introduced in the preceding chapter means that a small number of such elements can often be used to obtain practically adequate solutions. Unfortunately the simple shapes of the elements so far derived restrict severely their application in the analysis of practical problems, where often quite complex geometrical boundaries have to be modeled. This restriction would be removed if we could "map" a simple element, such as the rectangle in the local (ξ, η) element coordinates, into a more complex shape in the global (x, y) coordinate system. By *mapping* we understand here a unique, one-to-one relationship between the coordinates (ξ, η) and (x, y). In Fig. 5.1 we show the essentials of such a map of a square element, indicating how the coordinate lines in the (ξ, η) plane become continuously distorted.

Most readers for instance will be familiar with the relation between cylindrical polar and Cartesian coordinates where (Fig. 5.2)

$$x = r \cos \theta$$

$$y = r \sin \theta$$

(5.1)

and this transformation is nothing other than a mapping by which a rectangular (r, θ) domain (or element) is mapped into the (x, y) space.

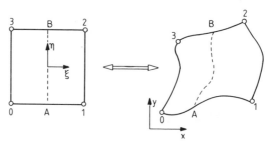

FIGURE 5.1.   A general mapping of a square (ξ, η) element.

A mapping is usually described in terms of some functional relationship between two coordinate systems which can be generally expressed as

$$x = f_1(\xi, \eta)$$

$$y = f_2(\xi, \eta)$$

(5.2)

Once a particular form of mapping is adopted and the coordinates are chosen for every element so that these map into contiguous spaces, then shape functions written in the local element (ξ, η) space can be used to represent the function variation over the element in the global (x, y) space without upsetting the interelement continuity requirements.

In the derivation of the element matrices for the various differential equations with which we have dealt so far, it has been necessary to establish

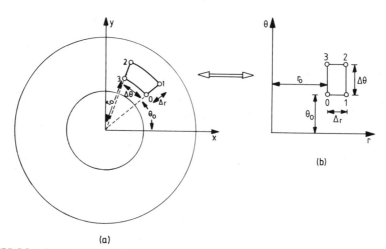

(a)

FIGURE 5.2.   A map between the Cartesian coordinates (x, y) and the cylindrical polar coordinates (r, θ).

the shape function derivatives with respect to the (x, y) coordinates. To obtain such derivatives we note that, in the previous chapter, we have generally defined the rectangular element shape functions in a local (ξ, η) element coordinate system, that is, $N_i^e = N_i^e(\xi, \eta)$. However, by the chain rule of differentiation we can always write*

$$\frac{\partial N_i^e}{\partial \xi} = \frac{\partial N_i^e}{\partial x}\frac{\partial x}{\partial \xi} + \frac{\partial N_i^e}{\partial y}\frac{\partial y}{\partial \xi}$$

$$\frac{\partial N_i^e}{\partial \eta} = \frac{\partial N_i^e}{\partial x}\frac{\partial x}{\partial \eta} + \frac{\partial N_i^e}{\partial y}\frac{\partial y}{\partial \eta}$$

(5.3)

and the required derivatives $\partial N_i^e/\partial x$ and $\partial N_i^e/\partial y$ can be obtained by inversion as

$$\begin{bmatrix} \dfrac{\partial N_i^e}{\partial x} \\[2ex] \dfrac{\partial N_i^e}{\partial y} \end{bmatrix} = \mathbf{J}^{-1} \begin{bmatrix} \dfrac{\partial N_i^e}{\partial \xi} \\[2ex] \dfrac{\partial N_i^e}{\partial \eta} \end{bmatrix}, \qquad \mathbf{J} = \begin{bmatrix} \dfrac{\partial x}{\partial \xi} & \dfrac{\partial y}{\partial \xi} \\[2ex] \dfrac{\partial x}{\partial \eta} & \dfrac{\partial y}{\partial \eta} \end{bmatrix}$$

(5.4)

provided that $\mathbf{J}$, the *Jacobian matrix* of the transformation, is nonsingular.

For the example involving the transformation from polar to Cartesian coordinates given in Eq. (5.1),

$$\mathbf{J} = \begin{bmatrix} \cos\theta & \sin\theta \\ -r\sin\theta & r\cos\theta \end{bmatrix}$$

(5.5a)

and so

$$\begin{bmatrix} \dfrac{\partial N_i^e}{\partial x} \\[2ex] \dfrac{\partial N_i^e}{\partial y} \end{bmatrix} = \mathbf{J}^{-1} \begin{bmatrix} \dfrac{\partial N_i^e}{\partial r} \\[2ex] \dfrac{\partial N_i^e}{\partial \theta} \end{bmatrix} = \begin{bmatrix} \cos\theta & -\dfrac{\sin\theta}{r} \\[2ex] \sin\theta & \dfrac{\cos\theta}{r} \end{bmatrix} \begin{bmatrix} \dfrac{\partial N_i^e}{\partial r} \\[2ex] \dfrac{\partial N_i^e}{\partial \theta} \end{bmatrix}$$

(5.5b)

In addition, the components of the element matrices [e.g., Eq. (3.36)] are normally expressed in terms of integrals involving the (x, y) coordinates, and it would be beneficial if these integrals could be reformulated in terms of the local (ξ, η) element coordinates only.

The element of area $dx\, dy$ has therefore to be replaced by an equivalent in the (ξ, η) coordinates. Without proof[1] we state the relationship

$$dx\, dy = \det(\mathbf{J})\, d\xi\, d\eta$$

(5.6)

---

*For three dimensions similar transformations are simply written by adding a further independent variable.

yielding in case of the polar coordinates simply by Eq. (5.5a) the well-known result

$$dx\, dy = r\, dr\, d\theta \tag{5.7}$$

It is now evident that the required objective can be achieved, and any integral of interest such as

$$I = \int_{\Omega^e} k \frac{\partial N_l^e}{\partial x} \frac{\partial N_m^e}{\partial x} dx\, dy \tag{5.8a}$$

can be recast in terms of integration over a square domain as

$$I = \int_{-1}^{1} \int_{-1}^{1} k \frac{\partial N_l^e}{\partial x} \frac{\partial N_m^e}{\partial x} \det(\mathbf{J})\, d\xi\, d\eta \tag{5.8b}$$

where $\partial N_l^e/\partial x$, $\partial N_m^e/\partial x$ can be expressed in terms of ξ and η by means of Eq. (5.4). In this way all the element matrices required for a finite element approximation can be evaluated if the integrals are sufficiently simple.

It should now be obvious how practical use may be made of mapping in finite element analyses. The elements can be defined by their coordinates in the Cartesian (x, y) space. The mapping rule is then applied to each element, and the resulting element matrices are obtained by use of expressions of the type given in Eq. (5.8). Assembly and solution of the problem then follow precisely the same procedures as were used with the simple elements.

Example 5.1

In this example we return to the problem of heat conduction discussed in Chapter 3, which is governed by the general equation (3.31) and with the full element matrices as specified in Eq. (3.36). Suppose that the region to be analyzed lies between two concentric circles of radius $r = a$ and $r = b$. It is convenient now to use simple sectional elements to represent the domain in the (x, y) plane and in Fig. 5.2a we show a typical element bounded by $\theta = \theta_0$, $\theta = \theta_0 + \Delta\theta$, $r = r_0$, $r = r_0 + \Delta r$ and with corner nodes numbered 0, 1, 2, and 3. A rectangular element results from mapping this element, by the relations of Eq. (5.1), into the (r, θ) space as shown in Fig. 5.2b. The application of the linear mapping

$$\xi = \frac{2(r - r_0)}{\Delta r} - 1 \qquad \eta = \frac{2(\theta - \theta_0)}{\Delta\theta} - 1$$

then ensures that the mapped element is the square defined by $-1 \leqslant \xi, \eta \leqslant 1$. Simple bilinear (ξ, η) shape functions will be used over this square element. We shall now indicate how the terms in the first row of the reduced element matrix $\mathbf{k}^e$ can be derived. The entries in this row are given explicitly by [Eq. (3.44)]

$$k^e_{0m} = \int_{\Omega^e} \left(\frac{\partial N^e_0}{\partial x} k \frac{\partial N^e_m}{\partial x} + \frac{\partial N^e_0}{\partial y} k \frac{\partial N^e_m}{\partial y} \right) dx\, dy; \qquad m = 0, 1, 2, 3$$

and the relevant element shape functions over the square (ξ, η) element can be written

$$N^e_0 = \frac{(1 - \xi)(1 - \eta)}{4} \qquad N^e_1 = \frac{(1 + \xi)(1 - \eta)}{4}$$

$$N^e_2 = \frac{(1 + \xi)(1 + \eta)}{4} \qquad N^e_3 = \frac{(1 - \xi)(1 + \eta)}{4}$$

and hence

$$\frac{\partial N^e_0}{\partial r} = -\frac{(1 - \eta)}{2\Delta r} \qquad \frac{\partial N^e_0}{\partial \theta} = -\frac{(1 - \xi)}{2\Delta \theta}$$

$$\frac{\partial N^e_1}{\partial r} = \frac{(1 - \eta)}{2\Delta r} \qquad \frac{\partial N^e_1}{\partial \theta} = -\frac{(1 + \xi)}{2\Delta \theta}$$

$$\frac{\partial N^e_2}{\partial r} = \frac{(1 + \eta)}{2\Delta r} \qquad \frac{\partial N^e_2}{\partial \theta} = \frac{(1 + \xi)}{2\Delta \theta}$$

$$\frac{\partial N^e_3}{\partial r} = -\frac{(1 + \eta)}{2\Delta r} \qquad \frac{\partial N^e_3}{\partial \theta} = \frac{(1 - \xi)}{2\Delta \theta}$$

The relationship between the x and y derivatives appearing in the above integral and those with respect to r and θ follows from Eq. (5.5) and gives here, for example,

$$\frac{\partial N^e_0}{\partial x} = \cos\theta \frac{\partial N^e_0}{\partial r} - \frac{\sin\theta}{r} \frac{\partial N^e_0}{\partial \theta}$$

$$\frac{\partial N^e_0}{\partial y} = \sin\theta \frac{\partial N^e_0}{\partial r} + \frac{\cos\theta}{r} \frac{\partial N^e_0}{\partial \theta}$$

Finally, using Eq. (5.7), we have

$$
k_{00}^e = \frac{\Delta r \Delta \theta}{4} \int_{-1}^{1} \int_{-1}^{1} k \left[\left(\cos \theta \frac{\partial N_0^e}{\partial r} - \frac{\sin \theta}{r} \frac{\partial N_0^e}{\partial \theta} \right)^2 \right.
$$

$$
\left. + \left(\sin \theta \frac{\partial N_0^e}{\partial r} + \frac{\cos \theta}{r} \frac{\partial N_0^e}{\partial \theta} \right)^2 \right] r \, d\xi \, d\eta
$$

$$
k_{01}^e = \frac{\Delta r \Delta \theta}{4} \int_{-1}^{1} \int_{-1}^{1} k \left[\left(\cos \theta \frac{\partial N_0^e}{\partial r} - \frac{\sin \theta}{r} \frac{\partial N_0^e}{\partial \theta} \right) \left(\cos \theta \frac{\partial N_1^e}{\partial r} - \frac{\sin \theta}{r} \frac{\partial N_1^e}{\partial \theta} \right) \right.
$$

$$
\left. + \left(\sin \theta \frac{\partial N_0^e}{\partial r} + \frac{\cos \theta}{r} \frac{\partial N_0^e}{\partial \theta} \right) \left(\sin \theta \frac{\partial N_1^e}{\partial r} + \frac{\cos \theta}{r} \frac{\partial N_1^e}{\partial \theta} \right) \right] r \, d\xi \, d\eta
$$

$$
k_{02}^e = \frac{\Delta r \Delta \theta}{4} \int_{-1}^{1} \int_{-1}^{1} k \left[\left(\cos \theta \frac{\partial N_0^e}{\partial r} - \frac{\sin \theta}{r} \frac{\partial N_0^e}{\partial \theta} \right) \left(\cos \theta \frac{\partial N_2^e}{\partial r} - \frac{\sin \theta}{r} \frac{\partial N_2^e}{\partial \theta} \right) \right.
$$

$$
\left. + \left(\sin \theta \frac{\partial N_0^e}{\partial r} + \frac{\cos \theta}{r} \frac{\partial N_0^e}{\partial \theta} \right) \left(\sin \theta \frac{\partial N_2^e}{\partial r} + \frac{\cos \theta}{r} \frac{\partial N_2^e}{\partial \theta} \right) \right] r \, d\xi \, d\eta
$$

$$
k_{03}^e = \frac{\Delta r \Delta \theta}{4} \int_{-1}^{1} \int_{-1}^{1} k \left[\left(\cos \theta \frac{\partial N_0^e}{\partial r} - \frac{\sin \theta}{r} \frac{\partial N_0^e}{\partial \theta} \right) \left(\cos \theta \frac{\partial N_3^e}{\partial r} - \frac{\sin \theta}{r} \frac{\partial N_3^e}{\partial \theta} \right) \right.
$$

$$
\left. + \left(\sin \theta \frac{\partial N_0^e}{\partial r} + \frac{\cos \theta}{r} \frac{\partial N_0^e}{\partial \theta} \right) \left(\sin \theta \frac{\partial N_3^e}{\partial r} + \frac{\cos \theta}{r} \frac{\partial N_3^e}{\partial \theta} \right) \right] r \, dr \, d\eta
$$

where $r = r_0 + (\xi + 1)\Delta r/2$ and $\theta = \theta_0 + (\eta + 1)\Delta\theta/2$.

Although the algebra is now more involved, it will be seen in Section 5.2 how expressions of this type may be simply evaluated using a computer, and sectional elements of this kind find considerable practical use.

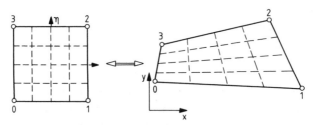

FIGURE 5.3.   A linear isoparametric map.

5.1.2. Parametric Mapping

A very convenient form of element mapping is one of the parametric kind in which the relationship between the local element coordinates (ξ, η) and the global coordinates (x, y) is written using the same kind of interpolation as we use to approximate the unknown function ϕ. If $N_l^e(\xi, \eta)$ represents a standard type of finite element shape function for an $(M + 1)$-noded element in the local domain, we can write the mapping relationship of Eq. (5.2) for each element as

$$x = N_0^e x_0 + N_1^e x_1 + \cdots + N_M^e x_M$$
$$N_l^e = N_l^e(\xi, \eta) \qquad (5.9)$$
$$y = N_0^e y_0 + N_1^e y_1 + \cdots + N_M^e y_M$$

where (x_l, y_l) are the global (x, y) coordinates of the point into which we wish to map node l of the element in the (ξ, η) space.

It is clear that, if the global shape functions N_l used possess the property of C^0 interelement continuity, the coordinate maps will be similarly continuous, even if different local origins are used in each element. In Figs. 5.3 and 5.4 we show how square elements in the (ξ, η) space can be mapped in this fashion into progressively more complex shapes in the global space by increasing the number of element nodes $(M + 1)$.

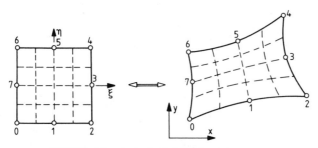

FIGURE 5.4.   A quadratic isoparametric map.

If, as here, the element mapping and the approximation $\hat{\phi}$ over the element are defined by means of the same shape functions, then the mapping and usage are termed *isoparametric*.

However, if required, the mapping can be performed by using only a selected number of the element shape functions. Generally, when mapping a square element, it suffices to use only four or at most eight nodes to produce a mapped element of reasonably general shape—the first gives an element with straight sides, while the second gives an element with sides in the form of arbitrary quadratics.

Care must be taken when using parametric mapping to avoid nonproper mapping forms. These will occur if the determinant of the Jacobian matrix changes sign in the transformation domain. It is found that, with quadrilateral elements, such a nonproper mapping is produced if an interior angle of the element exceeds 180° or if, in quadratic mapping, the distance between a center node and a corner node is less than one-third of the length of a quadrilateral side.

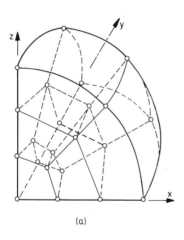

(a)

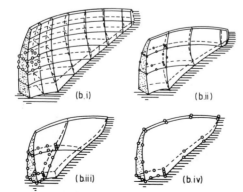

(b.i)   (b.ii)

(b.iii)   (b.iv)

FIGURE 5.5. Use of distorted hexahedral elements in some three-dimensional problems. (*a*) Representation of a portion of a sphere using seven quadratic elements. (*b*) Various representations of an arch dam. (*c*) Representation of a portion of a pressure vessel.

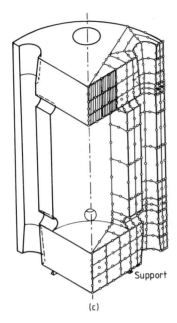

(c)   **FIGURE 5.5.** (*continued*).

The previously derived relations of Eqs. (5.4) and (5.6) can be used in the parametric context, although now the algebra involved is such that the inversion of the Jacobian, the calculation of its determinant, and also the integration involved in the evaluation of the coefficients of the element matrices have to be performed numerically. We discuss possible numerical integration processes in the next section. Elements of isoparametric type are popular in practical applications. The problem domain can be divided into elements of appropriate shape and size and the nodal coordinates specified. The satisfaction of interelement continuity is immediate, and the element integrals are evaluated as described above. Curved boundaries can be accurately approximated by isoparametric elements since the element sides are described in terms of polynomials, of degree appropriate to the type of element used, passing through the chosen nodal points. In the interior of the domain there is generally no need for such curvilinear interfaces between elements, and it is usual then to make the element interfaces straight lines. If such a choice is made, only the coordinates of the corner nodes need to be given, and the coordinates of intermediate nodes can be calculated automatically by interpolation.

The parametric mapping can also be used for triangular elements (see Exercises 5.8–5.11) as well as for all the three-dimensional element forms which were discussed in the last chapter. Indeed, in three dimensions its use is most common, and in Fig. 5.5 we illustrate how distorted hexahedral elements can be used to represent the domain of interest in some three-dimensional problems.

We should mention that the parametric mapping here has been discussed in the context of standard, rather than hierarchical, shape function forms. It is of course possible to use the latter if the geometry of a distorted element can be represented by, say, position of the corner nodes, departure of the midpoints from a straight line, and so on. This becomes, however, overly complex for practical usage.

Once an element is distorted by a general mapping, the approximation $\hat{\phi}$ to the unknown function is only a polynomial over the element in terms of the local coordinates (ξ, η) and need no longer be a polynomial in terms of the global (x, y) coordinates. The remarks on convergence made at the start of Chapter 4 now only apply with reference to the local coordinates rather than to the global system. However, it is often of interest in finite element solutions to ensure that the method be capable of reproducing exactly a solution which is a polynomial in the (x, y) coordinates. It is fortunate that the isoparametric expansion is able to reproduce exactly a linear polynomial in the (x, y) coordinates,[2] and it has recently been shown that quadratic elements, derived on the basis of a full Lagrangian expansion (nine-noded or equivalent hierarchical form), can give an exact quadratic expression, provided the mapping is carried out only by means of the bilinear corner-node functions which distort the element into an arbitrary quadrilateral. This feature is not available in the eight-noded serendipity quadratic elements which suffer under distortion.

EXERCISES

5.1. The nodal coordinates of a four-noded element Ω^e are $(0, 0)$, $(1, 0)$, $(0.4, 0.4)$, $(0, 1)$ with respect to an (x, y) coordinate system. Construct an isoparametric map between Ω^e and the square bilinear element defined by $-1 \leqslant \xi, \eta \leqslant 1$. Show that the mapping is not acceptable, since $\det(\mathbf{J})$ changes sign in the transformation domain, and identify the reason for the failure of the mapping process in this case.

5.2. Consider a square quadratic Lagrangian element in the local (ξ, η) coordinate system as shown in Fig. 4.5b. Relative to an (x, y) coordinate system the nodes (r, s) are defined by the following coordinates:

r	s	x	y
0	0	20	20
0	1	20	50
0	2	0	70
1	0	50	40
1	1	60	55
1	2	70	67
2	0	100	30
2	1	110	60
2	2	120	80

and an isoparametric mapping is employed. Plot in the (x, y) space the element sides (i.e., the lines $\xi = \pm 1, \eta = \pm 1$) and several intermediate contours of equal ξ and η.

5.3.   A certain eight-noded element Ω^e has its midside nodes in the (x, y) space placed so as to lie at the midpoints of the straight lines joining the corner nodes. Show that the resulting isoparametric mapping between Ω^e and the eight-noded square serendipity element defined by $-1 \leqslant \xi, \eta \leqslant 1$ involves only the bilinear shape functions corresponding to the corner nodes.

5.4.   In the analysis of a certain problem the domain of interest Ω is defined by $y^4 \leqslant 5x$. Following a finite element discretization of Ω, an element Ω^e, which is adjacent to the boundary curve $y^4 = 5x$, is to have eight nodes with coordinates as shown in the figure. Construct an isoparametric map between Ω^e and the eight-noded square serendipity element in the local (ξ, η) coordinate system and compare the shape of the curved side of Ω^e with that of the actual domain Ω.

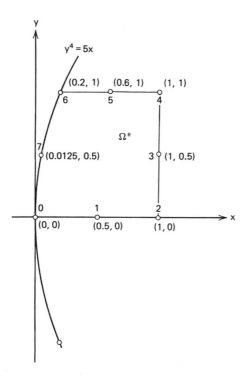

5.5.   Frequently in finite element computations boundary integrals must be evaluated along with domain integrals of the type considered in Eq. (5.8). Let Ω^e be an eight-noded element which is adjacent to the boundary Γ in a certain two-dimensional problem and suppose Γ^e denotes that side of Ω^e which is assumed to be along Γ. An isoparametric mapping is made between Ω^e and the square eight-noded serendipity element over $-1 \leqslant \xi, \eta \leqslant 1$ and is such that Γ^e is mapped into the line $\eta = -1$. If the nodes on Γ^e are numbered 0, 1, and 2, show that, along Γ^e,

$$x = x(\xi) = x_0 N_0^e(\xi) + x_1 N_1^e(\xi) + x_2 N_2^e(\xi)$$

$$y = y(\xi) = y_0 N_0^e(\xi) + y_1 N_1^e(\xi) + x_2 N_2^e(\xi)$$

where (x_j, y_j) denotes the coordinates of node j and $N_0^e(\xi)$, $N_1^e(\xi)$, $N_2^e(\xi)$ are the standard Lagrange one-dimensional quadratic shape functions of Eq. (4.12b). Hence prove that

$$\int_{\Gamma^e} f(x, y) \, d\Gamma = \int_{-1}^{1} f(x(\xi), y(\xi)) \sqrt{\left(\frac{dx}{d\xi}\right)^2 + \left(\frac{dy}{d\xi}\right)^2} \, d\xi$$

and verify the result by evaluating both integrals exactly for the element and conditions shown in the figure.

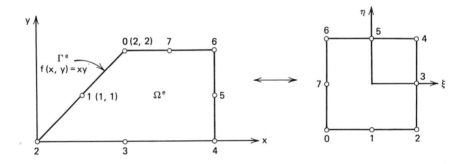

5.6. Show that the inverse of the Jacobian matrix $\mathbf{J}$ of Eq. (5.4) can be expressed as

$$\mathbf{J}^{-1} = \frac{1}{\det(\mathbf{J})} \begin{bmatrix} \dfrac{\partial y}{\partial \eta} & -\dfrac{\partial y}{\partial \xi} \\ -\dfrac{\partial x}{\partial \eta} & \dfrac{\partial x}{\partial \xi} \end{bmatrix}$$

and also prove the relationships

$$\frac{\partial \xi}{\partial x} = \frac{1}{\det(\mathbf{J})} \frac{\partial y}{\partial \eta}, \qquad \frac{\partial \eta}{\partial x} = -\frac{1}{\det(\mathbf{J})} \frac{\partial y}{\partial \xi}$$

$$\frac{\partial \xi}{\partial y} = -\frac{1}{\det(\mathbf{J})} \frac{\partial x}{\partial \eta}, \qquad \frac{\partial \eta}{\partial y} = \frac{1}{\det(\mathbf{J})} \frac{\partial x}{\partial \xi}$$

5.7. A single four-noded element is used to represent the region Ω^e shown in the figure. Construct an isoparametric mapping between Ω^e and the square bilinear element over $-1 \leqslant \xi, \eta \leqslant 1$. Then determine the steady-state distribution of temperature ϕ through Ω^e when the region is subjected to the boundary conditions shown.

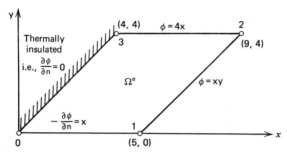

5.8. Consider the three-noded isosceles right-angled triangular element with nodes numbered 0, 1, and 2 at the points $(0,0)$, $(1,0)$, and $(0,1)$, respectively, in an (ξ, η) coordinate system. Show that the element shape functions can be written as $N_0^e = 1 - \xi - \eta = L_0$, $N_1^e = \xi = L_1$, $N_2^e = \eta = L_2$ and indicate how a general three-noded triangular element can be mapped into this element.

5.9. Return to Section 3.8.1 and obtain integral expressions for the element matrices $\mathbf{k}^e$ and $\mathbf{f}^e$ by mapping the general three-noded triangular element e into the isosceles right-angled triangular element described in Exercise 5.8.

5.10. An isosceles right-angled six-noded quadratic triangular element is illustrated in the figure. Show that the element shape functions can be

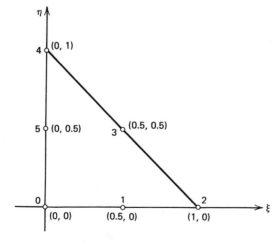

written

$$N_0^e = (1 - \xi - \eta)(1 - 2\xi - 2\eta),$$

$$N_2^e = \xi(2\xi - 1), \qquad N_4^e = \eta(2\eta - 1)$$

$$N_1^e = 4\xi(1 - \xi - \eta), \qquad N_3^e = 4\xi\eta, \qquad N_5^e = 4\eta(1 - \xi - \eta)$$

A six-noded element Ω^e has nodes numbered 0, 1, 2, 3, 4, and 5 with coordinates $(1, 1)$, $(5, 1)$, $(8, 2)$, $(7.5, 4)$, $(4, 5)$, and $(2, 3)$, respectively, in an (x, y) coordinate system. Construct an isoparametric mapping between Ω^e and the right-angled six-noded element described above and plot the corresponding curved sides of the domain Ω^e.

5.11.   Construct an isosceles right angled ten-noded triangular element and produce expressions for the element shape functions.

5.2.   NUMERICAL INTEGRATION

5.2.1.   General Remarks

In the derivation of the element matrices for higher order elements the complexity of the terms involved under the integration increases, making algebraic manipulation tedious. If, in addition, a mapping of the domain is used to distort its shape, the evaluation of the derivatives appearing in these expressions involves inversion of the Jacobian matrix [see Eq. (5.4)], and the integrals then become so complex that their exact evaluation is almost impossible. For such cases an approximate numerical evaluation of the integrals is necessary and then a typical integral, such as occurs in Eq. (5.8), will be replaced by a summation.

In general, we can approximate such integrals, over one-, two-, and three-dimensional domains, by using a simple summation of terms involving the integrand, evaluated at specified points in the domain, multiplied by suitable weightings, namely,

$$I = \int_{-1}^{1} G(\xi)\, d\xi \approx W_0 G(\xi_0) + W_1 G(\xi_1) + \cdots + W_n G(\xi_n) \qquad (5.10a)$$

$$I = \int_{-1}^{1} \int_{-1}^{1} G(\xi, \eta)\, d\xi\, d\eta$$

$$\approx W_0 G(\xi_0, \eta_0) + W_1 G(\xi_1, \eta_1) + \cdots + W_n G(\xi_n, \eta_n) \qquad (5.10b)$$

To develop methods of this type in one dimension we could choose the *sampling points* $\xi_0, \xi_1, \ldots, \xi_n$ at the outset and then determine the polynomial $F_n(\xi)$ of degree n which is exactly equal to $G(\xi)$ at each of these points. To do this we write

$$F_n(\xi) = \alpha_0 + \alpha_1 \xi + \cdots + \alpha_n \xi^n \qquad (5.11)$$

and the coefficients are given uniquely by solution of the equations

$$G(\xi_0) = \alpha_0 + \alpha_1\xi_0 + \cdots + \alpha_n\xi_0^n$$

$$G(\xi_1) = \alpha_0 + \alpha_1\xi_1 + \cdots + \alpha_n\xi_1^n$$

$$\vdots \qquad \vdots$$

$$G(\xi_n) = \alpha_0 + \alpha_1\xi_n + \cdots + \alpha_n\xi_n^n$$

(5.12)

By writing

$$I = \int_{-1}^{1} G(\xi)\,d\xi \approx \int_{-1}^{1} F_n(\xi)\,d\xi = 2\alpha_0 + \frac{2\alpha_2}{3} + \cdots + \frac{\alpha_n}{n+1}\left[1 - (-1)^{n+1}\right]$$

(5.13)

we can approximate to the value of the required integral and, if we substitute into this expression the values for the coefficients given by solution of Eq. (5.12), an approximation formula of the type given in Eq. (5.10a) results. This is illustrated in Fig. 5.6, which shows the use of the familiar trapezoidal approximation. Here we take $\xi_0 = -1$, $\xi_1 = 1$, and we can identify

$$F_1(\xi) = \frac{G(\xi_1) + G(\xi_0)}{2} + \frac{G(\xi_1) - G(\xi_0)}{2}\xi$$

(5.14)

as the linear function which is exactly equal to $G(\xi)$ at these two sampling points.

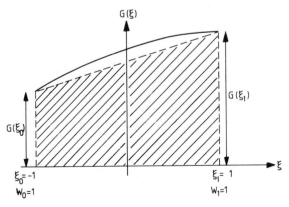

FIGURE 5.6.   Weights and sampling points for the trapezoidal approximation (shown shaded) to $\int_{-1}^{1} G(\xi)\,d\xi$.

The required integral is then approximated by

$$I = \int_{-1}^{1} G(\xi) \, d\xi \approx \int_{-1}^{1} F_1(\xi) \, d\xi = G(\xi_0) + G(\xi_1) \qquad (5.15)$$

which is of the general form of Eq. (5.10a) with $n = 1$.

Performing a detailed error analysis[3] on the approximation involved in Eq. (5.13) shows that if n is odd, this method will integrate exactly any polynomial up to degree n, while if n is even, the method is exact for any polynomial up to degree $n + 1$.

With the sampling points equally spaced, formulas of this type are known as Newton–Cotes methods, and they require generally $n + 1$ evaluations of the function $G(\xi)$. As we shall not recommend the use of such methods in the finite element context, the explicit derivation of the higher order formulas of this type is left to the reader.

5.2.2. Gaussian Quadrature for One-Dimensional Integration

Instead of proceeding as in the previous section, a better approach would be not to assign, a priori, the positions at which the function $G(\xi)$ is to be evaluated, but to attempt to determine these coordinates so that the approximation gives the exact value of the integral whenever $G(\xi)$ is a polynomial of degree less than or equal to p, where $p(\geqslant n)$ is also to be determined. Writing such a polynomial as

$$F_p(\xi) = \alpha_0 + \alpha_1 \xi + \cdots + \alpha_p \xi^p \qquad (5.16)$$

this can be integrated using the approximation of Eq. (5.10a) to obtain

$$I = \int_{-1}^{1} F_p(\xi) \, d\xi$$

$$\approx W_0 \big(\alpha_0 + \alpha_1 \xi_0 + \cdots + \alpha_p \xi_0^p \big) + W_1 \big(\alpha_0 + \alpha_1 \xi_1 + \cdots + \alpha_p \xi_1^p \big) \quad (5.17)$$

$$+ \cdots + W_n \big(\alpha_0 + \alpha_1 \xi_n + \cdots + \alpha_p \xi_n^p \big)$$

while, from Eq. (5.13), we know that the exact value of this integral is given by

$$I = 2\alpha_0 + \frac{2\alpha_2}{3} + \cdots + \frac{\alpha_p}{p+1} \big[1 - (-1)^{p+1} \big] \qquad (5.18)$$

By comparing coefficients, it follows that the formula of Eq. (5.10a) gives the exact value of the integral for any such polynomial $F_p(\xi)$, provided that

$$W_0 + W_1 + \cdots + W_n = 2$$

$$W_0\xi_0 + W_1\xi_1 + \cdots + W_n\xi_n = 0$$

$$\vdots \quad \vdots \tag{5.19}$$

$$W_0\xi_0^p + W_1\xi_1^p + \cdots + W_n\xi_n^p = \frac{1}{p+1}\left[1 - (-1)^{p+1}\right]$$

In the above set of $p + 1$ equations the quantities $\{W_i, \xi_i; i = 0, 1, \ldots, n\}$ are as yet unknown, and a solution will only be possible when the number of equations matches the number of unknowns, that is, when

$$p + 1 = 2(n + 1) \tag{5.20}$$

Since n is an integer, it can be seen that p will always be an odd number, and we tabulate below the variation of p with the first few values of n.

Number of Sampling Points $(n + 1)$	Degree of Polynomial Integrated Exactly (p)
1	1
2	3
3	5
4	7

In other words, with, for example, three function evaluations we can integrate exactly a polynomial of degree 5, while with the Newton–Cotes formulas, discussed in the previous section, only a polynomial of degree 3 could be integrated exactly with three sampling points. This method of integral evaluation is usually termed Gauss–Legendre quadrature.

Proceeding further, we can evaluate the sampling coordinates and the weights for particular values of n. Thus for $n = 0$ we write Eq. (5.19) simply as

$$W_0 = 2$$

$$W_0\xi_0 = 0 \tag{5.21}$$

Obviously with the single sampling point the best efficiency results from placing the point at the center of the region, and this method will exactly

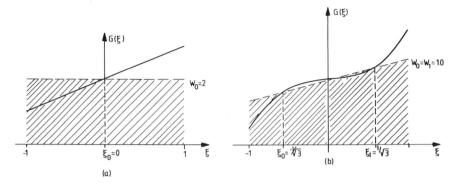

FIGURE 5.7. Gauss–Legendre quadrature. (a) One sampling point, exact for $p = 1$. (b) Two sampling points, exact for $p = 3$. In both cases the approximation to the area under the curve is shaded.

integrate a polynomial of first degree (Fig. 5.7a). Similarly for $n = 1$ we have

$$W_0 + W_1 = 2$$

$$W_0\xi_0 + W_1\xi_1 = 0 \qquad (5.22)$$

$$W_0\xi_0^2 + W_1\xi_1^2 = \tfrac{2}{3}$$

This nonlinear set of equations is more difficult to solve, but the reader can verify by back substitution that the solution is

$$\xi_1 = -\xi_0 = \frac{1}{\sqrt{3}} = 0.577\,350\,259\,1$$

$$\qquad (5.23)$$

$$W_0 = W_1 = 1.0$$

and this method can be used to integrate exactly a polynomial of degree 3, as shown in Fig. 5.7b.

The six equations produced when $n = 2$ result, by exactly similar analysis, in the solution

$$\xi_2 = -\xi_0 = \sqrt{\tfrac{3}{5}} = 0.774\,596\,669\,2, \qquad \xi_1 = 0.0$$

$$\qquad (5.24)$$

$$W_0 = W_2 = \tfrac{5}{9} = 0.555\,555\,555\,5, \qquad W_1 = \tfrac{8}{9} = 0.888\,888\,888\,9$$

Many texts dealing with numerical analysis present such values of the Gauss–Legendre sampling or integration points and the corresponding weightings to high values of n.[4]

Other possibilities for numerical integration or quadrature exist which the reader could explore. For instance, it may well be decided to fix at the outset some values of ξ_i and leave only the others to be determined. In such a case, and with a specified number of sampling points, the degree of the polynomial integrated exactly by the method would be no more than that achieved by the corresponding Gauss–Legendre quadrature and no less than that achieved by the corresponding Newton–Cotes method. In particular, it may sometimes be useful to ensure that the function is sampled at the two ends of the domain (i.e., $\xi_n = -\xi_0 = 1$), but to preserve the freedom of choice for interior points. Such an approach results in the so-called Gauss–Lobatto quadrature rules.

As an example consider a three-point rule with $\xi_2 = -\xi_0 = 1$. We thus have ξ_1, W_0, W_1, and W_2 as four unknowns and four equations of the form (5.19). Now we can write

$$W_0 + W_1 + W_2 + W_3 = 2$$

$$- W_0 + W_1\xi_1 + W_2 = 0$$

$$W_0 + W_1\xi_1^2 + W_2 = \tfrac{2}{3} \qquad (5.25)$$

$$- W_0 + W_1\xi_1^3 + W_2 = 0$$

and the solution is

$$\xi_1 = 0, \qquad W_0 = W_2 = \tfrac{1}{3}, \qquad W_1 = \tfrac{4}{3} \qquad (5.26)$$

This indeed is precisely the same as the Newton–Cotes integration formula with three equally spaced points, and we know that a third-degree polynomial will be exactly integrated.

A second member of this series would need the evaluation of ξ_1, ξ_2, W_0, W_1, W_2, and W_3, thus requiring the specification of six equations, and a rule which integrates exactly a polynomial of degree 5 results.

The reader can verify, as shown in Fig. 5.8, that the solution in this case is

$$\xi_3 = -\xi_0 = 1, \qquad \xi_2 = -\xi_1 = 1/\sqrt{5}$$

$$W_3 = W_0 = \tfrac{1}{6}, \qquad W_2 = W_1 = \tfrac{5}{6}$$

Many other quadrature formulas exist in the literature, but for finite element computation the Gauss–Legendre approach is particularly convenient as it requires the least number of computations to evaluate exactly integrals involving polynomials.

5.2.3.   Gauss Quadrature in Two and Three Dimensions

In a two-dimensional context we are faced with the problem of evaluating a double integral of the form [see Eq. (5.10b)]

$$I = \int_{-1}^{1}\int_{-1}^{1} G(\xi, \eta)\, d\xi\, d\eta \qquad (5.28)$$

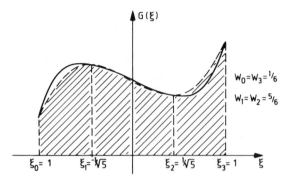

FIGURE 5.8. Gauss–Lobatto quadrature with four sampling points, exact for $p = 5$. The approximation to the area under the curve is shaded.

As the integration extends over a rectangular domain, the simplest procedure here is to perform two numerical integrations in the ξ and η directions independently. Thus we begin by evaluating the inner integral, using the formulas of the preceding section, and obtain

$$\int_{-1}^{1} G(\xi, \eta) \, d\xi \approx \sum_{i=0}^{n} W_i G(\xi_i, \eta) \tag{5.29}$$

We can follow this by a similar integration in the η direction so that

$$I \approx \int_{-1}^{1} \left[\sum_{i=0}^{n} W_i G(\xi_i, \eta) \right] d\eta \approx \sum_{j=0}^{n} \left[W_j \sum_{i=0}^{n} W_i G(\xi_i, \eta_j) \right] \tag{5.30}$$

and resulting finally in the approximation

$$I \approx \sum_{i=0}^{n} \sum_{j=0}^{n} \overline{W}_{ij} G(\xi_i, \eta_j), \qquad \overline{W}_{ij} = W_i W_j \tag{5.31}$$

Now (ξ_i, η_j) denotes the position of the sampling points whose exact location is determined by the type of the integration formula employed.

If the integrals in the ξ and η directions separately are exact for polynomials of degree p, then the expression of Eq. (5.31) will integrate exactly all terms such as $\xi^{p_1} \eta^{p_2}$ where $p_1, p_2 \leqslant p$. Typical Gauss–Legendre quadrature rules of this type are illustrated in Fig. 5.9. The three-dimensional extension to the evaluation of integrals over the prismatic three-dimensional region $-1 \leqslant$

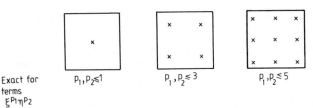

| Exact for terms $\xi^{P_1}\eta^{P_2}$ | $P_1, P_2 \leqslant 1$ | $P_1, P_2 \leqslant 3$ | $P_1, P_2 \leqslant 5$ |

FIGURE 5.9. Sampling points for Gauss–Legendre quadrature over quadrilaterals.

$\xi, \eta, \zeta \leqslant 1$ is trivial, resulting in

$$\int_{-1}^{1}\int_{-1}^{1}\int_{-1}^{1} G(\xi, \eta, \zeta)\, d\xi\, d\eta\, d\zeta \approx \sum_{i=0}^{n}\sum_{j=0}^{n}\sum_{k=0}^{n} \overline{W}_{ijk} G(\xi_i, \eta_j, \zeta_k),$$

$$\overline{W}_{ijk} = W_i W_j W_k \quad (5.32)$$

In two- and three-dimensional finite element computations product formulas of this type are commonly used, but the reader will observe that the integration process described above will be exact for terms additional to those appearing in a complete polynomial of degree p in the independent variables ξ, η, ζ. It should therefore be possible to establish formulas which integrate exactly complete polynomials of a certain degree and require fewer sampling

Number of sampling points	Figure	Degree of polynomial integrated exactly	Points	Triangular coordinates	Weights
1		1	0	$\frac{1}{3}, \frac{1}{3}, \frac{1}{3}$	1
3		2	0	$\frac{1}{2}, \frac{1}{2}, 0$	$\frac{1}{3}$
			1	$0, \frac{1}{2}, \frac{1}{2}$	$\frac{1}{3}$
			2	$\frac{1}{2}, 0, \frac{1}{2}$	$\frac{1}{3}$
4		3	0	$\frac{1}{3}, \frac{1}{3}, \frac{1}{3}$	$-\frac{27}{48}$
			1	0.6, 0.2, 0.2	
			2	0.2, 0.6, 0.2	$\frac{25}{48}$
			3	0.2, 0.2, 0.6	
7		4	0	$\frac{1}{3}, \frac{1}{3}, \frac{1}{3}$	0.225 000 000 0
			1	$\alpha_1, \beta_1, \beta_1$	
			2	$\beta_1, \alpha_1, \beta_1$	0.132 394 152 7
			3	$\beta_1, \beta_1, \alpha_1$	
			4	$\alpha_2, \beta_2, \beta_2$	
			5	$\beta_2, \alpha_2, \beta_2$	0.125 939 180 5
			6	$\beta_2, \beta_2, \alpha_2$	

with
$\alpha_1 = 0.059\ 715\ 871\ 7$
$\beta_1 = 0.470\ 142\ 064\ 1$
$\alpha_2 = 0.797\ 426\ 985\ 3$
$\beta_2 = 0.101\ 286\ 507\ 3$

FIGURE 5.10. Some quadrature formulas for triangles.

points than the product formulas, and this has been accomplished in certain circumstances.[5]

For triangular regions in two dimensions again a direct approach can be adopted. Thus with a single sampling point ($n = 0$) we have three unknowns, defining the position (ξ_0, η_0) and the weighting (W_0), and, clearly, a complete polynomial of degree 1 in ξ and η can be exactly integrated. With $n = 2$ a complete polynomial of degree 3 can be integrated exactly, and so on.

The value of the weighting coefficients and the position of the sampling points for $n = 0, 1, 2, 3$ are shown in Fig. 5.10.[6] Again the extension to three dimensions is possible, and integration rules for regions of tetrahedral shape have been produced.[7]

EXERCISES

5.12. Develop Simpson's rule for approximating $\int_{-1}^{1} G(\xi)\, d\xi$ using the sampling points $\xi_0 = -1, \xi_1 = 0, \xi_2 = 1$. By taking $G(\xi) = \xi^3$, show that the rule integrates a cubic exactly.

5.13. Produce the Newton–Cotes method for approximating $\int_{-1}^{1} G(\xi)\, d\xi$ which uses four equally spaced sampling points.

5.14. In finite element computations it is frequently necessary to evaluate element integrals such as $\int_{\Omega^e} (\partial N_i^e / \partial x)(\partial N_j^e / \partial x)\, dx\, dy$ and $\int_{\Omega^e} N_i^e N_j^e\, dx\, dy$. If numerical integration is used, determine the number of sampling points required for an exact evaluation of these integrals when Ω^e is (a) the four-noded bilinear element, (b) the eight-noded serendipity element, and (c) the nine-noded Lagrange element. Consider both Newton–Cotes and Gauss–Legendre methods.

5.15. Return to Exercise 5.9 and indicate how the components of the matrices $\mathbf{k}^e$ and $\mathbf{f}^e$ can be obtained by the use of Gaussian quadrature.

5.16. Derive an integration rule which will integrate exactly a general quadratic function over the eight-noded square serendipity element and which uses the element nodes as the sampling points. Show that the same requirement applied to the nine-noded square Lagrange element produces a family of possible integration formulas.

5.3. MORE ON MAPPING

5.3.1. General Remarks

In Section 5.1 we discussed element mapping in general and introduced a very simple form of so-called parametric mapping, using as its base the element trial functions. Many other forms of mapping have been employed and here we indicate three interesting possibilities, all of which have been used with good effect in finite element analyses.

The first of these, the so-called blending process, enables any shape of domain to be mapped simply into a quadrilateral region by direct interpolation. The second establishes a similar map of an arbitrary region by solving auxiliary differential equations while the third concerns itself with semi-infinite regions, such as may be found in many real problems, and maps such domains into finite regions. This type of mapping has been shown to be of great practical value in the analysis of many problems of engineering and physics.

5.3.2. Mapping by Blending Function

We begin by considering a quite general problem which requires that the complex shaped region A of Fig. 5.11a be mapped into the standard (ξ, η) square of Fig. 5.11b. A priori we decide arbitrarily upon a correspondence between the boundary points of A and the boundary points of B, that is, each point P of the boundary of A is associated with a unique image P' on the boundary of B. In particular the images of the corner nodes are fixed as indicated.

This decision leads to a one-to-one correspondence between points on the boundaries of the (ξ, η) space and the (x, y) boundary points of the real domain A. Consider Fig. 5.12 where we have displayed a perspective view of the (ξ, η) space and plotted the variation of x all along the boundary. The complete mapping will be achieved if a smooth surface can be defined between these specified edge values.

The *blending technique* of defining such a surface is graphically illustrated in Fig. 5.13 and consists of four stages.

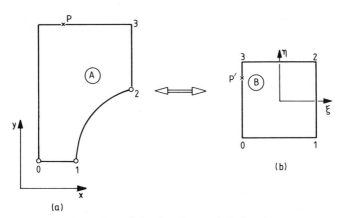

FIGURE 5.11. Mapping of an analysis domain.

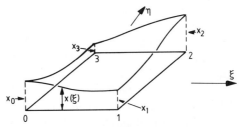

FIGURE 5.12. The variation of x along the boundary in (ξ, η) space.

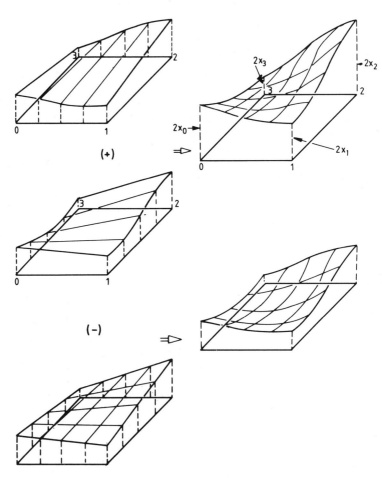

FIGURE 5.13. Construction of a mapping function by linear blending.

216

1. Produce a function which interpolates linearly with respect to η between the values of x specified on the lines $\eta = \pm 1$.

2. Produce a function which interpolates linearly with respect to ξ between the values of x specified on the lines $\xi = \pm 1$.

3. Add these two functions—the result is a continuous well-defined function which differs from the required edge values by a linear function.

4. Subtract the function produced by standard bilinear interpolation between the correct corner values.

The result is the required smooth curve $f_1(\xi, \eta)$ of Eq. (5.2) which defines the coordinate x at every point of the (ξ, η) space.

We do not give any mathematical expressions for this mapping function as the algebra involved is trivial if the concept is well understood (see Exercises 5.19 and 5.20).

Similar interpolation for the y coordinate results in the function $f_2(\xi, \eta)$ of Eq. (5.2) and completes the mapping.

This process may be extended and higher order interpolations used, for example, if we insist that a particular set of (x, y) points should lie on the axes of the (ξ, η) domain, in addition to the requirement along the edges of the domain, then the linear interpolations used above must be replaced by quadratic interpolations.

This mapping procedure enables us to map, in principle, an arbitrarily shaped domain (possibly subdivided into arbitrarily shaped subdomains or elements) into a single square (or family of square elements). To achieve this mapping it is of course necessary to define the external boundaries and element interfaces by suitable functions relating the behavior of the coordinates x and y on the sides $\xi = \pm 1, \eta = \pm 1$. This process is very useful in mapping a single domain with boundaries of complex form into a square in which solutions can be made using global functions or, if desired, using regular element subdivisions of a standard type. Indeed mapping of this kind allows the use of finite difference procedure, of the type introduced in Chapter 1, in the regular simple (ξ, η) space. By this procedure one of the drawbacks of the finite difference method may be removed, as a problem involving a domain of complex shape may be solved by straightforward application of the finite difference method to a suitably transformed governing equation over the mapped square in (ξ, η) space.

5.3.3. Mapping by Solving Auxiliary Equations

The problem of mapping, reduced in the previous section to determining a unique set of x or y coordinates (Fig. 5.12) in a square (ξ, η) domain with known values along the lines $\xi = \pm 1$ and $\eta \pm 1$, can be solved in many ways.

One approach, for instance, defines the variation of x and y by the equations

$$\frac{\partial^2 x}{\partial \xi^2} + \frac{\partial^2 x}{\partial \eta^2} = 0$$

$$-1 \leqslant \xi, \eta \leqslant 1 \qquad (5.33)$$

$$\frac{\partial^2 y}{\partial \xi^2} + \frac{\partial^2 y}{\partial \eta^2} = 0$$

The solutions of Eq. (5.33) over the square region with the values of x and y defined on the boundaries can be achieved by simple finite difference computation (or indeed by the equivalent finite element form), and results can be generated with speed. The form of Eq. (5.33) ensures that the generated solutions for x and y will be smooth and well ordered.

It may sound surprising that the method of mapping outlined here would ever be used in practice, requiring as it does the additional simultaneous solution of a pair of auxiliary equations before the main solution is attempted. However, these additional computations are made on a very simple domain, and for some complex shapes and equations advantages may accrue. All the merits of the blending function mapping described earlier are again available, and it is of interest to note that mappings of this and similar type are frequently used in finite difference solutions of problems involving complex domains.[8] With the x and y coordinates calculated on a regular mesh in the (ξ, η) domain, simple finite difference formulas can be used to evaluate quantities such as $\partial x / \partial \xi$, and hence the Jacobian matrices [Eq. (5.4)] necessary for modeling the original problem can be determined in straightforward fashion.

EXERCISES

5.17. A complex domain of the type shown in Fig. 5.2 is mapped onto the unit square $-1 \leqslant \xi, \eta \leqslant 1$, using the relationships of Eq. (5.33). The x and y coordinates are determined numerically at every point of a regular square mesh defined by $\Delta \xi = \Delta \eta = 0.1$. It is required to solve the linear heat conduction equation $k(\partial^2 T / \partial x^2) + k(\partial^2 T / \partial y^2) + Q = 0$ over the original domain. Show how the finite difference method can be applied directly in the mapped domain and investigate how a boundary condition of prescribed heat flux [Eq. (1.6b)] can be handled.

5.18. Repeat Exercise 5.17 using the finite element method and four-noded elements. (Finite difference expressions may be used in the evaluation of the Jacobian.)

5.19. Consider the process of mapping by the blending function illustrated in Fig. 5.13. Define

$$N_0(\xi) = \frac{1-\xi}{2}, \qquad N_1(\xi) = \frac{1+\xi}{2}$$

$$P(\xi, \eta) = x(\xi)|_{\eta=-1} N_0(\eta) + x(\xi)|_{\eta=1} N_1(\eta)$$

$$R(\xi, \eta) = x|_{\xi=\eta=-1} N_0(\xi) N_0(\eta)$$

$$+ x|_{\xi=1, \eta=-1} N_1(\xi) N_0(\eta) + x|_{\xi=-1, \eta=1} N_0(\xi) N_1(\eta)$$

$$+ x|_{\xi=\eta=1} N_1(\xi) N_1(\eta)$$

and show that

$$x = P(\xi, \eta) + P(\eta, \xi) - R(\xi, \eta)$$

is the required mapping.

5.20. It is required to determine the distribution of temperature ϕ through a rectangular four-noded element Ω^e of unit conductivity which is subjected to the boundary conditions shown in the figure. If a blending function representation for $\hat{\phi}$ is adopted over Ω^e in the form

$$\hat{\phi} = \left(1 - \frac{y}{h_y}\right)\phi|_{y=0} + \left(1 - \frac{x}{h_x}\right)\phi|_{x=0}$$

$$- \left(1 - \frac{x}{h_x}\right)\left(1 - \frac{y}{h_y}\right)\phi|_{x=y=0} + \phi_2 N_2^e$$

where N_2^e is the standard bilinear element shape function associated with node 2, show that $\hat{\phi}$ satisfies exactly the required boundary

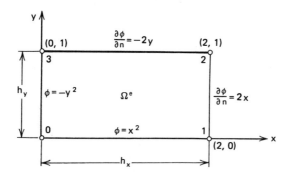

condition on the lines $x = 0$ and $y = 0$. Use this representation for $\hat{\phi}$ and the Galerkin weighted residual method to obtain an approximation to the value of ϕ_2. Compare the result with the value of ϕ_2 produced by the use of standard finite element bilinear interpolation for $\hat{\phi}$ over Ω^e and with the exact value of $\phi_2 = 3$.

5.3.4.   Infinite Elements

One of the most interesting, and practically useful, forms of mapping is that which converts an infinite domain into a finite space.

In many situations, such as the one illustrated in Fig. 5.14, the problem domain extends to infinity. In the case illustrated we are concerned with the

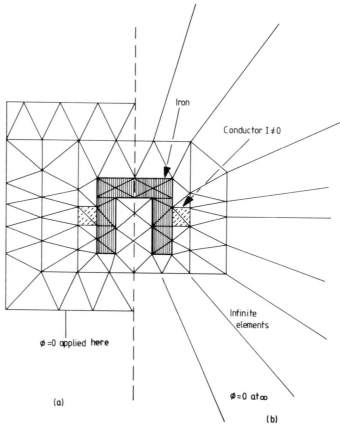

FIGURE 5.14.   Finite element mesh for the analysis of the magnetic field in an air–iron domain. (*a*) With the boundary condition applied at a large, but finite, distance. (*b*) With the boundary condition applied at infinity by using infinite elements.

determination of a magnetic field in an air–iron domain induced by conductors carrying a current of density I. The relevant equation to be solved is similar to the heat conduction equation and takes the form

$$\frac{\partial}{\partial x}\left(k\frac{\partial \phi}{\partial x}\right) + \frac{\partial}{\partial y}\left(k\frac{\partial \phi}{\partial y}\right) + I = 0 \tag{5.34}$$

where ϕ denotes potential and with a boundary condition requiring ϕ to be zero at infinity.*

Two main approaches for achieving an approximate numerical solution exist. In the first a pragmatic view is taken and the exterior boundary is fixed at a large, but finite, distance and the domain is only discretized up to this exterior boundary (see Fig. 5.14a). Such a procedure requires a large number of node or grid points and begs the question of specifying a value for the sufficiently large distance. (This would normally require some numerical experimentation.)

In the second approach we attempt to deal with the domain extending to infinity directly. Many devices exist here, ranging from the use of an analytical solution valid at large distances to the simplest method in which the infinite domain is mapped into a finite region. It is this latter possibility that we discuss in the present section.

At the outset let us consider the possibility of mapping a one-dimensional domain (or element), which extends from a point with coordinate x_0 through an intermediate point with coordinate x_Q to infinity at x_1, as shown in Fig. 5.15a, into the domain $-1 \leqslant \xi \leqslant 1$ of Fig. 5.15b.

We shall define the mapping by

$$x = \tilde{N}_P^e(\xi)x_P + \tilde{N}_Q^e(\xi)x_Q \tag{5.35}$$

where

$$\tilde{N}_P^e = -\frac{\xi}{1-\xi}, \qquad \tilde{N}_Q^e = 1 + \frac{\xi}{1-\xi} \tag{5.36}$$

The expressions are similar in form to the parametric mapping of Eq. (5.9), but the functions have been specially chosen to become infinite at point 1, (i.e., $\xi = 1$), and as yet the coordinate x_P is not determined.

Examining the above we note that for $\xi = 1$,

$$x \equiv \frac{\xi}{1-\xi}(x_Q - x_P) + x_Q = \infty, \qquad x_P \neq x_Q \tag{5.37a}$$

and for $\xi = 0$,

$$x = x_Q \tag{5.37b}$$

*Such a problem is well defined only if $\int_\Omega I\, d\Omega \equiv 0$, that is, when the currents are exactly in balance.

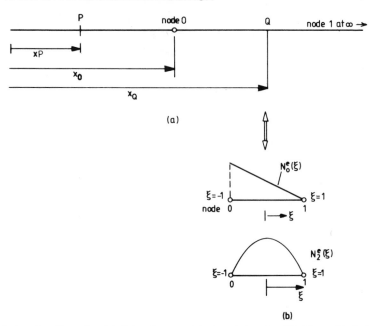

FIGURE 5.15.   One-dimensional infinite element. (a) Actual domain. (b) Mapped infinite element with hierarchical element shape functions.

as required, and finally for $\xi = -1$, which we want to map into the coordinate x_0, we must have

$$x \equiv x_0 = \tfrac{1}{2}x_P + \tfrac{1}{2}x_Q \qquad (5.37c)$$

This last relation determines the coordinate x_P in terms of x_0 and x_Q, and we observe immediately that the point x_0 lies at the center of the interval $[x_P, x_Q]$. Indeed the mapping of Eq. (5.35) could now be directly written in terms of x_0 and x_Q as

$$x = \left(2x_0 - x_Q\right)\tilde{N}_P^e + x_Q\tilde{N}_Q^e = \frac{2(x_Q - x_0)\xi}{1 - \xi} + x_Q \qquad (5.38)$$

and the required mapping has thus been completely defined.

Although many other function forms could be used in Eq. (5.36), it is important that the mapping functions chosen should satisfy the relation

$$\tilde{N}_P^e + \tilde{N}_Q^e = 1 \qquad (5.39)$$

This condition is identically satisfied by the functions specified in Eq. (5.36). The need for this condition follows from the fairly obvious requirement that the map should remain unchanged if the origin of the x coordinate is altered. Thus if we shift the origin of x by an amount Δx and write

$$X_P = x_P + \Delta x, \qquad X_Q = x_Q + \Delta x \qquad (5.40)$$

we require that $X = x + \Delta x$ for a given ξ.

Now using Eq. (5.35) we have

$$X = X_P \tilde{N}_P^e + X_Q \tilde{N}_Q^e \qquad (5.41)$$

and, substituting from Eq. (5.40), we see that the above requirement holds if

$$\Delta x = \Delta x \left(\tilde{N}_P^e + \tilde{N}_Q^e \right) \qquad (5.42)$$

which is only true if condition (5.39) is satisfied.*

It is now of interest to examine what happens if standard polynomial functions in ξ are used to describe the variation of an unknown function $\hat{\phi}$ in the mapped domain.

By using, for instance, hierarchical-type shape functions as shown in Fig. 5.15b, the condition that $\hat{\phi}^e = 0$ if $\xi = 1$ (or $x = \infty$) is automatically achieved by setting the nodal value ϕ_1 to zero, and the expansion over the element can be written as a polynomial,

$$\hat{\phi}^e = \phi_0 N_0^e + \sum_{l=2}^{p} a_l^e N_l^e(\xi) = \alpha_0 + \alpha_1 \xi + \alpha_2 \xi^2 + \cdots + \alpha_p \xi^p \qquad (5.43)$$

with the degree p determined by the number of shape functions used.

Now Eq. (5.35) can be solved for ξ in terms of x, thus defining the inverse map by

$$\xi = 1 - \frac{2(x_Q - x_P)}{r} \qquad (5.44)$$

where $r \, (= x - x_P)$ represents the distance from the point P of Fig. 5.15a.

Substituting this expression for ξ into Eq. (5.43) results in an expansion for $\hat{\phi}^e$, in terms of the global coordinates, of the form

$$\hat{\phi}^e = \beta_0 + \frac{\beta_1}{r} + \frac{\beta_2}{r^2} + \cdots + \frac{\beta_p}{r^p} \qquad (5.45)$$

with the number of terms present again depending on the degree of the polynomial expansion used. [The reader will observe that the requirement

---

*This condition was implicitly satisfied by the standard finite element shape functions we have used previously; see, for example, Eq. (3.48).

$\beta_0 = 0$ is implied if the condition $\hat{\phi}^e = 0$ at ∞ has been applied as in Eq. (5.43).]

An expression of the form given in Eq. (5.45) is typical of exact far-field solutions for exterior regions, and it can be used to represent "decay" functions to any desired degree of accuracy.

Clearly as the choice of the coordinate x_Q (or x_0) is arbitrary, a knowledge of the behavior of the far-field solution and of the approximate origin of the decay will be required if accurate solutions are to be produced in this way by the finite element method.

With the one-dimensional mapping of an infinite element achieved, the extension to two (or three) dimensions presents little difficulty.

Consider for instance a domain of finite elements shown shaded in Fig. 5.16a and, in particular, the face of a typical linear triangular element defined by nodes 0 and 3 to which an infinite element e is to be joined.

At first let us give attention to the mapping of the line through node 0 and point Q which defines a side of the infinite element. Clearly, we can extend the

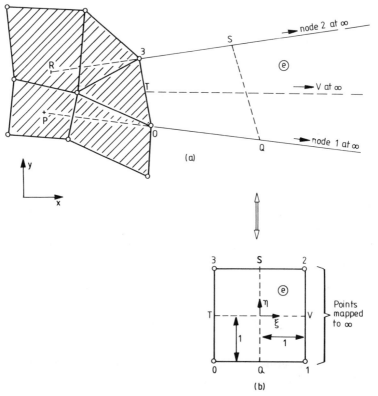

FIGURE 5.16. Infinite elements in two dimensions (a) Actual domain. (b) Mapped infinite element.

one-dimensional map here and write [using the form of Eq. (5.38)]

$$x = (2x_0 - x_Q)\tilde{N}_P^e + x_Q\tilde{N}_Q^e \equiv x_P\tilde{N}_P^e + x_Q\tilde{N}_Q^e$$

$$y = (2y_0 - y_Q)\tilde{N}_P^e + y_Q\tilde{N}_Q^e \equiv y_P\tilde{N}_P^e + y_Q\tilde{N}_Q^e$$

(5.46)

where the coordinates (x_P, y_P) of the point P are defined as before.

It is easy to see that, with respect to a coordinate system (X, Y) with origin at 0, we have

$$\frac{X}{Y} = \frac{\tilde{N}_Q^e X_Q}{\tilde{N}_Q^e Y_Q} = \frac{X_Q}{Y_Q}$$

(5.47)

so that the straight line through points P, 0, and Q corresponds to the direction of the mapped ξ coordinate.

Now if the position of point S is defined (with corresponding origin of decay R), it becomes apparent that we can write a complete map of the infinite element e

$$x = N_0^e(\eta)\left[(2x_0 - x_Q)\tilde{N}_P^e(\xi) + x_Q\tilde{N}_Q^e(\xi)\right]$$

$$+ N_3^e(\eta)\left[(2x_3 - x_S)\tilde{N}_P^e(\xi) + x_S\tilde{N}_Q^e(\xi)\right]$$

(5.48)

with y similarly defined and where $N_0^e(\eta)$ and $N_3^e(\eta)$ are the standard linear one-dimensional shape functions defined by

$$N_0^e(\eta) = \frac{1 - \eta}{2}, \qquad N_3^e(\eta) = \frac{1 + \eta}{2}$$

(5.49)

Again, in the (ξ, η) space polynomial trial functions can be used to represent the variation of any unknown function and we observe that along lines 0–1 and 3–2 (and therefore along all lines η = constant) expressions of the form given by Eq. (5.45) are obtained with r being the distance from a suitably defined pole.

Returning to the example of Fig. 5.14 we find that such a pole, if fixed near the center of the region, in fact defines a form identical with the exact solution at large distances, and an excellent approximation can then be achieved by the use of these elements.

Example 5.2

Consider the solution of the equation $d^2\phi/dx^2 = 2/x^3$ subject to the conditions $\phi = 1$ at $x = 1$ and $\phi = 0$ at $x = \infty$. We will produce a solution using a single infinite element. Then $x_0 = 1$, and if we choose $x_P = 0$, it follows from Eq. (5.37c) that $x_Q = 2$. The required mapping is then given by Eq. (5.38) as

$$x = \frac{2\xi}{1 - \xi} + 2 = \frac{2}{1 - \xi}$$

and the inverse map of Eq. (5.44) can be seen to be

$$\xi = 1 - \frac{2}{x}$$

Using a quadratic hierarchical representation for the approximation $\hat{\phi}$ over the element means that

$$\hat{\phi} = \phi_0 N_0^e + a_2 N_2^e$$

where $N_0^e = (1 - \xi)/2$, $N_2^e = 1 - \xi^2$ and $\phi_0 = 1$. The Galerkin weighted residual statement which determines a_2 is then

$$\int_1^\infty \frac{d\hat{\phi}}{dx} \frac{dN_2^e}{dx} dx + 2 \int_1^\infty \frac{N_2^e}{x^3} dx = 0$$

which can be written, making use of the mapping relationship between x and ξ, as

$$\phi_0 \int_{-1}^1 \frac{dN_0^e}{d\xi} \frac{dN_2^e}{d\xi} \frac{d\xi}{dx} d\xi + a_2 \int_{-1}^1 \left(\frac{dN_2^e}{d\xi} \right)^2 \frac{d\xi}{dx} d\xi + \frac{1}{4} \int_{-1}^1 (1 - \xi)^3 N_2^e \frac{dx}{d\xi} d\xi = 0$$

On evaluation of the integrals, this yields $a_2 = 0$, and the approximation is thus

$$\hat{\phi} = N_0^e = \frac{1 - \xi}{2} = \frac{1}{x}$$

which the reader can verify to be the exact solution.

We have been fortunate here in making the correct choice for the value of x_P at the outset. To investigate the effect of using a different value of x_P on the accuracy of the resulting approximation we can consider again the solution of the same differential equation, but subject now to the conditions $\phi = \frac{1}{2}$ at $x = 2$ and $\phi = 0$ at $x = \infty$. The exact solution is again $\phi = 1/x$ as before, but now, in constructing the approximation, we choose $x_P = 1$. The mapping becomes

$$x = \frac{3 - \xi}{1 - \xi}$$

since $x_0 = 2$ and $x_Q = 3$, and the inverse map is defined by

$$\xi = \frac{x - 3}{x - 1}$$

The approximation for $\hat{\phi}$ is as assumed above with $\phi_0 = \frac{1}{2}$, and the weighted residual statement is now

$$\phi_0 \int_{-1}^{1} \frac{dN_0^e}{d\xi} \frac{dN_2^e}{d\xi} \frac{d\xi}{dx} d\xi + a_2 \int_{-1}^{1} \left(\frac{dN_2^e}{d\xi} \right)^2 \frac{d\xi}{dx} d\xi + 2 \int_{-1}^{1} \frac{(1 - \xi)^3}{(3 - \xi)^3} N_2^e \frac{dx}{d\xi} d\xi = 0$$

Evaluation of the integrals produces the result $a_2 = 0.071\,542\,3$, and the approximation in this case is

$$\hat{\phi} = 0.32154 - 0.25 \frac{x - 3}{x - 1} - 0.07154 \left(\frac{x - 3}{x - 1} \right)^2$$

A comparison between this approximation and the exact solution is given in the table.

x	Exact	Approximate
2	0.5	0.5
3	0.3333	0.32154
4	0.25	0.23026
5	0.2	0.17866
10	0.1	0.08382
30	0.03333	0.02677

EXERCISES

5.21. Solve the equation $d^2\phi/dx^2 = e^{-x}$, subject to the conditions $\phi = e^{-1}$ at $x = 1$, $\phi = 0$ at $x = \infty$, by using a single infinite element with the pole defined by $x_P = 0$. Compare the hierarchical quadratic and cubic approximations with the exact solution.

5.22. Solve the equation $d^2\phi/dx^2 = 2/x^3$ subject to the conditions $\phi = \frac{1}{2}$ at $x = 2$ and $\phi = 0$ at $x = \infty$. Use the finite element mesh shown in the figure, consisting of two quadratic hierarchical elements, and compare the resulting approximation with that produced in Example 5.2.

5.4.   MESH GENERATION AND CONCLUDING REMARKS

In this and the previous chapter we have shown the following.

1. How any high-order polynomial shape functions can be easily generated for a variety of rectangular and triangular finite elements.
2. How mapping may be used to ensure that a global domain of almost any complex shape can be accurately represented by an assembly of such elements.
3. How all element matrices can be simply evaluated by use of numerical integration.

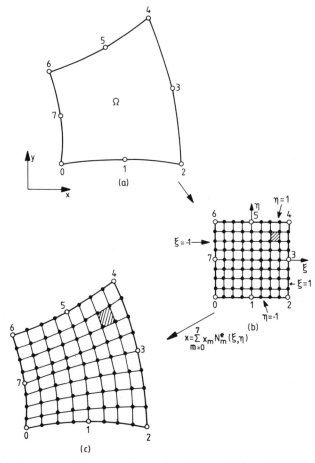

FIGURE 5.17.   Mesh generation by means of mapping. (*a*) Actual domain Ω. (*b*) Subdivided mapped domain in (ξ, η) space. (*c*) Generated mesh on Ω.

Such concepts as mapping and numerical integration have, obviously, many uses outside the finite element approximation field, and the subject matter expounded is therefore of general applicability. For instance, mapping and interpolation of complex surfaces by blending (Section 5.3.2) was in fact first developed to describe the complicated shapes of motor car surfaces, and numerical integration has been with us long before finite elements or other weighted residual forms were devised. However, there is one additional point which the reader may find of interest and in which the ideas of mapping are used not for deriving a finite element approximation, but rather for generating simply a mesh subdivision for a region of general shape, which obviously would need much computational effort if performed manually.

We have succeeded, by various procedures, in mapping arbitrary domains into squares. This has resulted in mapping functions giving the actual space coordinates (x, y) as

$$
\begin{aligned}
x &= f_1(\xi, \eta), \\
&\qquad\qquad -1 \leqslant \xi, \eta \leqslant 1 \qquad\qquad (5.50) \\
y &= f_2(\xi, \eta),
\end{aligned}
$$

Clearly, the generation of a mesh of equally spaced points in the (ξ, η) square domain is quite trivial, and if we observe that for each such mesh point, the coordinates of an equivalent point in the mapped space can be immediately computed, a means for deriving a mesh of elements of an arbitrary shape is at hand. Figure 5.17 illustrates such a process with a very simple parametric type map, but all forms of mapping developed here have been widely used in practical automatic mesh generation programs.

REFERENCES

[1] R. Courant, *Differential and Integral Calculus*, Vol. II, Blackie and Son, London, 1936.

[2] O. C. Zienkiewicz, *The Finite Element Method*, 3rd ed., McGraw-Hill, New York, 1977.

[3] C. E. Fröberg, *Introduction to Numerical Analysis*, Addison-Wesley, Reading, Mass., 1965.

[4] V. I. Krylov, *Approximate Calculation of Integrals*, Macmillan, New York, 1962.

[5] B. M. Irons, Quadrature rules for brick based finite elements, *Int. J. Num. Meth. Eng.* **3**, 293–284 (1971).

[6] E. R. Cowper, Gaussian quadrature formulae for triangles, *Int. J. Num. Meth. Eng.* **7**, 405–408 (1973).

[7] P. C. Hammer, O. P. Marlowe, and A. H. Stroud, Numerical integration over simplexes and cones, *Math. Tables Aids Comp.* **10**, 130–137 (1956).

[8] Examples will be found in (a) J. F. Thompson, F. C. Thames, and C. W. Mastin, TOMCAT—a code for numerical generation of boundary fixed curvilinear coordinate systems on fields containing any number of arbitrary two-dimensional bodies, *J. Comput. Phys.* **24**, 274–302 (1977); (b) S. B. Pope, The calculating of turbulent recirculating flows in general orthogonal coordinates, *J. Comput. Phys.* **26**, 197–217 (1978).

SUGGESTED FURTHER READING

E. B. Becker, G. F. Carey, and J. T. Oden, *Finite Elements: An Introduction*, Vol. 1, Prentice-Hall, Englewood Cliffs, N.J., 1981.

P. Bettess, Infinite elements, *Int. J. Num. Meth. Eng.* **11**, 53–64 (1977).

A. J. Davies, *The Finite Element Method*, Clarendon, Oxford, 1980.

W. J. Gordon, Blending-function methods of bivariate and multivariate interpolation and approximation, *SIAM J. Num. Anal.* **8**, 158–177 (1971).

W. J. Gordon and C. A. Hall, Construction of curvilinear coordinate systems and application to mesh generation, *Int. J. Num. Meth. Eng.* **7**, 461–477 (1973).

E. Hinton and D. R. J. Owen, *An Introduction to Finite Element Computations*, Pineridge Press, Swansea, 1979.

R. E. Smith (Editor), *Numerical Grid Generation Techniques*, NASA Conference Publication 2166, 1980.

CHAPTER SIX

Variational Methods

6.1. INTRODUCTION

It has been seen in the previous chapters that the analysis of many physical problems frequently requires the determination of a function which is the solution of a differential equation governing the behavior of the system under consideration. In some circumstances, however, it may be possible to determine a *natural variational principle* for the particular problem of interest, and then an alternative solution approach may be adopted, which consists of determining the function which makes a certain integral statement (or *functional*) stationary. Clearly, if such variational principles can be found, then immediately we have new methods available for constructing approximate solutions, for we can use the trial function or finite element methods of the previous chapters and attempt to make the functional stationary with respect to variations in the unknown parameters.

Some physical problems can be stated directly in the form of a variational principle—an obvious example being the requirement of minimization of total potential energy for the equilibrium of a mechanical system. However, the form of the natural variational principle is not always obvious and, indeed, such a principle does not exist for many continuous problems for which well-defined differential equations may be formulated. We shall therefore begin by considering under what circumstances a natural variational principle can be derived from a differential equation, and we shall then investigate how special contrived variational forms can be constructed if no natural variational principle exists. These special forms will be seen to use either the standard Lagrange multiplier approach, which introduces additional variables into the analysis, the penalty function method, or the method of least squares.

6.2. VARIATIONAL PRINCIPLES

Suppose we are given a functional Π in the integral form

$$\Pi(\phi) = \int_\Omega F\left(\phi, \frac{\partial \phi}{\partial x}, \cdots\right) d\Omega + \int_\Gamma G\left(\phi, \frac{\partial \phi}{\partial x}, \cdots\right) d\Gamma \qquad (6.1)$$

where F and G are functions of $\phi(x, \dots)$ and its derivatives, and Γ is the curve bounding the closed region Ω. We will now attempt to make Π stationary with respect to variations in ϕ among the *admissible set of functions* satisfying general boundary conditions, of the type introduced in Eq. (2.25),

$$B_1(\phi) = 0 \qquad \text{on } \Gamma_1 \qquad (6.2a)$$

$$B_2(\phi) = 0 \qquad \text{on } \Gamma_2 \qquad (6.2b)$$

where $\Gamma_1 + \Gamma_2 = \Gamma$. For a small admissible variation from ϕ to $\phi + \delta\phi$, we define the corresponding first variation in Π by

$$\delta\Pi = \int_\Omega \left[\frac{\partial F}{\partial \phi} \delta\phi + \frac{\partial F}{\partial(\partial\phi/\partial x)} \delta\left(\frac{\partial\phi}{\partial x}\right) \cdots \right] d\Omega$$

$$+ \int_\Gamma \left[\frac{\partial G}{\partial \phi} \delta\phi + \frac{\partial G}{\partial(\partial\phi/\partial x)} \delta\left(\frac{\partial\phi}{\partial x}\right) \cdots \right] d\Gamma \qquad (6.3)$$

and the stationarity of Π at ϕ then requires[1] that

$$\delta\Pi = 0 \qquad (6.4)$$

If, by suitable manipulation, Eq. (6.3) can be written

$$\delta\Pi = \int_\Omega A(\phi) \, \delta\phi \, d\Omega \qquad (6.5)$$

then the stationarity requirement of Eq. (6.4) shows that

$$A(\phi) = 0 \qquad \text{in } \Omega \qquad (6.6)$$

since $\delta\phi$ is arbitrary. We now have a natural variational principle for the problem of determining the solution ϕ of the differential equation (6.6) subject to the boundary conditions of Eq. (6.2). The required function makes the functional $\Pi(\phi)$ of Eq. (6.1) stationary with respect to variations in ϕ among the set of functions satisfying the boundary conditions. Equation (6.6) is termed the Euler equation corresponding to the variational principle which requires the stationarity of Π. It can be shown that for any variational principle of this form a corresponding Euler equation can be established, but the converse is unfortunately not true as only certain forms of differential equations can be identified as Euler equations of a variational functional.

In certain circumstances the manipulation of Eq. (6.3) will produce not Eq. (6.5), but an equation such as

$$\delta\Pi = \int_{\Omega} A(\phi)\,\delta\phi\,d\Omega + \int_{\Gamma_2} B_2(\phi)\,\delta\phi\,d\Gamma \qquad (6.7)$$

and Π is then stationary at ϕ provided that

$$A(\phi) = 0 \qquad \text{in } \Omega \qquad (6.8a)$$

and that

$$B_2(\phi) = 0 \qquad \text{on } \Gamma_2 \qquad (6.8b)$$

The boundary condition of Eq. (6.2b) or (6.8b) on Γ_2 is now a *natural boundary condition* as it is automatically satisfied by the function ϕ which makes Π stationary. In this case, the admissible set of functions is larger, since our only requirement is that any function belonging to this set should satisfy the *essential boundary condition* of Eq. (6.2a). This definition of a natural boundary condition might appear different to that introduced in Chapter 2 but, in fact, both are identical.

Example 6.1

Suppose that we are given the functional

$$\Pi(\phi) = \int_0^{L_x} \left[\frac{T}{2}\left(\frac{d\phi}{dx}\right)^2 - w(x)\phi \right] dx$$

where T is a constant, and that we want to determine the function ϕ which makes Π stationary. The admissible functions will be taken to be those satisfying the conditions $\phi = 0$ at $x = 0$ and $\phi = 0$ at $x = L_x$.

Performing a variation from ϕ to $\phi + \delta\phi$, we have

$$\delta\Pi = \int_0^{L_x} \left[T\frac{d\phi}{dx}\delta\left(\frac{d\phi}{dx}\right) - w(x)\,\delta\phi \right] dx$$

and since $\delta(d\phi/dx) = (d/dx)(\delta\phi)$, integration by parts can be used to give

$$\delta\Pi = \left[\frac{d\phi}{dx}\delta\phi \right]_0^{L_x} - \int_0^{L_x} \left[T\frac{d^2\phi}{dx^2} + w(x) \right]\delta\phi\,dx$$

Now both ϕ and $\phi + \delta\phi$ are admissible functions, and so $\delta\phi = 0$ at $x = 0$ and at $x = L_x$ and it follows that

$$\delta\Pi = -\int_0^{L_x}\left[T\frac{d^2\phi}{dx^2} + w(x)\right]\delta\phi\,dx$$

Stationarity of Π at ϕ then requires that $\delta\Pi = 0$ for any admissible variation $\delta\phi$, and this can only be true if

$$T\frac{d^2\phi}{dx^2} + w(x) = 0, \qquad 0 \leqslant x \leqslant L_x$$

which is thus the Euler equation for the given variational principle. This is the equation governing the small vertical deflection of an elastic string at tension T and subjected to a vertical loading $w(x)$. It can be observed that the total potential energy of such a system consists of the strain energy U and the load energy V defined by

$$U = \int_0^{L_x}\frac{T}{2}\left(\frac{d\phi}{dx}\right)^2 dx$$

$$V = -\int_0^{L_x}w(x)\phi\,dx$$

so that here we can write

$$\Pi = U + V$$

In this case it is therefore possible to attach a physical significance to Π, and the required function ϕ makes the total potential energy of the system stationary.

It can be seen from the above that if the admissible functions are only required to satisfy the condition $\phi = 0$ at $x = 0$, then

$$\delta\Pi = \frac{d\phi}{dx}\delta\phi\bigg|_{x=L_x} - \int_0^{L_x}\left(T\frac{d^2\phi}{dx^2} + w(x)\right)\delta\phi\,dx$$

and Π is therefore stationary, provided

$$T\frac{d^2\phi}{dx^2} + w(x) = 0, \qquad 0 \leqslant x \leqslant L_x$$

and

$$\frac{d\phi}{dx}\bigg|_{x=L_x} = 0$$

A condition of this type at $x = L_x$ is therefore a natural boundary condition for this problem.

Example 6.2

Consider now an example in two space dimensions where the functional is defined by

$$\Pi(\phi) = \int_{\Omega} \left[\frac{k}{2} \left(\frac{\partial \phi}{\partial x} \right)^2 + \frac{k}{2} \left(\frac{\partial \phi}{\partial y} \right)^2 - Q\phi \right] d\Omega + \int_{\Gamma_q} \bar{q}\phi \, d\Gamma$$

in which k and Q are functions of position only, and where the boundary curve $\Gamma = \Gamma_q + \Gamma_\phi$. The admissible functions in the variational process will consist of those functions satisfying the condition $\phi = \bar{\phi}$ on Γ_ϕ.

Now

$$\delta\Pi = \int_{\Omega} \left[k\frac{\partial \phi}{\partial x} \frac{\partial}{\partial x}(\delta\phi) + k\frac{\partial \phi}{\partial y} \frac{\partial}{\partial y}(\delta\phi) - Q\delta\phi \right] d\Omega + \int_{\Gamma_q} \bar{q}\delta\phi \, d\Gamma$$

which may be rewritten, using Green's lemma, to give

$$\delta\Pi = \int_{\Gamma_\phi + \Gamma_q} k\frac{\partial \phi}{\partial n} \delta\phi \, d\Gamma - \int_{\Omega} \left[\frac{\partial}{\partial x}\left(k\frac{\partial \phi}{\partial x}\right) + \frac{\partial}{\partial y}\left(k\frac{\partial \phi}{\partial y}\right) + Q \right]\delta\phi \, d\Omega$$

$$+ \int_{\Gamma_q} \bar{q}\delta\phi \, d\Gamma$$

and since ϕ and $\phi + \delta\phi$ are admissible functions, $\delta\phi = 0$ on Γ_ϕ, so that

$$\delta\Pi = -\int_{\Omega} \left[\frac{\partial}{\partial x}\left(k\frac{\partial \phi}{\partial x}\right) + \frac{\partial}{\partial y}\left(k\frac{\partial \phi}{\partial y}\right) + Q \right]\delta\phi \, d\Omega + \int_{\Gamma_q} \left(k\frac{\partial \phi}{\partial n} + \bar{q}\right)\delta\phi \, d\Gamma$$

Thus Π is stationary at ϕ provided that

$$\frac{\partial}{\partial x}\left(k\frac{\partial \phi}{\partial x}\right) + \frac{\partial}{\partial y}\left(k\frac{\partial \phi}{\partial y}\right) + Q = 0 \quad \text{in } \Omega$$

which is the Euler equation for the variational principle, and

$$k\frac{\partial \phi}{\partial n} = -\bar{q} \quad \text{on } \Gamma_q$$

which is therefore a natural boundary condition for this problem.

We see that the use of this functional enables us to approach the problem of two-dimensional steady heat conduction in an alternative manner—the Euler equation is the governing differential equation; the essential and natural boundary conditions on Γ_ϕ and Γ_q, respectively, are typical of the boundary conditions of interest.

EXERCISES

6.1. Consider the functional

$$\Pi(\phi) = \int_0^{L_x} \left[\frac{T}{2} \left(\frac{d\phi}{dx} \right)^2 + \frac{k}{2}\phi^2 - w(x)\phi \right] dx$$

where k and T are constants. Determine the Euler equation and identify typical essential and natural boundary conditions at $x = 0$ and $x = L_x$. The Euler equation describes the small deflection of a loaded cable resting on an elastic foundation of stiffness k. Show that Π as defined here is just equal to the total potential energy of the system.

6.2. Consider the functional

$$\Pi(\phi) = \int_0^{L_x} \left[\frac{EI}{2} \left(\frac{d^2\phi}{dx^2} \right)^2 - w(x)\phi \right] dx$$

where EI is a function of position only. Show that the Euler equation is the equation governing the deflection of a loaded beam of flexural rigidity EI, and identify typical essential and natural boundary conditions.

6.3. Consider the functional

$$\Pi(\phi) = \int_\Omega \left[\frac{k}{2} \left(\frac{\partial\phi}{\partial x} \right)^2 + \frac{k}{2} \left(\frac{\partial\phi}{\partial y} \right)^2 - Q\phi \right] d\Omega - \int_{\Gamma_q} \left(\frac{\alpha}{2}\phi^2 - \bar{q}\phi \right) d\Gamma$$

where k, Q, α, and $\bar{q}$ are functions of position only. Obtain the Euler equation and identify the natural boundary condition on Γ_q when the admissible functions satisfy $\phi = \bar{\phi}$ on $\Gamma_\phi = \Gamma - \Gamma_q$.

6.3.   THE ESTABLISHMENT OF NATURAL VARIATIONAL PRINCIPLES

6.3.1.   The Symmetric Operator

It has been shown in the last section that if we are presented with a variational principle in the form of a functional, then the corresponding Euler equation can always be determined. Normally, however, the behavior of a physical system will be described in terms of a differential equation, and it is of interest to attempt to determine if a variational formulation of the problem is possible. We shall restrict our attention to the case of linear differential equations, as general rules for nonlinear equations are complicated.[2] As we have seen, a general linear differential equation may be written in the form

$$\mathcal{L}\phi + p = 0 \quad \text{in } \Omega \tag{6.9}$$

where $\mathcal{L}$ is a linear operator and p is a known function, and the solution is required subject to the general boundary condition

$$\mathfrak{M}\phi + r = 0 \quad \text{on } \Gamma \tag{6.10}$$

where $\mathfrak{M}$ is a linear operator and r a given function of position.

Consider the set of functions θ which satisfy the homogeneous form of this boundary condition on Γ, that is

$$\mathfrak{M}\theta = 0 \quad \text{on } \Gamma \tag{6.11}$$

The operator $\mathcal{L}$ this is said to be *symmetric* (or *self-adjoint*) over the domain Ω with respect to this set of functions if, for any two members θ and v of this set, we have that

$$\int_\Omega \theta \mathcal{L} v \, d\Omega = \int_\Omega v \mathcal{L} \theta \, d\Omega \tag{6.12}$$

A symmetric operator $\mathcal{L}$ is said to be *positive definite* over Ω with respect to this set of functions if, for any member θ of the set,

$$\int_\Omega \theta \mathcal{L} \theta \, d\Omega \geqslant 0 \tag{6.13}$$

with equality if and only if θ is identically zero in Ω.

Example 6.3

Consider the operator $-d^2/dx^2$ on that portion of the x axis defined by $0 \leqslant x \leqslant L_x$ with an associated set of functions which are zero at $x = 0$ and at $x = L_x$. If v and θ are two such functions, then by using integration by parts, it is possible to write

$$\int_0^{L_x} -\theta \frac{d^2 v}{dx^2} \, dx = \left[-\theta \frac{dv}{dx} \right]_0^{L_x} + \int_0^{L_x} \frac{d\theta}{dx} \frac{dv}{dx} \, dx$$

and the first term on the right-hand side vanishes since $\theta = 0$ at $x = 0$ and at

$x = L_x$. Integrating by parts a second time, it is easy to show that

$$\int_0^{L_x} -\theta \frac{d^2 v}{dx^2} dx = \int_0^{L_x} -v \frac{d^2 \theta}{dx^2} dx$$

and it follows, from the definition of Eq. (6.12), that the operator $-d^2/dx^2$ is symmetric over $0 \leqslant x \leqslant L_x$ with respect to functions vanishing at $x = 0$ and $x = L_x$.

By taking $\theta = v$ in the above,

$$\int_0^{L_x} -\theta \frac{d^2 \theta}{dx^2} dx = \int_0^{L_x} \left(\frac{d\theta}{dx} \right)^2 dx \geqslant 0$$

and hence,

$$\int_0^{L_x} -\theta \frac{d^2 \theta}{dx^2} dx = 0$$

only if $d\theta/dx = 0$. Since $\theta = 0$ at $x = 0$ and at $x = L_x$, it follows that this integral is zero only when $\theta \equiv 0$ over $0 \leqslant x \leqslant L_x$ and the operator $-d^2/dx^2$ is thus positive definite over this range with respect to functions vanishing at $x = 0$ and at $x = L_x$.

EXERCISES

Prove that the following operators are symmetric and positive definite.

6.4. The operator $\mathcal{L} = -d^2/dx^2$ with respect to functions satisfying $d\phi/dx + a\phi = 0$ at $x = 0$ and $d\phi/dx + b\phi = 0$ at $x = L_x$. Here a and b are given constants such that $a < 0, b > 0$.

6.5. The operator $\mathcal{L}$ defined by

$$\mathcal{L}\phi = \frac{d^2}{dx^2} \left[a(x) \frac{d^2 \phi}{dx^2} \right] + b(x)\phi$$

with respect to functions satisfying $\phi = d\phi/dx = 0$ at $x = 0$ and at $x = L_x$. Here $a(x), b(x)$ are given nonnegative functions of x. If the conditions at $x = L_x$ are now changed to the requirement that

$$\frac{d^2 \phi}{dx^2} = \frac{d}{dx} \left[a(x) \frac{d^2 \phi}{dx^2} \right] = 0 \qquad \text{at } x = L_x$$

is the result still true?

6.6. The operator

$$-\frac{\partial}{\partial x} \left[a(x, y) \frac{\partial}{\partial x} \right] - \frac{\partial}{\partial y} \left[b(x, y) \frac{\partial}{\partial y} \right]$$

over a two-dimensional region Ω bounded by a closed curve Γ with respect to functions which vanish on Γ. Here $a(x, y)$, $b(x, y)$ are given nonnegative functions of x and y.

6.7. The operator

$$\frac{\partial^4}{\partial x^4} + 2\frac{\partial^4}{\partial x^2 \partial y^2} + \frac{\partial^4}{\partial y^4}$$

over a two-dimensional region Ω bounded by a closed curve Γ with respect to functions satisfying $\phi = \partial\phi/\partial n = 0$ on Γ.

6.3.2. The Variational Principle for Symmetric Operators

Following the definition of these properties of linear operators it is possible to produce the required variational principle.[3] Suppose $\mathcal{L}$ is a symmetric operator with respect to functions θ satisfying the homogeneous boundary condition (6.11) and let ψ be any function that satisfies the required condition on Γ, that is,

$$\mathfrak{M}\psi + r = 0 \qquad \text{on } \Gamma \tag{6.14}$$

Then the functional

$$\Pi(\phi) = \int_\Omega \left[(\phi - \psi)\{\tfrac{1}{2}\mathcal{L}(\phi - \psi) + \mathcal{L}\psi + p\} \right] d\Omega \tag{6.15}$$

is stationary, with respect to variations in functions ϕ satisfying Eq. (6.10), when ϕ is the solution of the differential equation (6.9) subject to the boundary condition of Eq. (6.10).

We can demonstrate that the Euler equation of this variational principle is the required differential equation by making a small admissible variation from ϕ to $\phi + \delta\phi$. For since $\mathcal{L}$ is linear,

$$\delta\Pi = \int_\Omega \left[\delta\phi\left\{ \frac{1}{2}\mathcal{L}(\phi - \psi) + \mathcal{L}\psi + p \right\} + \frac{(\phi - \psi)}{2}\mathcal{L}\delta\phi \right] d\Omega \tag{6.16}$$

and we note that

$$\int_\Omega (\phi - \psi)\mathcal{L}\delta\phi \, d\Omega = \int_\Omega \delta\phi\mathcal{L}(\phi - \psi) \, d\Omega \tag{6.17}$$

as $\mathcal{L}$ is symmetric with respect to functions satisfying Eq. (6.11). This means that

$$\delta\Pi = \int_\Omega \delta\phi(\mathcal{L}\phi + p) \, d\Omega \tag{6.18}$$

and, since $\delta\phi$ is arbitrary, stationarity of Π demands that

$$\mathcal{L}\phi + p = 0 \quad \text{in } \Omega \qquad (6.19)$$

which is just the required differential equation.

So far in this chapter we have not discussed the nature of the stationarity in Π (that is whether it is a maximum, minimum or merely a saddle point), but here it is possible to show that, for any admissible function χ, Eq. (6.15) can be written as

$$\Pi(\chi) = \int_\Omega (\chi - \phi)\mathcal{L}(\chi - \phi)\,d\Omega + \text{terms involving } \phi \text{ and } \psi \quad (6.20)$$

where ϕ is the function, satisfying Eq.(6.19), that makes Π stationary. If $\mathcal{L}$ is positive definite, then as χ varies, the terms involving ϕ and ψ in this equation remain constant, while the first term is always positive, unless $\chi = \phi$ when its value is zero. We therefore have in this case

$$\Pi(\chi) \geqslant \Pi(\phi) \qquad (6.21)$$

for all admissible functions χ, thus establishing the minimum of the functional Π at ϕ.

It should be noted that the functional of Eq. (6.15) can, upon expansion, be written as

$$\Pi(\phi) = \int_\Omega \left[\tfrac{1}{2}(\phi\mathcal{L}\phi - \psi\mathcal{L}\phi + \phi\mathcal{L}\psi) + p\phi\right] d\Omega + \text{terms involving } \psi \text{ only}$$

$$(6.22)$$

and it will sometimes be possible to write

$$\int_\Omega (\phi\mathcal{L}\psi - \psi\mathcal{L}\phi)\,d\Omega = 2\int_\Gamma \mathcal{N}\phi\,d\Gamma + \text{terms involving } \psi \text{ only} \quad (6.23)$$

where $\mathcal{N}$ is a linear operator. The above variational principle is thus equivalent to requiring that the new functional

$$\Pi(\phi) = \int_\Omega \left(\tfrac{1}{2}\phi\mathcal{L}\phi + p\phi\right) d\Omega + \int_\Gamma \mathcal{N}\phi\,d\Gamma \qquad (6.24)$$

be stationary among the set of functions ϕ satisfying Eq. (6.10), as the function ψ is constant. It should be observed that the variational principle now requires no direct knowledge of the function ψ of Eq. (6.14). This form of the variational principle is similar to that of Eq. (6.1), and, again, it may be possible to identify natural boundary conditions and so widen the admissible set of functions. As a general rule if $\mathcal{L}$ contains derivatives of order $2d$, then a boundary condition of the type specified in Eq. (6.10) will be a natural boundary condition if $\mathfrak{M}$ contains derivatives of order d or higher.

Example 6.4

Consider the problem of determining the small deflection of a loaded elastic string which is held fixed at both ends. The governing equation (Example 6.1) is

$$T\frac{d^2\phi}{dx^2} + w(x) = 0, \qquad 0 \leqslant x \leqslant L_x$$

with boundary conditions $\phi = 0$ at $x = 0$ and $\phi = 0$ at $x = L_x$. Writing this equation as

$$\mathcal{L}\phi + p = -T\frac{d^2\phi}{dx^2} - w(x) = 0$$

the operator $\mathcal{L}$ is positive definite with respect to functions which vanish at $x = 0$ and at $x = L_x$ (Example 6.3), and hence Eq. (6.15) can be used to obtain a variational formulation of the problem. Noting that $\psi \equiv 0$ satisfies all the problem boundary conditions, we have that

$$\Pi(\phi) = \int_0^{L_x}\left(-\frac{T}{2}\phi\frac{d^2\phi}{dx^2} - w\phi \right) dx$$

is the functional that must be minimized. Using integration by parts, we obtain

$$\Pi(\phi) = -\left[\phi\frac{d\phi}{dx} \right]_0^{L_x} + \int_0^{L_x}\left[\frac{T}{2}\left(\frac{d\phi}{dx}\right)^2 - w\phi \right] dx$$

which reduces to the functional introduced for this problem in Example 6.1 as $\phi = 0$ at $x = 0$ and $x = L_x$.

Example 6.5

Return to the problem of steady heat conduction in two dimensions described in Example 6.2. The governing equation is written here as

$$\mathcal{L}\phi + p = -\frac{\partial}{\partial x}\left(k\frac{\partial\phi}{\partial x} \right) - \frac{\partial}{\partial y}\left(k\frac{\partial\phi}{\partial y} \right) - Q = 0$$

and the boundary conditions are

$$\phi = \bar{\phi} \quad \text{on } \Gamma_\phi, \qquad k\frac{\partial \phi}{\partial n} = -\bar{q} \quad \text{on } \Gamma_q$$

Using the results of Exercise 6.6, since the thermal conductivity k is nonnegative, we see that the operator $\mathcal{L}$ is positive definite with respect to functions satisfying the homogeneous form of these conditions, namely,

$$\phi = 0 \quad \text{on } \Gamma_\phi, \qquad k\frac{\partial \phi}{\partial n} = 0 \quad \text{on } \Gamma_q$$

and so, by Eq. (6.22), a suitable functional for this problem is

$$\Pi(\phi) = \int_\Omega \left[-\frac{\phi}{2}\frac{\partial}{\partial x}\left(k\frac{\partial \phi}{\partial x}\right) - \frac{\phi}{2}\frac{\partial}{\partial y}\left(k\frac{\partial \phi}{\partial y}\right) - Q\phi \right] d\Omega$$

$$+ \frac{1}{2}\int_\Omega \left[\psi\frac{\partial}{\partial x}\left(k\frac{\partial \phi}{\partial x}\right) + \psi\frac{\partial}{\partial y}\left(k\frac{\partial \phi}{\partial y}\right) - \phi\frac{\partial}{\partial x}\left(k\frac{\partial \psi}{\partial x}\right) - \phi\frac{\partial}{\partial y}\left(k\frac{\partial x}{\partial y}\right) \right] d\Omega$$

where ψ is any function that satisfies the full nonhomogeneous boundary conditions. Using Green's lemma, the second integral may be rewritten to give

$$\int_\Omega \left[\psi\frac{\partial}{\partial x}\left(k\frac{\partial \phi}{\partial x}\right) + \psi\frac{\partial}{\partial y}\left(k\frac{\partial \phi}{\partial y}\right) - \phi\frac{\partial}{\partial x}\left(k\frac{\partial \psi}{\partial x}\right) - \phi\frac{\partial}{\partial y}\left(k\frac{\partial \psi}{\partial y}\right) \right] d\Omega$$

$$= \int_{\Gamma_\phi} \left(\psi k\frac{\partial \phi}{\partial n} - \phi k\frac{\partial \psi}{\partial n} \right) d\Gamma + \int_{\Gamma_q} \left(\psi k\frac{\partial \phi}{\partial n} - \phi k\frac{\partial \psi}{\partial n} \right) d\Gamma$$

But $\phi = \psi = \bar{\phi}$ on Γ_ϕ while $k\,\partial\phi/\partial n = k\,\partial\psi/\partial n = -\bar{q}$ on Γ_q. Hence

$$\Pi(\phi) = \int_\Omega \left[-\frac{\phi}{2}\frac{\partial}{\partial x}\left(k\frac{\partial \phi}{\partial x}\right) - \frac{\phi}{2}\frac{\partial}{\partial y}\left(k\frac{\partial \phi}{\partial y}\right) - Q\phi \right] d\Omega$$

$$+ \frac{1}{2}\int_{\Gamma_\phi} \bar{\phi} k\frac{\partial \phi}{\partial n}\,d\Gamma + \frac{1}{2}\int_{\Gamma_q} \phi\bar{q}\,d\Gamma + \text{terms independent of } \phi$$

and using Green's lemma again in the first integral, it is found that, ignoring the constant terms independent of ϕ, the required function makes

$$\Pi(\phi) = \int_\Omega \left[\frac{k}{2}\left(\frac{\partial \phi}{\partial x}\right)^2 + \frac{k}{2}\left(\frac{\partial \phi}{\partial y}\right)^2 - Q\phi \right] d\Omega + \int_{\Gamma_q} \phi\bar{q}\,d\Gamma$$

stationary among the set of functions satisfying the full nonhomogeneous boundary conditions. In fact, this set of admissible functions may be widened for, on evaluation of the first variation $\delta\Pi$ (as in Example 6.2), the boundary condition on Γ_q can be identified as a natural boundary condition, and so the function ϕ can be sought among the set of functions which satisfy the boundary condition on Γ_ϕ. The fact that this boundary condition $k\,\partial\phi/\partial n = -\bar{q}$ is natural could also have been deduced from the closing remarks of the previous section, for the operator $\mathcal{L}$ here includes differentiation of order 2 (i.e., $d = 1$), and so a boundary condition involving differentiation of order 1 is natural.

EXERCISES

6.8. A light beam of length L_x is clamped at both ends and subjected to a load $w(x)$ per unit length. If $R = EI$ is the flexural rigidity of the beam, the deflection ϕ is given by the solution of the equation

$$\frac{d^2}{dx^2}\left[R(x)\frac{d^2\phi}{dx^2}\right] = w(x)$$

subject to $\phi = d\phi/dx = 0$ at $x = 0$ and at $x = L_x$. Show that ϕ minimizes

$$\Pi(\phi) = \int_0^{L_x}\left[R(x)\left(\frac{d^2\phi}{dx^2}\right)^2 - 2w(x)\phi\right]dx$$

with respect to functions satisfying $\phi = d\phi/dx = 0$ at $x = 0$ and $x = L_x$. If the end $x = L_x$ is free so that the boundary condition becomes $d^2\phi/dx^2 = d^3\phi/dx^3 = 0$ at $x = L_x$, what are the functional and the admissible set of functions in this case?

6.9. If the beam of the previous question rests on an elastic foundation of stiffness $k(> 0)$, the governing differential equation becomes

$$\frac{d^2}{dx^2}\left[R(x)\frac{d^2\phi}{dx^2}\right] + k\phi = w(x)$$

What is the appropriate variational formulation of this problem?

6.10. Obtain the variational formulation for the torsion problem specified in the form of Example 2.8, that is,

$$\frac{\partial^2\theta}{\partial x^2} + \frac{\partial^2\theta}{\partial y^2} = 0, \qquad -3 \leqslant x \leqslant 3, -2 \leqslant y \leqslant 2$$

with $\theta = \frac{1}{2}(x^2 + y^2)$ on the boundaries.

6.11. Show that the solution of the heat conduction equation of Example 6.5 subject to $\phi = 1$ on Γ_ϕ and $\partial\phi/\partial n = -h\phi$ on Γ_q, where h is a function of position, minimizes

$$\Pi(\phi) = \int_{\Omega}\left[\frac{k}{2}\left(\frac{\partial\phi}{\partial x}\right)^2 + \frac{k}{2}\left(\frac{\partial\phi}{\partial y}\right)^2 - Q\phi\right]d\Omega + \int_{\Gamma_q}h\phi^2\,d\Gamma$$

What is the admissible set of functions for the minimization process?

6.12. The deflection ϕ of a thin elastic plate is governed by the equation

$$\frac{\partial^4\phi}{\partial x^4} + 2\frac{\partial^4\phi}{\partial x^2\,\partial y^2} + \frac{\partial^4\phi}{\partial y^4} = w/D \qquad \text{in } \Omega$$

Here w is the load per unit area and D is the flexural rigidity of the plate. If the edge Γ of the plate is rigidly clamped so that $\phi = \partial\phi/\partial n = 0$ on Γ, show that ϕ minimizes

$$\Pi(\phi) = \int_{\Omega}\left[\left(\frac{\partial^2\phi}{\partial x^2}\right)^2 + 2\left(\frac{\partial^2\phi}{\partial x\,\partial y}\right)^2 + \left(\frac{\partial^2\phi}{\partial y^2}\right)^2 - 2\frac{w}{D}\phi\right]d\Omega$$

with respect to functions satisfying $\phi = \partial\phi/\partial n = 0$ on Γ.

6.4. APPROXIMATE SOLUTION OF DIFFERENTIAL EQUATIONS BY THE RAYLEIGH – RITZ METHOD

The results of the previous section may be used to develop a method for obtaining an approximate solution to the differential equation (6.9) subject to the general linear boundary conditions of Eq. (6.10), provided that $\mathcal{L}$ is a symmetric linear operator with respect to functions satisfying the condition of Eq. (6.11). The solution method to be adopted is similar to those introduced in Chapters 2 and 3 and was first used by Rayleigh and later generalised by Ritz. A suitable function ψ is defined satisfying the problem boundary conditions, that is,

$$\mathfrak{M}\psi + r = 0 \qquad \text{on } \Gamma \tag{6.25}$$

and a set of independent trial functions $\{N_m;\ m = 1, 2, 3, \dots\}$ is chosen such that

$$\mathfrak{M}N_m = 0 \qquad \text{on } \Gamma \tag{6.26}$$

The approximation is then written as

$$\phi \approx \hat{\phi} = \psi + \sum_{m=1}^{M} a_m N_m \tag{6.27}$$

and the boundary conditions on Γ are automatically satisfied for all values of the constants $a_1, a_2, \ldots, a_M$. The basis of the Rayleigh–Ritz method is now to evaluate [see Eq. (6.15)]

$$\Pi(\hat{\phi}) = \int_\Omega \left[(\hat{\phi} - \psi)\{\tfrac{1}{2}\mathcal{L}(\hat{\phi} - \psi) + \mathcal{L}\psi + p\} \right] d\Omega \qquad (6.28)$$

and then to make Π stationary with respect to the parameters $a_1, a_2, \ldots, a_M$.

It can be shown formally[3] that this process converges in the sense that $\lim_{M \to \infty} \int_\Omega (\phi - \hat{\phi})^2 \, d\Omega = 0$ if the operator $\mathcal{L}$ is positive definite, and it follows from Eq. (6.21) that $\Pi(\hat{\phi}) \geq \Pi(\phi)$ in this case.

Using Eq. (6.27), we can write the functional as

$$\Pi(\hat{\phi}) = \sum_{l=1}^{M} \sum_{m=1}^{M} \frac{a_l a_m}{2} \int_\Omega N_l \mathcal{L} N_m \, d\Omega + \sum_{l=1}^{M} a_l \int_\Omega N_l (\mathcal{L}\psi + p) \, d\Omega \quad (6.29)$$

and this expression attains its stationary value when

$$\frac{\partial \Pi}{\partial a_1} = \frac{\partial \Pi}{\partial a_2} = \cdots = \frac{\partial \Pi}{\partial a_M} = 0 \qquad (6.30)$$

Performing the differentiation in Eq. (6.29) leads to a set of linear equations

$$\sum_{m=1}^{M} a_m \int_\Omega N_l \mathcal{L} N_m \, d\Omega = -\int_\Omega N_l (\mathcal{L}\psi + p) \, d\Omega, \qquad l = 1, 2, \ldots, M \quad (6.31)$$

which may be solved for the constants $a_1, a_2, \ldots, a_M$. This equation set can be written in the familiar standard matrix form

$$\mathbf{Ka} = \mathbf{f} \qquad (6.32)$$

where now

$$K_{lm} = \int_\Omega N_l \mathcal{L} N_m \, d\Omega$$

$$f_l = -\int_\Omega N_l (\mathcal{L}\psi + p) \, d\Omega \qquad (6.33)$$

It should be noted that the matrix $\mathbf{K}$ is always symmetric because of the assumed properties of the operator $\mathcal{L}$. The reader will also observe that the approximation equations produced here are identical to those produced in Chapter 2 [Eq. (2.33)] for the Galerkin method of solution of this problem (i.e., $W_l = N_l$). We can therefore deduce that the approximations produced by the Rayleigh–Ritz and the Galerkin methods for the solution of Eqs. (6.9) and

(6.10) will be identical if the operator $\mathcal{L}$ is symmetric, that is if a variational form of the problem exists. The Galerkin method can, of course, be used whether or not such a variational form can be found and has, therefore, wider applicability.

Example 6.6

We return to the problem of Example 2.6 where we sought the solution of the equation

$$-\frac{d^2\phi}{dx^2} + \phi = 0, \qquad 0 \leqslant x \leqslant 1$$

subject to the conditions $\phi = 0$ at $x = 0$ and $d\phi/dx = 20$ at $x = 1$. The reader can show that the operator $\mathcal{L} = -d^2/dx^2 + 1$ appearing in this problem is positive definite with respect to functions θ satisfying $\theta = 0$ at $x = 0$, $d\theta/dx = 0$ at $x = 1$. Then, from Eq. (6.22), this problem is equivalent to finding the function ϕ which satisfies the boundary conditions and which minimizes

$$\Pi(\phi) = \int_0^1 \left[\phi\left(-\frac{d^2\phi}{dx^2} + \phi \right) - \psi\left(-\frac{d^2\phi}{dx^2} + \phi \right) + \phi\left(-\frac{d^2\psi}{dx^2} + \psi \right) \right] dx$$

where ψ is any function satisfying the problem boundary conditions. Following through a one-dimensional equivalent of the analysis of Example 6.5, this is equivalent to minimizing a new functional

$$\Pi(\phi) = \int_0^1 \left[\left(\frac{d\phi}{dx} \right)^2 + \phi^2 \right] dx - 2[20\phi]_{x=1}$$

The operator $\mathcal{L}$ for this problem contains differentiation of order 2, and thus, from the preceding theory, the condition at $x = 0$ is an essential boundary condition, while the condition at $x = 1$ is a natural boundary condition. Proceeding as in Example 2.6, we construct an approximation

$$\hat{\phi} = a_1 x + a_2 x^2 + \cdots + a_M x^M$$

thus automatically satisfying the essential condition at $x = 0$, but not the natural condition at $x = 1$. The constants are now determined by the Rayleigh–Ritz method of minimizing Π with respect to variations in $a_1, a_2, \ldots, a_M$.

For example, using a two-term approximation means that

$$\Pi(\hat{\phi}) = \int_0^1 \left[(a_1 + 2a_2 x)^2 + (a_1 x + a_2 x^2)^2 \right] dx - 40(a_1 + a_2)$$

and this attains its minimum value when

$$\frac{\partial \Pi}{\partial a_1} = 2\int_0^1 \left[(a_1 + 2a_2 x) + x(a_1 x + a_2 x^2) \right] dx - 40 = 0$$

and

$$\frac{\partial \Pi}{\partial a_2} = 2\int_0^1 \left[2x(a_1 + 2a_2 x) + x^2(a_1 x + a_2 x^2) \right] dx - 40 = 0$$

Performing the integrations produces the equation set

$$\tfrac{4}{3}a_1 + \tfrac{5}{4}a_2 = 20$$

$$\tfrac{5}{4}a_1 + \tfrac{23}{15}a_2 = 20$$

As would be expected from previous observations, these are just the equations produced when the Galerkin method of solution is applied to this problem, and therefore the solution $a_1 = 11.7579$, $a_2 = 3.4582$ of Example 2.6 is reproduced, and with the one and two-term approximations again illustrating convergence to the natural boundary condition on $d\phi/dx$ at $x = 1$.

EXERCISES

6.13. Obtain, by the Rayleigh–Ritz method, an approximate solution to a problem of steady one-dimensional heat conduction with a distributed heat source governed by the equation

$$\frac{d^2\phi}{dx^2} + e^{-x} = 0, \qquad 0 \leqslant x \leqslant 1$$

and the boundary conditions $\phi = 0$ at $x = 0$, $d\phi/dx = 1$ at $x = 1$. Use piecewise linear finite element trial functions with a distance of $\tfrac{1}{3}$ between successive nodes.

6.14. Return to the problem of Example 3.4 of a bar under the action of axial body forces b per unit volume, where the governing equation was written

$$\frac{d}{dx}\left(AE\frac{d\phi}{dx} \right) + Ab = 0, \qquad 0 \leqslant x \leqslant 1$$

subject to $\phi = 0$ at $x = 0$, $d\phi/dx = 0$ at $x = 1$, and where A is the cross section of the bar, E is Young's modulus for the bar, and ϕ is the displacement at any point. Show that the variational form of this problem is equivalent to minimizing the total potential energy of the system. When A, E, b are constant, obtain an approximate solution to the problem using suitable continuous trial functions.

6.15. The deflection of a certain light beam of unit length and unit flexural rigidity is governed by the equation

$$\frac{d^4\phi}{dx^4} = \sin \pi x, \qquad 0 \leqslant x \leqslant 1$$

Obtain a one-term approximation, using polynomial-type trial functions, by the Rayleigh–Ritz method for the case when the beam is rigidly clamped at both ends so that $\phi = d\phi/dx = 0$ at $x = 0$ and $x = 1$. Compare this approximation with the exact solution for this problem.

6.16. If the right-hand end of the beam in Exercise 6.15 is now free, so that $d^2\phi/dx^2 = d^3\phi/dx^3 = 0$ at $x = 1$, determine a one-term approximation using the Rayleigh–Ritz method and satisfying the essential boundary conditions only. Examine the accuracy to which the natural boundary conditions are satisfied by comparing the approximation with the exact solution for this problem.

6.5.   THE USE OF LAGRANGE MULTIPLIERS

In the previous sections it has been shown how the solution of a linear differential equation [in the form of Eq. (6.9)], subject to the general boundary conditions of Eq. (6.10), can be found by seeking the function ϕ, which makes a certain functional $\Pi(\phi)$ stationary among the set of functions that satisfy the appropriate boundary conditions for the problem. If, however, we view the boundary condition as an additional *constraint* on the problem of making $\Pi(\phi)$ stationary, then we can use an approach due to Lagrange in which the variation of a new functional $\Pi_1(\phi, \lambda)$ is considered. This new functional is constructed as

$$\Pi_1(\phi, \lambda) = \Pi(\phi) + \int_\Gamma \lambda(\mathfrak{M}\phi + r)\,d\Gamma \qquad (6.34a)$$

where λ, known as a Lagrange multiplier, is a function of the space coordinates. The first variation in Π_1 is then given by

$$\delta\Pi_1 = \delta\Pi + \int_\Gamma \delta\lambda(\mathfrak{M}\phi + r)\,d\Gamma + \int_\Gamma \lambda\mathfrak{M}\delta\phi\,d\Gamma \qquad (6.34b)$$

When Π_1 is stationary, that is, $\delta\Pi_1 = 0$ for all variations $\delta\phi$, $\delta\lambda$ we have

$$\mathfrak{M}\phi + r = 0 \qquad \text{on } \Gamma \qquad (6.35a)$$

and hence

$$\mathfrak{M}\delta\phi = 0 \qquad \text{on } \Gamma \qquad (6.35b)$$

and

$$\delta\Pi = 0 \qquad \text{in } \Omega \qquad (6.35c)$$

It has been seen previously that condition (6.35c) requires that

$$\mathcal{L}\phi + p = 0 \qquad \text{in } \Omega \qquad (6.36)$$

and so the function ϕ which makes Π_1 stationary is the solution of this equation which satisfies the boundary condition of Eq. (6.35a).

If approximations are constructed for both ϕ and λ in the usual manner as

$$\phi \approx \hat{\phi} = \sum_{m=1}^{M} a_m N_{m,1}$$

$$\lambda \approx \hat{\lambda} = \sum_{m=1}^{M} b_m N_{m,2} \qquad (6.37)$$

where $\hat{\phi}$ does not now necessarily satisfy any or all of the required boundary conditions on ϕ, then the constants $\{a_m, b_m; \ m = 1, 2, \ldots, M\}$ can be determined by the requirement that $\Pi_1(\hat{\phi}, \hat{\lambda})$ be stationary. Inserting the approximations of Eq. (6.37) into Eq. (6.34a), gives

$$\Pi_1(\hat{\phi}, \hat{\lambda}) = \Pi(\hat{\phi}) + \int_{\Gamma} \left[\sum_{m=1}^{M} b_m N_{m,2} \right] \left[\left(\sum_{m=1}^{M} a_m \mathfrak{M} N_{m,1} \right) + r \right] d\Gamma \quad (6.38)$$

and this is stationary with respect to variations in a_l, b_l, provided that

$$\frac{\partial \Pi_1}{\partial a_l} = \frac{\partial \Pi}{\partial a_l} + \int_{\Gamma} \left[\sum_{m=1}^{M} b_m N_{m,2} \right] \mathfrak{M} N_{l,1} \, d\Gamma = 0 \qquad (6.39a)$$

$$\frac{\partial \Pi}{\partial b_l} = \int_{\Gamma} N_{l,2} \left[\left(\sum_{m=1}^{M} a_m \mathfrak{M} N_{m,1} \right) + r \right] d\Gamma = 0 \qquad (6.39b)$$

The first term on the right-hand side of Eq. (6.39a) is given by the original variational principle, and we can write

$$\frac{\partial \Pi}{\partial a_l} = \sum_{m=1}^{M} K_{lm} a_m - f_l \qquad (6.40)$$

with suitable K_{lm} and f_l. The equation set (6.39) then becomes, in matrix form,

$$\begin{bmatrix} \mathbf{K} & \mathbf{K}_1 \\ \mathbf{K}_1^T & \mathbf{0} \end{bmatrix} \begin{bmatrix} \mathbf{a} \\ \mathbf{b} \end{bmatrix} = \begin{bmatrix} \mathbf{f} \\ \mathbf{f}_1 \end{bmatrix} \tag{6.41}$$

where

$$[\mathbf{K}_1]_{lm} = \int_\Gamma N_{m,2} \mathfrak{M} N_{l,1} \, d\Gamma$$

$$[\mathbf{f}_1]_l = -\int_\Gamma r N_{l,2} \, d\Gamma \tag{6.42}$$

It is immediately apparent that this formulation results in a larger number of unknown parameters than the original formulation of the preceding sections. Moreover, the matrix which has to be inverted now possesses zeros on its main diagonal, and allowance must be made for this in any numerical computation of the solution.

The Lagrange multiplier can be applied to the enforcement of constraints other than boundary conditions, and such use is at times convenient.

EXERCISES

6.17. Obtain a solution to the problem of Example 2.4 by a Lagrange multiplier method.

6.18. Return to Example 2.5 and obtain a two-term solution by a Lagrange mutiplier method.

6.19. For Example 2.4 construct a solution using two linear finite elements and a Lagrange multiplier method. Compare the values of λ at $x = 0$ and $x = 1$ with the exact values of $-\partial\phi/\partial n$ at these points. (See the next section for the significance of this comparison.)

6.5.1. Physical Identification of the Lagrange Multiplier. Modified Variational Principles.

Although the Lagrange multipliers have been introduced as a device for the enforcement of certain boundary constraints which are necessary for the satisfaction of the original variational principle, it is found that in many physical problems they can often be identified with physical quantities of importance in the original mathematical model. We can best illustrate this by reference to a particular example, and return again to the solution of the two-dimensional steady problem of heat conduction in a region Ω. The governing equation is as given in Example 6.5, and it will be assumed that the boundary condition to be applied is one in which the value of the temperature ϕ is specified on the boundary curve Γ, that is,

$$\phi = \bar{\phi} \qquad \text{on } \Gamma \qquad (6.43)$$

If this boundary condition is treated as a constraint on ϕ, then the previous theory (e.g., Example 6.5) gives that the solution sought makes

$$\Pi_1(\phi, \lambda) = \int_\Omega \left[\frac{k}{2}\left(\frac{\partial \phi}{\partial x}\right)^2 + \frac{k}{2}\left(\frac{\partial \phi}{\partial y}\right)^2 - Q\phi \right] d\Omega + \int_\Gamma \lambda(\phi - \bar{\phi})\, d\Gamma$$

$$(6.44)$$

stationary with respect to variations in ϕ, λ. Performing the variation gives

$$\delta\Pi_1 = \int_\Omega \left[k\frac{\partial \phi}{\partial x}\frac{\partial}{\partial x}(\delta\phi) + k\frac{\partial \phi}{\partial y}\frac{\partial}{\partial y}(\delta\phi) - Q\delta\phi \right] d\Omega$$

$$+ \int_\Gamma \delta\lambda(\phi - \bar{\phi})\, d\Gamma + \int_\Gamma \lambda\delta\phi\, d\Gamma \qquad (6.45)$$

and the use of Green's lemma produces the result

$$\delta\Pi_1 = -\int_\Omega \delta\phi \left[\frac{\partial}{\partial x}\left(k\frac{\partial \phi}{\partial x}\right) + \frac{\partial}{\partial y}\left(k\frac{\partial \phi}{\partial y}\right) + Q \right] d\Omega + \int_\Gamma k\frac{\partial \phi}{\partial n}\delta\phi\, d\Gamma$$

$$+ \int_\Gamma \delta\lambda(\phi - \bar{\phi})\, d\Gamma + \int_\Gamma \lambda\delta\phi\, d\Gamma \qquad (6.46)$$

Then if Π_1 is to be stationary for all variations $\delta\phi, \delta\lambda$, it follows that

$$\phi - \bar{\phi} = 0 \qquad \text{on } \Gamma \qquad (6.47a)$$

$$\lambda + k\frac{\partial \phi}{\partial n} = 0 \qquad \text{on } \Gamma \qquad (6.47b)$$

$$\frac{\partial}{\partial x}\left(k\frac{\partial \phi}{\partial x}\right) + \frac{\partial}{\partial y}\left(k\frac{\partial \phi}{\partial y}\right) + Q = 0 \qquad \text{in } \Omega \qquad (6.47c)$$

Equations (6.47a) and (6.47c) show that the function ϕ which makes Π_1 stationary is the required solution, while Eq. (6.47b) defines the appropriate value of λ. Noting that $-k\,\partial\phi/\partial n$ on Γ is equal to the boundary flux of heat, a physical interpretation of the value of the Lagrange multiplier in this case has been produced. This identification of the Lagrange multiplier leads to the possible establishment of a modified variational principle in which λ is replaced at the outset by its identification. We could thus replace Eq. (6.44) with a new principle for this example and seek to make

$$\Pi(\phi) = \int_\Omega \left[\frac{k}{2} \left(\frac{\partial \phi}{\partial x} \right)^2 + \frac{k}{2} \left(\frac{\partial \phi}{\partial y} \right)^2 - Q\phi \right] d\Omega - \int_\Gamma k \frac{\partial \phi}{\partial n} (\phi - \bar{\phi}) \, d\Gamma$$

$$(6.48)$$

stationary, where again ϕ is not constrained to satisfy any boundary conditions. Constructing an approximate solution $\hat{\phi}$ in the usual form of an expansion in terms of trial functions and seeking to make $\Pi(\hat{\phi})$ stationary with respect to variations in the parameters in the expansion, it can be seen that the use of this new principle restores the problem to the original number of unknown parameters, and that this method may be computationally advantageous.

Principles of this form have been used in many different areas and have been widely used in the field of structural mechanics.[4]

Example 6.7

In Example 2.4 the equation $-d^2\phi/dx^2 + \phi = 0$ was solved, subject to the boundary conditions $\phi = 0$ at $x = 0$, $\phi = 1$ at $x = 1$, by using a weighted residual method and a trial function expansion which did not satisfy, a priori, the problem boundary conditions. Now we shall attempt to produce a solution using the same trial function expansion and a variational method of the form just described.

By following through the steps leading from Eq. (6.44) to Eq. (6.48), the reader should first show that this problem is equivalent to finding the function ϕ which makes

$$\Pi(\phi) = \int_0^1 \frac{1}{2} \left[\left(\frac{d\phi}{dx} \right)^2 + \phi^2 \right] dx + \left[\frac{d\phi}{dx} \phi \right]_{x=0} - \left[\frac{d\phi}{dx} (\phi - 1) \right]_{x=1}$$

stationary. Using the same three-term approximation

$$\hat{\phi} = a_1 + a_2 x + a_3 x^2$$

as used in Example 2.4, the constants a_1, a_2, a_3 are now chosen so as to make $\Pi(\hat{\phi})$ stationary with respect to variations in these quantities, that is, we require that

$$\frac{\partial \Pi}{\partial a_1} = \frac{\partial \Pi}{\partial a_2} = \frac{\partial \Pi}{\partial a_3} = 0$$

Now

$$\Pi(\hat{\phi}) = \int_0^1 \tfrac{1}{2}\Big[(a_2 + 2a_3x)^2 + (a_1 + a_2x + a_3x^2)^2\Big]\, dx$$

$$+ \big[(a_2 + 2a_3x)(a_1 + a_2x + a_3x^2)\big]_{x=0}$$

$$- \big[(a_2 + 2a_3x)(a_1 + a_2x + a_3x^2 - 1)\big]_{x=1}$$

which means that

$$\frac{\partial\Pi}{\partial a_1} = \int_0^1 (a_1 + a_2x + a_3x^2)\, dx + a_2 - (a_2 + 2a_3) = 0$$

$$\frac{\partial\Pi}{\partial a_2} = \int_0^1 \big[(a_2 + 2a_3x) + x(a_1 + a_2x + a_3x^2)\big]\, dx + a_1$$

$$- (a_1 + a_2 + a_3 - 1) - (a_2 + 2a_3) = 0$$

$$\frac{\partial\Pi}{\partial a_3} = \int_0^1 \big[2x(a_2 + 2a_3x) + x^2(a_1 + a_2x + a_3x^2)\big]\, dx$$

$$- 2(a_1 + a_2 + a_3 - 1) - (a_2 + 2a_3) = 0$$

Performing the integrations produces a linear equation system **Ka = f** where

$$\mathbf{K} = \begin{bmatrix} 1 & \tfrac{1}{2} & -\tfrac{5}{3} \\ \tfrac{1}{2} & -\tfrac{2}{3} & -\tfrac{7}{4} \\ -\tfrac{5}{3} & -\tfrac{7}{4} & -\tfrac{37}{15} \end{bmatrix}$$

$$\mathbf{f} = \begin{bmatrix} 0 \\ -1 \\ -2 \end{bmatrix}$$

The solution of the equation set for the three-term approximation is

$$a_1 = -0.0448, \qquad a_2 = 0.8598, \qquad a_3 = 0.2311$$

and this is compared with the exact solution and the corresponding one- and two-term approximations in Fig. 6.1.

EXERCISES

6.20. Return to the problem of the displacement of an axially loaded bar which was solved in Example 3.4. Obtain a solution using a Lagrange multiplier method with a suitable set of continuous trial functions.

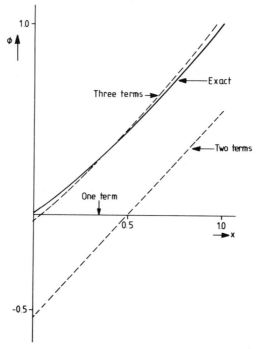

FIGURE 6.1. Comparison of approximate and exact solutions to the problem of Example 6.7.

6.21. Identify the physical significance of the Lagrange multiplier λ in Exercise 6.20 and then re-solve the problem by constructing a modified variational principle in which λ is replaced by this identification.

6.22. Return to Example 2.5 and obtain an approximate solution by using the same polynomial trial function set together with the modified variational principle of Eq. (6.48).

6.6. GENERAL VARIATIONAL PRINCIPLES

The Lagrangian multiplier technique leads to an obvious procedure for creating a variational principle for a general differential equation of the form

$$\mathcal{L}\phi + p = 0 \quad \text{in } \Omega \tag{6.49}$$

where the operator $\mathcal{L}$ may not possess the necessary properties for the existence of a natural variational principle. By regarding the above equation as a constraint, we can obtain a variational formulation for this equation simply by putting

$$\Pi(\phi, \lambda) = \int_\Omega \lambda(\mathcal{L}\phi + p) \, d\Omega \tag{6.50}$$

and requiring Π to be stationary for all variations $\delta\phi$ and $\delta\lambda$ with respect to functions ϕ satisfying the problem boundary conditions.

For, performing the variation, results in

$$\delta\Pi = \int_\Omega \delta\lambda(\mathcal{L}\phi + p)\, d\Omega + \int_\Omega \lambda\mathcal{L}\delta\phi\, d\Omega \qquad (6.51)$$

and introducing the adjoint operator $\mathcal{L}^*$ defined by

$$\int_\Omega \lambda\mathcal{L}\delta\phi\, d\Omega = \int_\Omega \delta\phi\mathcal{L}^*\lambda\, d\Omega \qquad (6.52)$$

it is possible to deduce that Π is stationary for all such variations $\delta\phi, \delta\lambda$ provided that

$$\mathcal{L}\phi + p = 0 \qquad \text{in } \Omega \qquad (6.53)$$

as required, and

$$\mathcal{L}^*\lambda = 0 \qquad \text{in } \Omega \qquad (6.54)$$

The adjoint operator[5] as defined here exists only for linear problems, and λ is known as the adjoint function to ϕ.

Again we can see that this variational principle has been introduced at the expense of doubling the number of variables in any approximate solution. If we use the approximations

$$\phi \approx \hat{\phi} = \psi + \sum_{m=1}^{M} a_m N_{m,1}$$

$$\lambda \approx \hat{\lambda} = \sum_{m=1}^{M} b_m N_{m,2} \qquad (6.55)$$

and follow the steps involved in producing Eqs. (6.38)–(6.42), it is found that, in this case, the final system of equations takes the form

$$\begin{bmatrix} \mathbf{K} & \mathbf{0} \\ \mathbf{0} & \mathbf{K}^T \end{bmatrix} \begin{bmatrix} \mathbf{a} \\ \mathbf{b} \end{bmatrix} = \begin{bmatrix} \mathbf{f} \\ \mathbf{0} \end{bmatrix} \qquad (6.56)$$

where

$$K_{lm} = \int_\Omega N_{l,2}\mathcal{L}N_{m,1}\, d\Omega$$

$$f_l = -\int_\Omega N_{l,2}(\mathcal{L}\psi + p)\, d\Omega \qquad (6.57)$$

It can be seen that this final system is completely decoupled and the first set can be solved independently for the parameters a_m, which describe the unknowns in the approximation $\hat{\phi}$ in which we are primarily interested, without consideration of the parameters b_m. This first set of equations, it will be observed, is identical to the equations which result from a general weighted residual process, and thus we have completed a full circle and obtained the weighted residual forms of Chapter 2 from a general variational principle.

6.7.   PENALTY FUNCTIONS

In Section 6.5 it was shown how the general process of introducing Lagrange multipliers produces variational principles in which, normally, the total number of unknowns is increased. It was also demonstrated that the solution of the resulting matrix equation is complicated by the appearance of zero diagonal terms in the matrix to be inverted, namely, Eq. (6.41). In this section an alternative method of imposing constraints will be considered, and this method will be shown to be free of these drawbacks associated with the standard Lagrange multiplier approach.

To illustrate the method, we return again to the problem considered in Section 6.5, which is that of minimizing a certain functional $\Pi(\phi)$ with respect to functions satisfying the boundary condition constraint of

$$\mathcal{M}\phi + r = 0 \qquad \text{on } \Gamma \qquad (6.58)$$

Defining a modified functional $\Pi_1(\phi)$ by

$$\Pi_1(\phi) = \Pi(\phi) + \alpha \int_\Gamma (\mathcal{M}\phi + r)^2 \, d\Gamma \qquad (6.59)$$

where α is a large positive *penalty number*, it is clear that the function which makes Π_1 stationary will simultaneously approximately satisfy the differential equation and the constraint of Eq. (6.58). The larger the value of α adopted, the better will be the satisfaction of the boundary constraint, and the better the approximation to ϕ. The reader will observe that a problem involving a constraint formulated in this manner introduces no additional unknown parameters. Further, if the original variational principle is such that $\Pi(\phi)$ is minimized, the penalty function method will always result in strongly positive definite matrices.

Example 6.8

The application of the penalty function method will be illustrated by considering again the problem of Example 6.7. The equation to be solved is $-d^2\phi/dx^2$

$+ \phi = 0$ subject to the boundary conditions $\phi = 0$ at $x = 0$ and $\phi = 1$ at $x = 1$. Now by Eq. (6.59), we consider the variation of

$$\Pi_1(\phi) = \int_0^1 \frac{1}{2}\left[\left(\frac{d\phi}{dx}\right)^2 + \phi^2\right] dx + \alpha[\phi^2]_{x=0} + \alpha[(\phi - 1)^2]_{x=1}$$

where α is a penalty number. Using, as previously, a three-term approximation

$$\hat{\phi} = a_1 + a_2 x + a_3 x^2$$

the constants a_1, a_2, a_3 are then chosen so as to make Π_1 stationary with respect to variations in a_1, a_2, a_3, that is, we require that

$$\frac{\partial \Pi}{\partial a_1} = \frac{\partial \Pi}{\partial a_2} = \frac{\partial \Pi}{\partial a_3} = 0$$

Inserting the approximation and performing the differentiation produces the equations

$$\frac{\partial \Pi}{\partial a_1} = \int_0^1 (a_1 + a_2 x + a_3 x^2)\, dx + 2\alpha a_1 + \alpha(2a_1 + 2a_2 + 2a_3 - 2) = 0$$

$$\frac{\partial \Pi}{\partial a_2} = \int_0^1 (a_2 + 2a_3 x + a_1 x + a_2 x^2 + a_3 x^3)\, dx$$

$$+ \alpha(2a_1 + 2a_2 + 2a_3 - 2) = 0$$

$$\frac{\partial \Pi}{\partial a_3} = \int_0^1 (2a_2 x + 4a_3 x^2 + a_1 x^2 + a_2 x^3 + a_3 x^4)\, dx$$

$$+ \alpha(2a_1 + 2a_2 + 2a_3 - 2) = 0$$

which, on evaluation of the integrals, leads to the linear equation system $\mathbf{Ka} = \mathbf{f}$ where

$$\mathbf{K} = \begin{bmatrix} 2 + 4\beta & 1 + 2\beta & \frac{2}{3} + 2\beta \\ 1 + 2\beta & \frac{8}{3} + 2\beta & \frac{5}{2} + 2\beta \\ \frac{2}{3} + 2\beta & \frac{5}{2} + 2\beta & \frac{46}{15} + 2\beta \end{bmatrix}$$

$$\mathbf{f} = \begin{bmatrix} 2\beta \\ 2\beta \\ 2\beta \end{bmatrix}$$

and $\beta = 2\alpha$.

This equation set can be solved once the value of β has been specified. In Fig. 6.2 the approximations resulting from taking $\beta = 1, 10, 1000$ in turn are

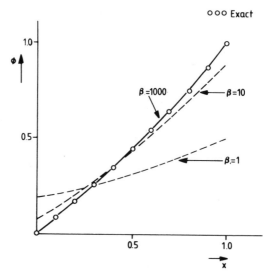

FIGURE 6.2. Effect of increasing the penalty parameter on the approximate solution to the problem of Example 6.8.

compared with the exact solution. The improvement of the solution with increasing β is apparent, and this is further demonstrated in the following table.

β	$\hat{\phi}|_{x=0}$	$\hat{\phi}|_{x=1}$
1	0.1841	0.4999
10	0.67×10^{-1}	0.8889
10^2	0.83×10^{-2}	0.9871
10^3	0.85×10^{-3}	0.9987

This table demonstrates the convergence of the three-term approximations to the prescribed boundary conditions at $x = 0$ and $x = 1$.

EXERCISES

6.23. Obtain, by a penalty function method, the distribution of bending moment in the beam of Exercise 1.2 by using an approximation which satisfies the condition at $x = 0$ but does not satisfy automatically the condition at $x = 1$. Investigate the effect of increasing the penalty parameter α.

6.24. Use a penalty function method to solve the torsion problem of Example 2.5 by using a polynomial approximation which satisfies the required condition on $y = \pm 2$ but does not satisfy automatically the condition on $x = \pm 3$. Investigate the effect of increasing the penalty parameter α on the estimate of the twisting moment and of the maximum value of the shear stress.

6.25. Return to Exercise 2.7 and obtain the distribution of temperature by using a penalty function method and an approximation which satisfies only the boundary conditions on the sides $x = \pm 1$.

6.8. LEAST-SQUARES METHOD

In Section 6.6 we have shown how a Lagrange multiplier technique could be used to construct a general variational principle by regarding the differential equation which is to be solved as a constraint to be applied over the whole domain. Clearly, the same procedure can be used in the context of the penalty function approach of the previous section, for if the equation to be solved is

$$\mathcal{L}\phi + p = 0 \qquad \text{in } \Omega \qquad (6.60)$$

then we can determine the solution by making

$$\Pi(\phi) = \alpha \int_\Omega (\mathcal{L}\phi + p)^2 \, d\Omega \qquad (6.61)$$

stationary, where the functions ϕ being considered are assumed to satisfy the required boundary conditions on Γ. It is interesting that since the parameter α is now just a multiplier, it need not be considered further and can be removed from this equation. Inserting an approximation $\hat{\phi}$ of the usual form, this statement may then be recognized as just the least-squares statement of Chapter 2 in which we required the sum of the squares of the residual over Ω to be a minimum.
    Writing

$$\hat{\phi} = \psi + \sum_{m=1}^{M} a_m N_m \qquad (6.62)$$

then

$$\Pi(\hat{\phi}) = \int_\Omega \left[\mathcal{L}\psi + \left(\sum_{m=1}^{M} a_m \mathcal{L} N_m \right) + p \right]^2 d\Omega \qquad (6.63)$$

and Π takes its minimum value when

$$2\int_{\Omega}\left[\mathcal{L}\psi + \left(\sum_{m=1}^{M} a_m\mathcal{L}N_m\right) + p\right]\mathcal{L}N_l\, d\Omega = 0, \qquad l = 1, 2, \ldots, M \qquad (6.64)$$

This approach therefore immediately yields an equation of the usual form

$$\mathbf{Ka} = \mathbf{f} \qquad (6.65)$$

where

$$K_{lm} = \int_{\Omega}(\mathcal{L}N_l)(\mathcal{L}N_m)\, d\Omega$$

$$f_l = -\int_{\Omega}\left[(\mathcal{L}\psi)(\mathcal{L}N_l) + p\mathcal{L}N_l\right] d\Omega \qquad (6.66)$$

Least-squares methods of this kind are a very powerful alternative procedure for obtaining integral forms which can be used to construct approximate solutions. A least-squares variational principle can be written for any differential equation without introducing additional variables, but the reader will observe, by performing a variation for a particular example, that the Euler equation is some derivative of the original equation. This introduces the possibility of spurious solutions if incorrect boundary conditions are used. In addition, it will be noted from Eq. (6.66) that the least-squares method will normally require higher order continuity of the trial functions used, and this is important if a finite element approximation is attempted. Although this may be a serious drawback in some cases, it can frequently be overcome by adopting a mixed formulation and replacing the original equation by a set of lower order equations. This, however, again necessitates the introduction of additional variables.

Example 6.9

Consider again the problem of linear steady heat conduction in one dimension, discussed in Example 3.3. The governing equation is

$$k\frac{d^2\phi}{dx^2} + Q = 0$$

where $k = 1$, $Q = 1$ for $x < \frac{1}{2}$, $Q = 0$ for $x > \frac{1}{2}$, and the boundary conditions are $\phi = 0$ at $x = 0$, $d\phi/dx = 0$ at $x = 1$. Using a finite element least-squares method, with an approximation

$$\hat{\phi} = \sum_{m=1}^{M} \phi_m N_m$$

we see that the matrix **K** of Eq. (6.65) has coefficients

$$K_{lm} = \int_0^1 \frac{d^2 N_l}{dx^2} \frac{d^2 N_m}{dx^2} dx$$

The trial functions must therefore exhibit C^1 continuity in order that the integrals appearing here be defined. If we wish to use linear elements, which possess only C^0 continuity, then we must use the mixed approach adopted in Example 3.3 and replace the above equation by the two first-order equations

$$k \frac{d\phi}{dx} + q = 0$$

$$\frac{dq}{dx} = Q$$

It has been observed previously that this pair of equations may be written as a single equation by introducing the vector

$$\boldsymbol{\phi} = \begin{bmatrix} \phi \\ q \end{bmatrix}$$

and then we have

$$\mathcal{L}\boldsymbol{\phi} + \mathbf{p} = \mathbf{0}$$

where

$$\mathcal{L} = \begin{bmatrix} \dfrac{d}{dx} & 1 \\ 0 & \dfrac{d}{dx} \end{bmatrix} \qquad \mathbf{p} = \begin{bmatrix} 0 \\ Q \end{bmatrix}$$

Although the least-squares analysis has only been described for a single equation, its extension to the case of a coupled equation system is straightforward, and here we would require that

$$\Pi(\boldsymbol{\phi}) = \int_0^1 (\mathcal{L}\boldsymbol{\phi} + \mathbf{p})^T (\mathcal{L}\boldsymbol{\phi} + \mathbf{p}) \, d\Omega$$

be a minimum. Using the mesh of four equal linear elements shown in Fig.

3.7a, finite element approximations can be constructed as

$$\phi \simeq \hat{\phi} = \sum_{m=1}^{5} \phi_m N_m = \sum_{m=1}^{5} \begin{bmatrix} \phi_m \\ q_m \end{bmatrix} N_m$$

where $\phi_0 = q_5 = 0$ by the boundary conditions. When these expansions are substituted into the above expression, it is found that $\Pi(\hat{\phi})$ is a minimum when

$$\mathbf{K\Phi} = \mathbf{f}$$

where

$$\mathbf{\Phi}^T = (\phi_1, \phi_2, \phi_3, \phi_4, \phi_5)$$

and a typical submatrix $\mathbf{K}_{lm}$ of $\mathbf{K}$ and a typical subvector $\mathbf{f}_l$ of $\mathbf{f}$ are defined by

$$\mathbf{K}_{lm} = \int_0^1 \begin{bmatrix} \dfrac{dN_l}{dx}\dfrac{dN_m}{dx} & \dfrac{dN_l}{dx}N_m \\[3mm] N_l\dfrac{dN_m}{dx} & \dfrac{dN_l}{dx}\dfrac{dN_m}{dx} + N_l N_m \end{bmatrix} dx$$

$$\mathbf{f}_l = \int_0^1 \begin{bmatrix} 0 \\[2mm] Q\dfrac{dN_l}{dx} \end{bmatrix} dx$$

As expected, it can now be observed that it is sufficient that the shape functions adopted be C^0 continuous. The reduced element matrices for the typical element e of Fig. 3.7b become

$$\mathbf{k}^e = \int_{\Omega^e} \begin{bmatrix} \left(\dfrac{dN_i^e}{dx}\right)^2 & \dfrac{dN_i^e}{dx}N_i^e & \dfrac{dN_i^e}{dx}\dfrac{dN_j^e}{dx} & \dfrac{dN_i^e}{dx}N_j^e \\[3mm] N_i^e\dfrac{dN_i^e}{dx} & \left(\dfrac{dN_i^e}{dx}\right)^2 + (N_i^e)^2 & N_i^e\dfrac{dN_j^e}{dx} & \dfrac{dN_i^e}{dx}\dfrac{dN_j^e}{dx} + N_i^e N_j^e \\[3mm] \dfrac{dN_i^e}{dx}\dfrac{dN_j^e}{dx} & N_i^e\dfrac{dN_j^e}{dx} & \left(\dfrac{dN_j^e}{dx}\right)^2 & \dfrac{dN_j^e}{dx}N_j^e \\[3mm] N_j^e\dfrac{dN_i^e}{dx} & \dfrac{dN_i^e}{dx}\dfrac{dN_j^e}{dx} + N_i^e N_j^e & N_j^e\dfrac{dN_j^e}{dx} & \left(\dfrac{dN_j^e}{dx}\right)^2 + (N_j^e)^2 \end{bmatrix} dx$$

$$\mathbf{f}^e = \int_{\Omega^e} \begin{bmatrix} 0 \\[2mm] Q\dfrac{dN_i^e}{dx} \\[2mm] 0 \\[2mm] Q\dfrac{dN_j^e}{dx} \end{bmatrix} dx$$

and evaluating these entries and performing the assembly process results in

$$
\begin{bmatrix}
1/h & -1/2 & -1/h & -1/2 & 0 & 0 & 0 & 0 & 0 & 0 \\
-1/2 & 1/h+h/3 & 1/2 & -1/h+h/6 & 0 & 0 & 0 & 0 & 0 & 0 \\
-1/h & 1/2 & 2/h & 0 & -1/h & -1/2 & 0 & 0 & 0 & 0 \\
-1/2 & -1/h+h/6 & 0 & 2/h+2h/3 & 1/2 & -1/h+h/6 & 0 & 0 & 0 & 0 \\
0 & 0 & -1/h & 1/2 & 2/h & 0 & -1/h & -1/2 & 0 & 0 \\
0 & 0 & -1/2 & -1/h+h/6 & 0 & 2/h+2h/3 & 1/2 & -1/h+h/6 & 0 & 0 \\
0 & 0 & 0 & 0 & -1/h & 1/2 & 2/h & 0 & -1/h & -1/2 \\
0 & 0 & 0 & 0 & -1/2 & -1/h+h/6 & 0 & 2/h+2h/3 & 1/2 & -1/h+h/6 \\
0 & 0 & 0 & 0 & 0 & 0 & -1/h & 1/2 & 2/h & 1/2 \\
0 & 0 & 0 & 0 & 0 & 0 & -1/2 & -1/h+h/6 & 1/2 & 1/h+h/3
\end{bmatrix}
\begin{bmatrix}
\phi_1 \\ q_1 \\ \phi_2 \\ q_2 \\ \phi_3 \\ q_3 \\ \phi_4 \\ q_4 \\ \phi_5 \\ q_5
\end{bmatrix}
=
\begin{bmatrix}
0 \\ -1 \\ 0 \\ 0 \\ 0 \\ 1 \\ 0 \\ 0 \\ 0 \\ 0
\end{bmatrix}
$$

The solution of this equation set is

$$\phi_2 = 0.09326, \quad \phi_3 = 0.1244, \quad \phi_4 = 0.1244, \quad \phi_5 = 0.1244$$

$$q_1 = -0.4974, \quad q_2 = -0.2487, \quad q_3 = 0, \quad q_4 = 0$$

which the reader should compare with the solution produced by the standard Galerkin method in Example 3.3.

EXERCISES

6.26. Use the C^1 continuous finite elements introduced in Exercise 4.12 to obtain directly a least-squares solution to the problem of Example 6.9.

6.27. Obtain a least-squares finite element solution to the beam deflection problem of Exercise 6.15.

6.9. CONCLUDING REMARKS

We have shown in this chapter how variational principles can be established for many classes of problems and how such variational methods lead readily to approximating, discrete equations. Indeed such equations are identical to those obtainable directly by the use of Galerkin processes previously described in Chapters 2 and 3. Is there any merit therefore in introducing the additional concepts described here? The answer is (weakly) in the affirmative, for the following reasons.

1. The existence of the functional guarantees that the matrix equation from which the problem parameters are to be determined is symmetric. This is computationally advantageous, as we have repeatedly remarked. Indeed it can be shown that symmetric discretized equations always imply the existence of a variational functional.

2. The functional often has a distinct physical meaning (such as, energy of the field). If minimization of such a functional leads to the exact solution, we are assured that the approximate solution will always be an overestimate of the functional. Bounds can thus be introduced at times on the error.

3. The understanding of the relations between differential equations and variational functionals does help in obtaining a new perspective on approximation.

REFERENCES

[1] R. Courant, *Differential and Integral Calculus*, Vol. II, Blackie and Son, London, 1936.

[2] M. M. Veinberg, *Variational Methods for the Study of Nonlinear Operators*, Holden-Day, San Francisco, 1964.

[3] S. G. Mikhlin, *Variational Methods in Mathematical Physics*, Macmillan, New York, 1964.

[4] K. Washizu, *Variational Methods in Elasticity and Plasticity*, 2nd ed., Pergamon, New York, 1975.

[5] I. Stakgold, *Boundary Value Problems of Mathematical Physics*, Macmillan, New York, 1967.

SUGGESTED FURTHER READING

L. Collatz, *The Numerical Treatment of Differential Equations*, Springer-Verlag, Berlin, 1960.

B. A. Finlayson, *The Method of Weighted Residuals and Variational Principles*, Academic, New York, 1972.

I. Fried, *Numerical Solution of Differential Equations*, Academic, New York, 1979.

S. G. Mikhlin and K. L. Smolitskiy, *Approximate Methods for Solution of Differential and Integral Equations*, Elsevier, New York, 1967.

T. H. Richards, *Energy Methods in Stress Analysis*, Ellis Horwood, Chichester, 1977.

G. Strang and G. J. Fix, *An Analysis of the Finite Element Method*, Prentice-Hall, Englewood Cliffs, N.J., 1973.

Partial Discretization and Time-Dependent Problems

7.1. INTRODUCTION

In the boundary value problems which have been analyzed in the previous chapters it has been assumed that steady-state conditions exist, that is, the solution sought has been taken to be independent of time. However, in a great number of practical problems the conditions are unsteady (i.e., time dependent), and the effects of the time dimension have to be considered. Typically we are given the state of a system at some time $t = 0$, and we require to determine the state of the system at subsequent times. Problems of this type are often referred to as being *initial value problems*, and they occur frequently in such fields as heat conduction, wave propagation, and the dynamic behavior of structures, for example, we saw in Section 1.2, for the case of heat conduction, how the time dependence arose naturally during the formulation of the problem, leading to the time-dependent governing equation (1.7). In this chapter we will be concerned with the development of solution methods for initial value problems. Although in such problems we could proceed with the discretization of the whole space–time domain, it is more convenient to use the so-called partial discretization procedure* which will replace the original partial differential equation by a set of ordinary differential equations. This set can then be solved by applying a second discretization in time alone or by other, often analytical, procedures. This process of partial discretization will be first described in the context of boundary value problems before proceeding to consider its use for time-dependent problems.

---

*This technique is termed Kantorovich's method in some texts.

7.2. PARTIAL DISCRETIZATION APPLIED TO BOUNDARY VALUE PROBLEMS

In the preceding chapters we have considered chiefly the problem of solving the general linear boundary value problem defined by an equation written in the form

$$\mathcal{L}\phi + p = 0 \quad \text{in } \Omega \tag{7.1}$$

subject to boundary conditions which were written as

$$\mathcal{M}\phi + r = 0 \quad \text{on } \Gamma \tag{7.2}$$

The approach adopted was to use a trial function expansion

$$\phi \simeq \hat{\phi} = \psi + \sum_{m=1}^{M} a_m N_m \tag{7.3}$$

where the trial functions N_m were chosen so as to include all the independent coordinates of the problem, and the quantities a_m were assumed to be constant. The final set of equations produced was then always of a linear algebraic form, namely,

$$\mathbf{Ka} = \mathbf{f} \tag{7.4}$$

from which a unique set of values for the constants $\{a_m;\ m = 1, 2, \ldots, M\}$ could be obtained.

However, a different approach may prove to be convenient for some problems. For example, if the independent variables are x, y, and z, we could allow the quantities a_m to be, say, functions of y and choose the trial functions only in the domain $\bar{\Omega}$ of x and z.

Thus in Eq. (7.3) we would have

$$\phi \simeq \hat{\phi} = \psi + \sum_{m=1}^{M} a_m(y) N_m(x, z) \tag{7.5}$$

where the functions ψ and N_m are now such that $\hat{\phi}$ satisfies the essential boundary conditions on $\bar{\Omega}$. Then, clearly, in the application of the weighted residual method the derivatives of a_m with respect to y will remain, and the resulting equations will be in the form of a set of coupled ordinary differential equations with y as the independent variable. Equation (7.4) will thus be replaced by the equation

$$\mathbf{Ka} + \mathbf{C}\frac{d\mathbf{a}}{dy} + \cdots = \mathbf{f} \tag{7.6}$$

with the order of the equation being governed by the highest derivative with respect to y appearing in the original equation (7.1).

This approach proves to be particularly useful when the domain $\overline{\Omega}$ is independent of y, that is, when the problem is *prismatic*. The coefficients in Eq. (7.6) are then independent of y, and the system may, in principle at least, be solved by direct analytic means. The detailed application of this method will now be illustrated by considering again the solution of the torsion problem introduced in Example 1.5.

Example 7.1

Recall that the problem is to solve the equation $\partial^2\phi/\partial x^2 + \partial^2\phi/\partial y^2 = -2$ in the rectangular region $-3 \leqslant x \leqslant 3$, $-2 \leqslant y \leqslant 2$ subject to the condition $\phi = 0$ on the boundaries.

For illustration we will use continuous trial functions over the region and attempt a solution in the form

$$\phi \simeq \hat{\phi} = \psi + \sum_{m=1}^{M} a_m(y)N_m(x)$$

The domain $\overline{\Omega}$ of x is then the region $-3 \leqslant x \leqslant 3$, and the boundary conditions at the two ends of this region can be satisfied by choosing $\psi = 0$ and requiring that the shape functions be such that $N_m = 0$ at $x = \pm 3$. It has been noted previously that this problem is symmetrical about $x = 0$, and so suitable trial functions could be defined by

$$N_m(x) = x^{2(m-1)}(9 - x^2), \qquad m = 1, 2, \ldots$$

This choice ensures satisfaction of the required conditions at $x = \pm 3$ and also takes account of the essential symmetry of the problem. The approximation then becomes

$$\hat{\phi} = \sum_{m=1}^{M} a_m(y)x^{2(m-1)}(9 - x^2)$$

and the boundary conditions on $y = \pm 2$ will be satisfied if the functions $a_m(y)$ are such that

$$a_m(y) = 0 \qquad \text{when } y = \pm 2, \qquad m = 1, 2, \ldots, M$$

If we restrict consideration to a one-term approximation, the residual $R_{\overline{\Omega}}$ can be calculated as

$$R_{\overline{\Omega}} = \frac{\partial^2\hat{\phi}}{\partial x^2} + \frac{\partial^2\hat{\phi}}{\partial y^2} + 2 = -2a_1 + (9 - x^2)\frac{d^2a_1}{dy^2} + 2$$

and the weighted residual method can be applied as before, noting, however, that the integration has now to be carried out only for the region $\bar{\Omega}$, that is, for $-3 \leqslant x \leqslant 3$. Thus the weighted residual statement here is (because of symmetry)

$$2\int_0^3 \left\{ -2a_1 + (9 - x^2)\frac{d^2a_1}{dy^2} + 2 \right\} W_1(x)\, dx = 0$$

and again the various forms for the weighting functions may be adopted. Choosing $W_1(x) = \delta(x - 1.5)$ produces point collocation with the residual made equal to zero at $x = \pm 1.5$, that is,

$$-2a_1 + \frac{27}{4}\frac{d^2a_1}{dy^2} + 2 = 0$$

This equation may be integrated exactly to give

$$a_1 = A \cosh\sqrt{\tfrac{8}{27}}\, y + B \sinh\sqrt{\tfrac{8}{27}}\, y + 1$$

where A and B are constants. The boundary conditions on a_1 then require that

$$A = -\operatorname{sech}\sqrt{\tfrac{32}{27}}, \qquad B = 0$$

and the one-term approximation is

$$\hat{\phi} = \left(1 - \operatorname{sech}\sqrt{\tfrac{32}{27}} \cosh\sqrt{\tfrac{8}{27}}\, y\right)(9 - x^2)$$

leading to the values 77.32 for the twisting moment and 3.901 for the maximum shear stress, compared to the exact values of 76.4 and 2.96, respectively.

EXERCISES

7.1. Repeat Example 7.1 using a one-term approximation and the Galerkin method. Compare the answers with those given by the point collocation method.

7.2. Solve the problem of steady-state heat conduction in the square region $0 \leqslant x, y \leqslant 1$ when the temperature ϕ is subjected to the boundary conditions shown. Divide the region $0 \leqslant x \leqslant 1$ into three equal linear finite elements and seek a solution in the form $\hat{\phi} = \sum_{m=1}^{4} a_m(y)N_m(x)$, where $N_m(x)$ denotes a piecewise linear trial function.

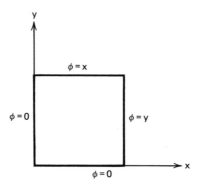

7.3. A finite difference grid is constructed over the region $0 \leqslant x \leqslant 1$ with equally spaced grid points $x_1, x_2, \ldots, x_M$ and with $x_1 = 0$, $x_M = 1$. This grid is to be used to analyze the problem of Exercise 7.2 with $\phi_j(y)$ denoting the value of the temperature on the line $x = x_j$. Show that the governing equation evaluated at $x = x_j$ can be expressed as

$$\frac{d^2\phi_j}{dy^2} + \frac{(\phi_{j+1} - 2\phi_j + \phi_{j-1})}{\Delta^2} = 0$$

where Δ is the grid spacing, and hence obtain a solution for the case $M = 4$.

7.3 TIME-DEPENDENT PROBLEMS VIA PARTIAL DISCRETIZATION

In many linear time-dependent physical problems the governing differential equation can be written in the form

$$\mathcal{L}\phi + p - \alpha\frac{\partial\phi}{\partial t} - \beta\frac{\partial^2\phi}{\partial t^2} = 0 \qquad \text{in } \Omega \tag{7.7}$$

where $\mathcal{L}$ is a linear operator involving differentiation with respect to the spatial coordinates only and p, α, and β are prescribed functions of position and time. Two typical examples of problems governed by this general formulation are as follows.

1. Transverse vibrations of a stretched string of density ρ and at tension T are governed by the equation

$$\frac{\partial^2\phi}{\partial x^2} - \frac{\rho}{T}\frac{\partial^2\phi}{\partial t^2} = 0 \tag{7.8}$$

which is Eq. (7.7) with $\alpha = p = 0$, $\beta = \rho/T$, and $\mathcal{L}\phi = \partial^2\phi/\partial x^2$.

2. Linear transient conduction of heat in a two-dimensional region of thermal capacity ρc and constant conductivity k is governed by Eq. (1.7), in the form

$$k\left(\frac{\partial^2 \phi}{\partial x^2} + \frac{\partial^2 \phi}{\partial y^2}\right) + Q - \rho c \frac{\partial \phi}{\partial t} = 0 \qquad (7.9)$$

which is Eq. (7.7) with $\beta = 0$, $\alpha = \rho c$, $p = Q$, and $\mathcal{L}\phi = k(\partial^2\phi/\partial x^2 + \partial^2\phi/\partial y^2)$.

In addition, typical associated conditions normally specify the value of the function ϕ and (if $\beta \neq 0$) $\partial\phi/\partial t$ at all points in the space domain $\bar{\Omega}$ at time $t = 0$ together with boundary conditions, valid for all $t \geq 0$, on the boundary $\bar{\Gamma}$ of the space domain. We have seen in the previous section that the method of partial discretization is particularly useful for prismatic problems, and so we might expect this method to be directly applicable to time-dependent problems if the space domain $\bar{\Omega}$ does not vary with time.

The application of the partial discretization method to the solution of Eq. (7.7) follows closely the description of the previous section. We use shape functions N_m which are functions of the space coordinates only and write the approximation as

$$\hat{\phi} = \psi + \sum_{m=1}^{M} a_m(t)N_m(x, y, z) \qquad (7.10)$$

where ψ and N_m are such that the essential boundary conditions on $\bar{\Gamma}$ are satisfied. The use of the weighted residual method then produces the matrix differential equation

$$\mathbf{M}\frac{d^2\mathbf{a}}{dt^2} + \mathbf{C}\frac{d\mathbf{a}}{dt} + \mathbf{Ka} = \mathbf{f} \qquad (7.11)$$

where the components of the individual matrices are defined by

$$M_{lm} = \int_{\Omega} \beta W_l N_m \, d\Omega$$

$$C_{lm} = \int_{\Omega} \alpha W_l N_m \, d\Omega$$

$$K_{lm} = -\int_{\Omega} W_l \mathcal{L} N_m \, d\Omega \qquad (7.12)$$

$$f_l = \int_{\Omega} \left(p + \mathcal{L}\psi - \alpha\frac{\partial\psi}{\partial t} - \beta\frac{\partial^2\psi}{\partial t^2} \right) W_l \, d\Omega$$

To complete the solution, this ordinary differential equation must be solved subject to the initial values of $\mathbf{a}$ and (if $\beta \neq 0$) $d\mathbf{a}/dt$ being given. This is a classical problem in the theory of ordinary differential equations, and its solution can, in principle, be obtained exactly.

Example 7.2

Consider the problem of unsteady heat conduction in the one-dimensional region defined by $0 \leqslant x \leqslant 1$. The thermal properties are assumed to be given by $k = \rho c = 1$, and the initial temperature distribution satisfies $\phi = x(1 - x)$. The temperature at $x = 0$ and at $x = 1$ is held fixed at zero for all time, and it is required to determine the resulting temperature distribution at later times. We therefore have to solve the one-dimensional unsteady heat conduction equation

$$\frac{\partial^2 \phi}{\partial x^2} - \frac{\partial \phi}{\partial t} = 0$$

subject to the conditions

$$\phi = x(1 - x), \qquad \text{at } t = 0$$

$$\phi = 0, \qquad \text{at } x = 0, 1 \text{ for } t \geqslant 0$$

The finite element method will be used, and the space domain (i.e, the region $0 \leqslant x \leqslant 1$) is divided into M linear elements, with the elements and nodes numbered as shown in Fig. 7.1a. Using the partial discretization method, the approximation $\hat{\phi}$ is then constructed in the finite element form as

$$\hat{\phi} = \sum_{m=1}^{M+1} \phi_m(t) N_m(x)$$

where ϕ_m is the time-dependent approximation to the temperature at node m and N_m denotes the standard linear shape function associated with this node (see Fig. 7.1b). We could immediately apply the essential boundary conditions at $x = 0$ and $x = 1$ in the form

$$\phi_1 = \phi_{M+1} = 0 \qquad \text{for all } t$$

but, as we have already seen, it is more convenient at this stage to regard ϕ_1 and ϕ_{M+1} as unknowns and to apply these essential conditions after the matrix assembly process has been completed.

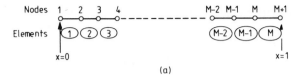

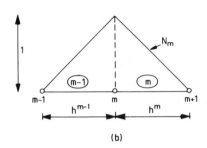

FIGURE 7.1. (*a*) Element and node numbering for Example 7.2 (*b*) Global piecewise linear shape function associated with node *m*.

The Galerkin weighted residual statement is

$$\int_0^1 \left(\frac{\partial^2 \hat{\phi}}{\partial x^2} - \frac{\partial \hat{\phi}}{\partial t} \right) N_l \, dx = 0, \qquad l = 1, 2, \ldots, M + 1$$

and, using integration by parts, the corresponding weak form is

$$\int_0^1 \left(\frac{\partial \hat{\phi}}{\partial x} \frac{dN_l}{dx} + \frac{\partial \hat{\phi}}{\partial t} N_l \right) dx - \left[N_l \frac{\partial \hat{\phi}}{\partial x} \right]_0^1 = 0, \qquad l = 1, 2, \ldots, M + 1$$

Inserting the approximation $\hat{\phi}$ leads to the equation set

$$\sum_{m=1}^{M+1} \left(\int_0^1 \frac{dN_l}{dx} \frac{dN_m}{dx} dx \right) \phi_m + \sum_{m=1}^{M+1} \left(\int_0^1 N_l N_m \, dx \right) \frac{d\phi_m}{dt} = \left[N_l \frac{\partial \hat{\phi}}{\partial x} \right]_0^1,$$

$$l = 1, 2, \ldots, M + 1$$

which may be written in the form of Eq. (7.11), namely,

$$\mathbf{C} \frac{d\boldsymbol{\phi}}{dt} + \mathbf{K} \boldsymbol{\phi} = \mathbf{f}$$

where typical matrix elements are now

$$C_{lm} = \int_0^1 N_l N_m \, dx$$

$$K_{lm} = \int_0^1 \frac{dN_l}{dx} \frac{dN_m}{dx} \, dx$$

$$f_l = \left[\frac{\partial \hat{\phi}}{\partial x} N_l \right]_0^1$$

and

$$\boldsymbol{\phi}^T = \left(\phi_1, \phi_2, \ldots, \phi_M, \phi_{M+1} \right)$$

With linear elements being used, most of the elements of the column vector **f** are zero, the only nonzero components being

$$f_1 = - \left. \frac{\partial \hat{\phi}}{\partial x} \right|_{x=0}$$

and

$$f_{M+1} = \left. \frac{\partial \hat{\phi}}{\partial x} \right|_{x=1}$$

The reduced element matrices can be evaluated, with the typical results

$$\mathbf{c}^e = h^e \begin{bmatrix} \frac{1}{3} & \frac{1}{6} \\ \frac{1}{6} & \frac{1}{3} \end{bmatrix}, \qquad \mathbf{k}^e = \frac{1}{h^e} \begin{bmatrix} 1 & -1 \\ -1 & 1 \end{bmatrix}$$

and the matrices **C** and **K** obtained by the standard assembly process. Finally the essential boundary conditions are applied to the system (by removing the equations corresponding to nodes 1 and $M + 1$ and replacing them by the conditions $\phi_1 = 0$, $\phi_{M+1} = 0$) and the variation in time of the nodal temperatures may then be obtained by solving the resulting equation set.

This process may be illustrated if we restrict consideration to two linear elements of equal length, that is, $M = 2$, $h^1 = h^2 = \frac{1}{2}$, for then the assembled form of the equation is

$$\frac{1}{2}\begin{bmatrix} \frac{1}{3} & \frac{1}{6} & 0 \\ \frac{1}{6} & \frac{2}{3} & \frac{1}{6} \\ 0 & \frac{1}{6} & \frac{1}{3} \end{bmatrix}\begin{bmatrix} \frac{d\phi_1}{dt} \\ \frac{d\phi_2}{dt} \\ \frac{d\phi_3}{dt} \end{bmatrix} + 2\begin{bmatrix} 1 & -1 & 0 \\ -1 & 2 & -1 \\ 0 & -1 & 1 \end{bmatrix}\begin{bmatrix} \phi_1 \\ \phi_2 \\ \phi_3 \end{bmatrix} = \begin{bmatrix} -\frac{\partial\hat{\phi}}{\partial x}\Big|_{x=0} \\ 0 \\ \frac{\partial\hat{\phi}}{\partial x}\Big|_{x=1} \end{bmatrix}$$

The imposition of the essential boundary conditions at $x = 0$ and $x = 1$ results in

$$\frac{1}{2}\begin{bmatrix} \frac{1}{3} & \frac{1}{6} & 0 \\ \frac{1}{6} & \frac{2}{3} & \frac{1}{6} \\ 0 & \frac{1}{6} & \frac{1}{3} \end{bmatrix}\begin{bmatrix} 0 \\ \frac{d\phi_2}{dt} \\ 0 \end{bmatrix} + 2\begin{bmatrix} 1 & -1 & 0 \\ -1 & 2 & -1 \\ 0 & -1 & 1 \end{bmatrix}\begin{bmatrix} 0 \\ \phi_2 \\ 0 \end{bmatrix} = \begin{bmatrix} -\frac{\partial\hat{\phi}}{\partial x}\Big|_{x=0} \\ 0 \\ \frac{\partial\hat{\phi}}{\partial x}\Big|_{x=1} \end{bmatrix}$$

and in this case the behavior of the unknown ϕ_2 is governed by the second equation, namely,

$$\frac{1}{3}\frac{d\phi_2}{dt} + 4\phi_2 = 0$$

The solution is

$$\phi_2 = Ae^{-12t}$$

where A is a constant, and fitting ϕ_2 to its known value at $t = 0$ gives $A = \frac{1}{4}$. As we have seen previously, the first and third equations can, if required, then be used to obtain approximations to the heat flux variation with time across the boundaries at $x = 0$ and $x = 1$.

EXERCISES

7.4. The displacement of a flexible string of uniform mass per unit length ρ and tension T is governed by the equation $\partial^2\phi/\partial x^2 = (\rho/T)\,\partial^2\phi/\partial t^2$. The ends of the string are held fixed at the points $x = 0$ and $x = 1$, and initially the string is at rest with $\phi = \sin \pi x$. Use polynomial trial functions in the x direction and the partial discretization method to produce a one-term approximate solution for the subsequent displacement.

7.5. Repeat Exercise 7.4 using three equal piecewise linear finite elements in the x direction.

7.6. Transverse vibrations of a beam, of flexural rigidity EI and mass per unit length ρ, are governed by the equation $\partial^4\phi/\partial x^4 + (\rho/EI)\,\partial^2\phi/\partial t^2 = 0$, where ϕ is the displacement. If the beam is of unit length and is clamped horizontally at both ends, determine a one-term approximate solution when initially the beam is at rest with $\phi = x^2(1 - x)^2$.

7.7. The region $-1 \leqslant x, y \leqslant 1$ is initially at a temperature $\phi = (1 - x^2)(1 - y^2)$, and the boundary of the region is held at zero temperature for all time. Determine a one-term approximate solution for the variation of temperature in the region with time.

7.8. In Exercise 7.7 a four-noded bilinear element is used to represent the region $0 \leqslant x, y \leqslant 1$. Determine the resulting approximation for the variation of temperature with time.

7.4. ANALYTICAL SOLUTION PROCEDURES

By using the method of partial discretization, it has been shown that many time-dependent problems can be reduced to a system of ordinary differential equations of the form

$$\mathbf{M}\frac{d^2\mathbf{a}}{dt^2} + \mathbf{C}\frac{d\mathbf{a}}{dt} + \mathbf{K}\mathbf{a} = \mathbf{f} \qquad (7.13)$$

where, in general, if the Galerkin method has been used, the matrices $\mathbf{M}, \mathbf{C}, \mathbf{K}$ are symmetric. We shall now examine possible solution methods for such a system of ordinary differential equations. For general problems, this system of equations can be nonlinear (e.g., the system arising from the problem of unsteady heat conduction in a medium with temperature-dependent thermal conductivity), but we shall concentrate in this section on linear systems only, as such systems can always, in principle, be solved analytically. However, although such analytic solutions are possible they may be so complex that further recourse has to be made to approximation procedures, and these are considered in the next section. The analytical approach is frequently useful in that it provides an insight into the behavior of the problem under consideration, which an investigator usually finds helpful.

7.4.1. Free Response of Second-Order Equations

If both $\mathbf{C}$ and $\mathbf{f}$ are zero, Eq. (7.13) reduces to

$$\mathbf{M}\frac{d^2\mathbf{a}}{dt^2} + \mathbf{K}\mathbf{a} = \mathbf{0} \qquad (7.14)$$

and we may seek a solution in the form

$$\mathbf{a} = \boldsymbol{\alpha} \cos(\omega t - \delta) \tag{7.15}$$

Direct substitution shows this to be a solution of Eq. (7.14) if

$$(-\omega^2 \mathbf{M} + \mathbf{K})\boldsymbol{\alpha} = \mathbf{0} \tag{7.16}$$

which is a standard *eigenvalue problem* and nonzero solutions are possible for $\boldsymbol{\alpha}$ only if ω is such that

$$\det(-\omega^2 \mathbf{M} + \mathbf{K}) = 0 \tag{7.17}$$

If the matrices $\mathbf{K}$ and $\mathbf{M}$ are of size $M \times M$, such a determinantal equation will have as solution M eigenvalues $\{\omega_m^2;\ m = 1, 2, \ldots, M\}$ with corresponding eigenvectors $\{\boldsymbol{\alpha}_m;\ m = 1, 2, \ldots, M\}$. Provided that the matrices $\mathbf{K}$ and $\mathbf{M}$ are positive definite,* the eigenvalues of Eq. (7.17) can be shown to be real distinct positive numbers, and these are termed the natural frequencies of the system. The general solution of Eq. (7.14) can then be expressed as a linear combination of M solutions of the type shown in Eq. (7.15) in the form

$$\mathbf{a} = \sum_{m=1}^{M} \boldsymbol{\alpha}_m \cos(\omega_m t - \delta_m) \tag{7.18}$$

where the constants $\delta_1, \delta_2, \ldots, \delta_M$ are arbitrary. It is worth noting that the eigenvectors are not uniquely defined by Eq. (7.16) and are only determined up to constant. They are frequently normalized by requiring that

$$\boldsymbol{\alpha}_m^T \mathbf{M} \boldsymbol{\alpha}_m = 1 \tag{7.19}$$

and this ensures that the eigenvectors are uniquely defined. The eigenvectors can also be shown to be orthogonal with respect to the matrices $\mathbf{M}$ and $\mathbf{K}$, that is,

$$\boldsymbol{\alpha}_l^T \mathbf{M} \boldsymbol{\alpha}_m = 0,$$
$$\qquad\qquad l \neq m \tag{7.20}$$
$$\boldsymbol{\alpha}_l^T \mathbf{K} \boldsymbol{\alpha}_m = 0,$$

To determine the actual eigenvalues in a particular problem it is seldom practicable to solve the determinantal equation (7.17) directly, and alternative techniques suitable for computer implementation have been developed.[1]

*The $M \times M$ matrix $\mathbf{A}$ is positive definite if $\mathbf{x}^T \mathbf{A} \mathbf{x} > 0$ for all $M \times 1$ vectors $\mathbf{x} \neq \mathbf{0}$.

If **f** is zero, but **C** is nonzero, the governing equation is

$$\mathbf{M}\frac{d^2\mathbf{a}}{dt^2} + \mathbf{C}\frac{d\mathbf{a}}{dt} + \mathbf{K}\mathbf{a} = \mathbf{0} \tag{7.21}$$

and a nontrivial solution of the form

$$\mathbf{a} = \boldsymbol{\alpha}e^{\lambda t} \tag{7.22}$$

exists if

$$\det(\lambda^2\mathbf{M} + \lambda\mathbf{C} + \mathbf{K}) = 0 \tag{7.23}$$

This eigenvalue problem is considerably more complicated than that considered previously in Eq. (7.17), as the eigenvalues and eigenvectors are, in general, complex.

7.4.2. Free Response of First-Order Equations

If both **M** and **f** are zero, then the appropriate form of Eq. (7.13) is

$$\mathbf{C}\frac{d\mathbf{a}}{dt} + \mathbf{K}\mathbf{a} = \mathbf{0} \tag{7.24}$$

A nontrivial solution of the form defined by Eq. (7.22) will exist if

$$\det(\lambda\mathbf{C} + \mathbf{K}) = 0 \tag{7.25}$$

If **K** and **C** are positive definite and of size $M \times M$, then the M solutions of this equation $\{\lambda_m; \ m = 1, 2, \ldots, M\}$ will all be negative and real. The general solution in this case is just a linear combination of exponential decay terms.

7.4.3. Transient Response by Modal Decomposition

The solution of the general problem of Eq. (7.13) will now be considered, where **f** is an arbitrary function of time. With **f** and **C** both equal to zero, we have seen that the solution is given by Eq. (7.18). For the general problem we could attempt to express the solution in terms of a linear combination of the eigenvectors of Eq. (7.16), that is, we seek a solution of the form

$$\mathbf{a} = \sum_{m=1}^{M} \boldsymbol{\alpha}_m y_m(t) \tag{7.26}$$

If the matrix **C** is such that

$$\boldsymbol{\alpha}_l^T \mathbf{C} \boldsymbol{\alpha}_m = 0, \qquad l \neq m \tag{7.27}$$

then substituting Eq. (7.26) into Eq. (7.13) and, with condition (7.19) satisfied, premultiplying the resulting equation by α_l^T, produces a set of scalar independent equations

$$\frac{d^2 y_l}{dt^2} + 2w_l c_l' \frac{dy_l}{dt} + \omega_l^2 y_l = f_l, \qquad l = 1, 2, \ldots, M \qquad (7.28)$$

where

$$c_l = \alpha_l^T C \alpha_l = 2\omega_l c_l' \qquad (7.29)$$
$$f_l = \alpha_l^T f$$

In general, the condition of Eq. (7.27) will not apply as the eigenvectors α_m used here are only orthogonal with respect to the matrices K and M [Eq. (7.20)], but if C can be expressed as a linear combination of K and M, then such conditions automatically result. (It should be noted that even if C cannot be expressed in this way, it may still be possible to decouple the system.[2])

Equation (7.28) is now in the form of a second-order nonhomogeneous linear equation with constant coefficients, and can be solved by standard techniques.

For the equation system (7.13) with $M = 0$, exactly analogous procedures can be adopted, but using now the real eigenvalues determined by Eq. (7.25) to construct the solution. With the eigenvectors suitably normalized, the resulting decomposition produces an uncoupled equation set of the form

$$\frac{dy_l}{dt} - \lambda_l y_l = f_l, \qquad l = 1, 2, \ldots, M \qquad (7.30)$$

which is again amenable to analytical solution.

Example 7.3

A problem of one-dimensional unsteady heat conduction with generation in the region $0 \leqslant x \leqslant 1$ is governed by the equation $\partial\phi/\partial t = k\, \partial^2\phi/\partial x^2 + e^{-t}$. The temperature ϕ is maintained at zero at $x = 0$ and $x = 1$ and initially $\phi = x(1 - x)/2$. It is required to determine an approximate solution for the resulting temperature variation using four equal piecewise linear finite elements in the x direction, when $k = 0.01$.

We proceed as in Example 7.2, using the finite element distribution and numbering system shown in Fig. 7.2. Application of the Galerkin weighted residual method produces the equation system

$$C\frac{d\phi}{dt} + kK\phi = f$$

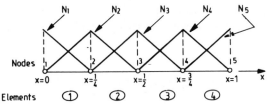

FIGURE 7.2.   Element and node numbering used in the solution of the problem of Example 7.3.

where the matrices $\mathbf{C}$ and $\mathbf{K}$ are as given in Example 7.2 and

$$
\mathbf{f} = e^{-t}
\begin{bmatrix}
\dfrac{1}{8} - k \left.\dfrac{\partial \hat{\phi}}{\partial x}\right|_{x=0} \\[12pt]
\dfrac{1}{4} \\[10pt]
\dfrac{1}{4} \\[10pt]
\dfrac{1}{4} \\[10pt]
\dfrac{1}{8} + k \left.\dfrac{\partial \hat{\phi}}{\partial x}\right|_{x=1}
\end{bmatrix}
$$

Applying the boundary conditions on ϕ_1 and ϕ_5 and deleting the first and fifth equations produces the set

$$
\frac{1}{4}
\begin{bmatrix}
\dfrac{2}{3} & \dfrac{1}{6} & 0 \\[8pt]
\dfrac{1}{6} & \dfrac{2}{3} & \dfrac{1}{6} \\[8pt]
0 & \dfrac{1}{6} & \dfrac{2}{3}
\end{bmatrix}
\begin{bmatrix}
\dfrac{d\phi_2}{dt} \\[8pt]
\dfrac{d\phi_3}{dt} \\[8pt]
\dfrac{d\phi_4}{dt}
\end{bmatrix}
+ 4k
\begin{bmatrix}
2 & -1 & 0 \\
-1 & 2 & -1 \\
0 & -1 & 2
\end{bmatrix}
\begin{bmatrix}
\phi_2 \\
\phi_3 \\
\phi_4
\end{bmatrix}
= \frac{e^{-t}}{4}
\begin{bmatrix}
1 \\
1 \\
1
\end{bmatrix}
$$

The solution of this equation set will be obtained by decoupling the equations. Following Eq. (7.25), we determine λ such that

$$
\det
\begin{vmatrix}
\dfrac{\lambda}{6} + 8k & \dfrac{\lambda}{24} - 4k & 0 \\[10pt]
\dfrac{\lambda}{24} - 4k & \dfrac{\lambda}{6} + 8k & \dfrac{\lambda}{24} - 4k \\[10pt]
0 & \dfrac{\lambda}{24} - 4k & \dfrac{\lambda}{6} + 8k
\end{vmatrix} = 0
$$

The solution of this determinantal equation produces eigenvalues

$$\lambda_1 = -\frac{48k}{7}(10 - 6\sqrt{2}) = -10 \cdot 3866k, \qquad \lambda_2 = -48k,$$

$$\lambda_3 = -\frac{48k}{7}(10 + 6\sqrt{2}) = -126.7562k$$

with corresponding eigenvectors

$$\boldsymbol{\alpha}_1 = \begin{bmatrix} 1 \\ \sqrt{2} \\ 1 \end{bmatrix}, \qquad \boldsymbol{\alpha}_2 = \begin{bmatrix} 1 \\ 0 \\ -1 \end{bmatrix}, \qquad \boldsymbol{\alpha}_3 = \begin{bmatrix} 1 \\ -\sqrt{2} \\ 1 \end{bmatrix}$$

Now using Eq. (7.26), we write

$$\boldsymbol{\phi} = \boldsymbol{\alpha}_1 y_1(t) + \boldsymbol{\alpha}_2 y_2(t) + \boldsymbol{\alpha}_3 y_3(t)$$

which is a solution provided that

$$\mathbf{C}\boldsymbol{\alpha}_1 \frac{dy_1}{dt} + \mathbf{C}\boldsymbol{\alpha}_2 \frac{dy_2}{dt} + \mathbf{C}\boldsymbol{\alpha}_3 \frac{dy_3}{dt} + k\mathbf{K}\boldsymbol{\alpha}_1 y_1 + k\mathbf{K}\boldsymbol{\alpha}_2 y_2 + k\mathbf{K}\boldsymbol{\alpha}_3 y_3 = \mathbf{f}$$

Premultiplying both sides of this equation by $\boldsymbol{\alpha}_1^T$, $\boldsymbol{\alpha}_2^T$, and $\boldsymbol{\alpha}_3^T$ in turn, leads to the set of three decoupled equations

$$\frac{2}{3}\left(1 + \frac{\sqrt{2}}{4}\right)\frac{dy_1}{dt} + 16(2 - \sqrt{2})ky_1 = \left(\frac{2 + \sqrt{2}}{4}\right)e^{-t}$$

$$\frac{1}{3}\frac{dy_2}{dt} + 16ky_2 = 0$$

$$\frac{2}{3}\left(1 - \frac{\sqrt{2}}{4}\right)\frac{dy_3}{dt} + 16(2 + \sqrt{2})ky_3 = \left(\frac{2 - \sqrt{2}}{4}\right)e^{-t}$$

and these can be solved by standard methods to give

$$y_1 = A_1 e^{\lambda_1 t} - 1.0555 e^{-t}$$

$$y_2 = A_2 e^{\lambda_2 t}$$

$$y_3 = A_3 e^{\lambda_3 t} + 1.2700 e^{-t}$$

The variation of the nodal temperatures with time follows as

$$\phi_2 = \left(A_1 e^{\lambda_1 t} - 1.0555 e^{-t}\right) + A_2 e^{\lambda_2 t} + \left(A_3 e^{\lambda_3 t} + 1.2700 e^{-t}\right)$$

$$\phi_3 = \sqrt{2}\left(A_1 e^{\lambda_1 t} - 1.0555 e^{-t}\right) - \sqrt{2}\left(A_3 e^{\lambda_3 t} + 1.2700 e^{-t}\right)$$

$$\phi_4 = \left(A_1 e^{\lambda_1 t} - 1.0555 e^{-t}\right) - A_2 e^{\lambda_2 t} + \left(A_3 e^{\lambda_3 t} + 1.2700 e^{-t}\right)$$

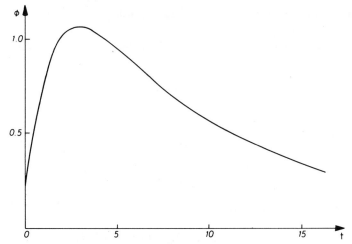

FIGURE 7.3.   Calculated variation of temperature with time at $x = \frac{1}{2}$ in Example 7.3.

where the constants A_1, A_2, A_3 can be determined by using the initial conditions. If we make the value of the approximation at $t = 0$ exactly equal to the initial temperature at the nodal points, this requires that

$$0.09375 = A_1 - 1.0555 + A_2 + A_3 + 1.2700$$

$$0.12500 = \sqrt{2}\,(A_1 - 1.0555 - A_3 - 1.2700)$$

$$0.09375 = A_1 - 1.0555 - A_2 + A_3 + 1.2700$$

with solution

$$A_1 = 1.147, \qquad A_2 = 0, \qquad A_3 = -1.267$$

The resulting variation of temperature ϕ_3 at the center of the region with time is shown in Fig. 7.3.

EXERCISES

7.9.   Return to Exercise 7.4 and obtain a three-term approximate solution for the displacement of the flexible string.

7.10.   Return to Exercise 7.4 and obtain the displacement using four piecewise linear finite elements.

7.11.   Return to Exercise 7.6 and obtain a three-term approximation for the displacement of the vibrating beam.

7.5.   FINITE ELEMENT SOLUTION PROCEDURES IN THE TIME DOMAIN

In the previous section it has been shown that analytical solution procedures may be applied to the system of ordinary differential equations resulting from the application of the method of partial discretization to many time-dependent physical problems. Clearly, although such techniques are available for linear problems, they are going to be difficult to apply in practice, and indeed such techniques cannot be used at all for nonlinear problems. In this section, therefore, we shall consider the solution of equation systems such as (7.13) by trial function techniques. Finite elements will be used to represent the time domain which is of infinite extent. Conditions at the end of the first element will be determined by use of the governing equations plus the initial conditions. This process is then repeated for subsequent elements, using the newly calculated information as the initial conditions for each element in turn.

Since trial functions in terms of the space coordinates have already been introduced and it is now proposed to introduce trial functions in time, it might be asked why trial functions in space and time were not introduced directly at the outset. There are many reasons why an approach of this form is not normally adopted and among these the following can be listed: (1) With a large time domain, the size of the problem becomes excessively large. (2) The resulting equation system is, in general, nonsymmetric, even if a Galerkin-type weighting is used. (3) The geometrically simple nature of the time domain offers little incentive for the use of an irregular subdivision of space–time elements. (4) With the use of trial functions of the form $N_m(x, y, z)\overline{N}_m(t)$ in the general domain, the result would be the same as that achieved by successive space and time discretization.

In this section it is convenient to note that the system of equations (7.13) can be written as the set of modally uncoupled equations (7.28) or, if $\mathbf{M} = 0$, Eqs. (7.30). In general, we shall work with the original matrix equations, but we shall consider the equations in their modally uncoupled form when we discuss certain features of the methods to be described. Initially we restrict consideration to equations that are first order in time and then develop the technique to deal with second-order equations.

7.5.1.   First-Order Equations

It has been seen in the preceding sections that when the method of partial discretization is applied to a general linear first-order equation in time of the form

$$\mathcal{L}\phi + p - \alpha\frac{\partial\phi}{\partial t} = 0 \qquad \text{in } \Omega \qquad (7.31)$$

with an assumed approximate solution $\hat{\phi}$, as detailed in Eq. (7.10), the result is a coupled system of equations

$$\mathbf{C}\frac{d\mathbf{a}}{dt} + \mathbf{Ka} = \mathbf{f} \tag{7.32}$$

The initial conditions, giving the value of ϕ everywhere at time $t = 0$, imply that the initial value of $\mathbf{a}$, say $\mathbf{a}^0$, is known.

Suppose that the time domain is divided into linear elements as shown in Fig. 7.4a. Then, proceeding in the standard finite element manner, we write

$$\mathbf{a} \simeq \hat{\mathbf{a}} = \sum_{m=1}^{\infty} \mathbf{a}^m N_m \tag{7.33}$$

where $\mathbf{a}^m$ denotes the value of the approximation to $\mathbf{a}$ at node m, that is, at time $t = t_m$. Since we use the same shape functions N_m to represent the variation of each component of $\mathbf{a}$, in Eq. (7.33) N_m is a scalar quantity. Consider the linear element n in time with nodes placed at $t = t_n$ and $t = t_{n+1}$, as shown in Fig. 7.4.b. On this element we have that

$$\hat{\mathbf{a}} = \mathbf{a}^n N_n^n + \mathbf{a}^{n+1} N_{n+1}^n \tag{7.34}$$

since all the other trial functions are zero on element n, and where

$$N_n^n = 1 - T, \qquad \frac{dN_n^n}{dt} = \frac{-1}{\Delta t_n}$$

$$N_{n+1}^n = T, \qquad \frac{dN_{n+1}^n}{dt} = \frac{1}{\Delta t_n} \tag{7.35}$$

$$T = \frac{t - t_n}{\Delta t_n}, \qquad \Delta t_n = t_{n+1} - t_n$$

The standard weighted residual technique applied to Eq. (7.32) leads to

$$\int_0^{\infty} \left(\mathbf{C}\frac{d\hat{\mathbf{a}}}{dt} + \mathbf{K}\hat{\mathbf{a}} - \mathbf{f} \right) W_n \, dt = 0, \qquad n = 0, 1, 2, \ldots \tag{7.36}$$

and, if we consider only weighting functions W_n which are such that

$$W_n = 0, \qquad t < t_n \quad \text{and} \quad t > t_{n+1} \tag{7.37}$$

this statement can be written

$$\int_{t_n}^{t_{n+1}} \left(\mathbf{C}\frac{d\hat{\mathbf{a}}}{dt} + \mathbf{K}\hat{\mathbf{a}} - \mathbf{f}(t) \right) W_n \, dt = 0, \qquad n = 0, 1, 2, \ldots \tag{7.38}$$

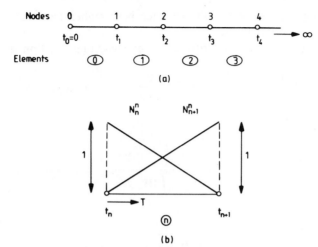

FIGURE 7.4. (*a*) Subdivision of the time domain by means of linear finite elements. (*b*) Element shape functions on element *n*.

The integration now extends over element n only, and so Eq. (7.34) can be used to replace $\hat{a}$ in this expression. Then, making use of Eq. (7.35), the weighted residual statement becomes

$$\int_0^1 \left[\frac{\mathbf{C}}{\Delta t_n}(-\mathbf{a}^n + \mathbf{a}^{n+1}) + \mathbf{K}\{\mathbf{a}^n(1 - T) + \mathbf{a}^{n+1}T\} - \mathbf{f}(t_n + \Delta t_n T)\right] W_n \, dT = 0,$$

$$n = 0, 1, 2, \ldots \quad (7.39)$$

and, if the matrices $\mathbf{C}$ and $\mathbf{K}$ can be assumed constant over the element, it is found that

$$\left\{ \frac{\mathbf{C}}{\Delta t_n} \int_0^1 W_n \, dT + \mathbf{K} \int_0^1 T W_n \, dT \right\} \mathbf{a}^{n+1}$$

$$+ \left\{ -\frac{\mathbf{C}}{\Delta t_n} \int_0^1 W_n \, dT + \mathbf{K} \int_0^1 (1 - T) W_n \, dT \right\} \mathbf{a}^n$$

$$= \int_0^1 \mathbf{f}(t_n + T\Delta t_n) W_n \, dT \quad\quad\quad (7.40)$$

A relation of this form is valid over each element n and can, with suitable choice of weighting function W_n, be used to generate successive values $\mathbf{a}^1, \mathbf{a}^2, \mathbf{a}^3, \ldots$, starting from the known initial conditions $\mathbf{a}^0$. A relationship of the form of Eq. (7.40) which just relates $\mathbf{a}^{n+1}$ to $\mathbf{a}^n$ is known as a *two-level time-stepping scheme.* We shall see later how higher level schemes can be produced in a similar fashion.

Equation (7.40) can be rewritten, for any weighting function, as

$$\left(\frac{\mathbf{C}}{\Delta t_n} + \gamma_n \mathbf{K}\right)\mathbf{a}^{n+1} + \left(-\frac{\mathbf{C}}{\Delta t_n} + (1 - \gamma_n)\mathbf{K}\right)\mathbf{a}^n = \bar{\mathbf{f}}^n \qquad (7.41)$$

where

$$\gamma_n = \frac{\int_0^1 W_n T \, dT}{\int_0^1 W_n \, dT} \qquad (7.42a)$$

and

$$\bar{\mathbf{f}}^n = \frac{\int_0^1 W_n \mathbf{f}(t_n + T\Delta t_n) \, dT}{\int_0^1 W_n \, dT} \qquad (7.42b)$$

If the known function $\mathbf{f}$ varies smoothly in time, it is sometimes convenient to represent its behavior over element n by using the same form of interpolation as that adopted for $\hat{\mathbf{a}}$; that is, if we write

$$\mathbf{f}(t_n + T\Delta t_n) \simeq \mathbf{f}^n N_n^n(T) + \mathbf{f}^{n+1} N_{n+1}^n(T), \qquad 0 \leqslant T \leqslant 1 \qquad (7.43)$$

then

$$\bar{\mathbf{f}}^n = (1 - \gamma_n)\mathbf{f}^n + \gamma_n \mathbf{f}^{n+1} \qquad (7.44)$$

and the relationship defining $\mathbf{a}^{n+1}$ in terms of $\mathbf{a}^n$ [Eq. (7.41)] becomes

$$\left\{\frac{\mathbf{C}}{\Delta t_n} + \gamma_n \mathbf{K}\right\}\mathbf{a}^{n+1} + \left\{-\frac{\mathbf{C}}{\Delta t_n} + (1 - \gamma_n)\mathbf{K}\right\}\mathbf{a}^n = (1 - \gamma_n)\mathbf{f}^n + \gamma_n \mathbf{f}^{n+1}$$

$$(7.45)$$

If the interpolation of Eq. (7.43) is not acceptable, because of the nature of the variation of the function $\mathbf{f}$ over the element n (as would be the case, for example, for a system subjected to a sudden impact), then Eq. (7.42b) should be evaluated exactly.

7.5.2. Particular Schemes for First-Order Equations

By choosing different forms for the weighting function W_n in Eq. (7.42a), we are able to produce a variety of solution schemes. If we take point collocation at $T = \theta$ for each element, that is, $W_n = \delta(T - \theta)$ for $n = 0, 1, 2, 3, \ldots$, then the integration in Eq. (7.42a) can be performed directly to give

$$\gamma_n = \theta$$

and Eq. (7.45) can be expressed as

$$\left\{\frac{\mathbf{C}}{\Delta t_n} + \theta \mathbf{K}\right\}\mathbf{a}^{n+1} + \left\{-\frac{\mathbf{C}}{\Delta t_n} + (1-\theta)\mathbf{K}\right\}\mathbf{a}^n = (1-\theta)\mathbf{f}^n + \theta \mathbf{f}^{n+1}$$

$$(7.46)$$

We have produced this relationship from Eq. (7.32) by using the finite element method. An alternative, widely used approach would be to apply the finite difference method of Chapter 1 directly to Eq. (7.32). Then marking off the discrete set of points $t_0, t_1, t_2, \ldots$ on the time axis, we obtain, exactly,

$$\mathbf{C}\left.\frac{d\mathbf{a}}{dt}\right|^{n+\theta} + \mathbf{K}\mathbf{a}^{n+\theta} = \mathbf{f}^{n+\theta} \tag{7.47}$$

where the superscript $n + \theta$ denotes evaluation at time $t = t_n + \theta \Delta t_n, 0 \leqslant \theta \leqslant 1$. Using Taylor series expansions as in Eqs. (1.14) and (1.19) gives

$$\mathbf{a}^n = \mathbf{a}^{n+\theta} - \theta \Delta t_n \left.\frac{d\mathbf{a}}{dt}\right|^{n+\theta} + \frac{\theta^2 \Delta t_n^2}{2} \left.\frac{d^2\mathbf{a}}{dt^2}\right|^{n+\theta} - \frac{\theta^3 \Delta t_n^3}{6} \left.\frac{d^3\mathbf{a}}{dt^3}\right|^{n+\theta_1}$$

$$(7.48)$$

$$\mathbf{a}^{n+1} = \mathbf{a}^{n+\theta} + (1-\theta) \Delta t_n \left.\frac{d\mathbf{a}}{dt}\right|^{n+\theta} + \frac{(1-\theta)^2 \Delta t_n^2}{2} \left.\frac{d^2\mathbf{a}}{dt^2}\right|^{n+\theta}$$

$$+ \frac{(1-\theta)^3 \Delta t_n^3}{6} \left.\frac{d^3\mathbf{a}}{dt^3}\right|^{n+\theta_2} \tag{7.49}$$

Subtracting these two equations produces the approximation

$$\left.\frac{d\mathbf{a}}{dt}\right|^{n+\theta} \approx \frac{\mathbf{a}^{n+1} - \mathbf{a}^n}{\Delta t_n} \tag{7.50}$$

with the error E in this approximation given by

$$E = \left\{(1-\theta)^2 - \theta^2\right\}\frac{\Delta t_n}{2} \left.\frac{d^2\mathbf{a}}{dt^2}\right|^{n+\theta}$$

$$+ \frac{\Delta t_n^2}{6}\left\{(1-\theta)^3 \left.\frac{d^3\mathbf{a}}{dt^3}\right|^{n+\theta_2} + \theta^3 \left.\frac{d^3\mathbf{a}}{dt^3}\right|^{n+\theta_1}\right\} \tag{7.51}$$

We note that E is $O(\Delta t_n^2)$ when $\theta = \frac{1}{2}$, whereas E is $O(\Delta t_n)$ otherwise. Multiplying Eq. (7.48) by $(1 - \theta)$, Eq. (7.49) by θ, and adding the resulting equations, gives

$$\mathbf{a}^{n+\theta} \approx (1 - \theta)\mathbf{a}^n + \theta\mathbf{a}^{n+1} \tag{7.52}$$

with the error here being of $O(\Delta t_n^2)$.

When the approximations of Eqs. (7.50) and (7.52) are used in Eq. (7.47), using the same representation for $\mathbf{f}^{n+\theta}$ as that adopted for $\mathbf{a}^{n+\theta}$ in Eq. (7.52), this gives

$$\left\{\frac{\mathbf{C}}{\Delta t_n} + \theta\mathbf{K}\right\}\mathbf{a}^{n+1} + \left\{-\frac{\mathbf{C}}{\Delta t_n} + (1 - \theta)\mathbf{K}\right\}\mathbf{a}^n = (1 - \theta)\mathbf{f}^n + \theta\mathbf{f}^{n+1}$$

$$\tag{7.53}$$

which is just the result obtained by the finite element method in Eq. (7.46).

By suitably choosing the collocation point θ within each element, we can thus make Eq. (7.46) correspond to some of the following well-known finite difference schemes for Eq. (7.32).

1. *Forward difference (Euler) scheme.* This results from putting $\theta = 0$ in Eq. (7.46) and is

$$\frac{\mathbf{C}}{\Delta t_n}\mathbf{a}^{n+1} + \left(-\frac{\mathbf{C}}{\Delta t_n} + \mathbf{K}\right)\mathbf{a}^n = \mathbf{f}^n \tag{7.54}$$

If the matrix $\mathbf{C}$ is diagonal, so that the elements of its inverse $\mathbf{C}^{-1}$ can be obtained trivially, this scheme is said to be *explicit*, as $\mathbf{a}^{n+1}$ is given directly in terms of $\mathbf{a}^n$ by

$$\mathbf{a}^{n+1} = \Delta t_n \mathbf{C}^{-1}\left\{\left(\frac{\mathbf{C}}{\Delta t_n} - \mathbf{K}\right)\mathbf{a}^n + \mathbf{f}^n\right\} \tag{7.55}$$

We shall see shortly that the computational advantage of an explicit scheme of this type is normally accompanied by the disadvantage that the time step Δt_n, being used should not exceed a certain magnitude.

The matrix $\mathbf{C}$ will be diagonal if a finite difference discretization of the space domain has been employed. However, if the space discretization has been performed by the finite element method, the matrix $\mathbf{C}$ will not normally be diagonal [see Eq. (7.12)], and the explicit nature of this scheme is lost. Frequently, in practical calculations this is overcome by diagonalizing or *lumping* the matrix $\mathbf{C}$. To illustrate this process we consider a problem involving a single space dimension, and we assume that the space discretization has been performed by using two-noded linear one-dimensional elements.

Then if the Galerkin method is used, the reduced element matrix $\mathbf{c}^e$, for a typical element e with nodes numbered i and j, is given by

$$\mathbf{c}^e = \int_{\Omega^e} \begin{bmatrix} \alpha(N_i^e)^2 & \alpha N_i^e N_j^e \\ \alpha N_j^e N_i^e & \alpha(N_j^e)^2 \end{bmatrix} dx \qquad (7.56)$$

If these matrix entries are evaluated by numerical integration, using a formula of the type introduced in Eq. (5.10a) with the sampling points located at the nodes, we see that

$$\int_{\Omega^e} \alpha N_i^e N_j^e \, dx = \frac{h^e}{2}\left\{ \left(\alpha N_i^e N_j^e\right)_{\text{node } i} + \left(\alpha N_i^e N_j^e\right)_{\text{node } j}\right\} = 0 \qquad (7.57)$$

and it follows that $\mathbf{c}^e$ and, hence, also the global matrix $\mathbf{C}$, are diagonalized. A description of alternative lumping procedures and a full discussion of the lumping process can be found in Reference 2.

2. *Crank–Nicolson (central difference) method.* With $\theta = \frac{1}{2}$, Eq. (7.46) becomes

$$\left(\frac{\mathbf{C}}{\Delta t_n} + \frac{1}{2}\mathbf{K}\right)\mathbf{a}^{n+1} + \left(-\frac{\mathbf{C}}{\Delta t_n} + \frac{1}{2}\mathbf{K}\right)\mathbf{a}^n = \frac{1}{2}\mathbf{f}^n + \frac{1}{2}\mathbf{f}^{n+1} \qquad (7.58)$$

and this scheme is said to be *implicit* as it requires the solution of a nondiagonal system of equations to determine $\mathbf{a}^{n+1}$.

3. *Backward difference scheme.* This results from putting $\theta = 1$ in Eq. (7.46) and is the implicit scheme

$$\left(\frac{\mathbf{C}}{\Delta t_n} + \mathbf{K}\right)\mathbf{a}^{n+1} - \frac{\mathbf{C}}{\Delta t_n}\mathbf{a}^n = \mathbf{f}^{n+1} \qquad (7.59)$$

An obvious alternative to point collocation is to use a Galerkin type method over each element, that is, by defining $W_n = N_n^n$ or $W_n = N_{n+1}^n$ over element n. These weightings just produce a special case of Eq. (7.46), and the reader could show that, with $W_n = N_{n+1}^n$, the resulting scheme is identical to that obtained by applying point collocation at $T = \frac{2}{3}$ on each element.

7.5.3. Higher Level Schemes for First-Order Equations

If we remove the restriction of Eq. (7.37) and allow weighting functions W_n, which are nonzero over more than one element, then we can produce schemes of any desired level. To illustrate this process, we may return to Eq. (7.36) and use the standard Galerkin approach directly, that is, we put

$$W_n = N_n \qquad (7.60)$$

and since

$$N_n = 0, \quad \text{for } t \leqslant t_{n-1} \text{ and } t \geqslant t_{n+1} \qquad (7.61)$$

the weighted residual statement becomes

$$\int_{t_{n-1}}^{t_{n+1}} \left(C \frac{d\hat{a}}{dt} + K\hat{a} - f(t) \right) N_n \, dt = 0, \qquad n = 1, 2, 3, \ldots \qquad (7.62)$$

If the form of interpolation used for $\hat{a}$ is also used for f [see Eq. (7.43)], we may write

$$\int_{t_{n-1}}^{t_n} \left\{ C \left(a^{n-1} \frac{dN_{n-1}}{dt} + a^n \frac{dN_n}{dt} \right) + K \left(a^{n-1} N_{n-1} + a^n N_n \right) \right\} N_n \, dt$$

$$+ \int_{t_n}^{t_{n+1}} \left\{ C \left(a^n \frac{dN_n}{dt} + a^{n+1} \frac{dN_{n+1}}{dt} \right) + K \left(a^n N_n + a^{n+1} N_{n+1} \right) \right\} N_n \, dt$$

$$= \int_{t_{n-1}}^{t_n} \left(f^{n-1} N_{n-1} + f^n N_n \right) N_n \, dt + \int_{t_n}^{t_{n+1}} \left(f^n N_n + f^{n+1} N_{n+1} \right) N_n \, dt$$

$$(7.63)$$

Then making use of Eq. (7.35) and performing the integrations produces the *three-level scheme*

$$\left(\frac{C}{2\,\Delta t} + \frac{K}{6} \right) a^{n+1} + \frac{2}{3} K a^n + \left(-\frac{C}{2\,\Delta t} + \frac{K}{6} \right) a^{n-1} = \frac{f^{n+1}}{6} + \frac{2f^n}{3} + \frac{f^{n-1}}{6}$$

$$(7.64)$$

where it has been assumed that $\Delta t_{n-1} = \Delta t_n = \Delta t$ and that the matrices C and K can be regarded as constant. Such a scheme can be used repeatedly, provided two initial values, a^0 and a^1, are known. Since only a^0 is given by the initial conditions, a^1 has to be calculated in some way, and this can be done by performing the initial step by one of the two-level schemes described earlier. Details of alternative schemes of this type are given in Reference 2.

Example 7.4

We will consider the application of the various schemes which have been introduced to the simple single equation $da/dt + \bar{\lambda} a = 0$ where $\bar{\lambda}$ is some positive constant, subject to the initial condition $a = 1$ at time $t = 0$.

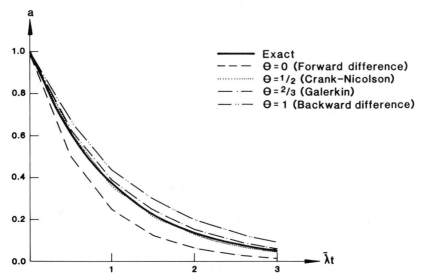

FIGURE 7.5. Result of applying some two-level schemes to the equation $da/dt + \bar{\lambda}a = 0$ with $\bar{\lambda}\Delta t_n = 0.5$.

In this case, the general expression for the two-level method of Eq. (7.46) reduces to

$$\left(\frac{1}{\bar{\lambda}\,\Delta t_n} + \theta\right)a^{n+1} + \left(-\frac{1}{\bar{\lambda}\,\Delta t_n} + (1-\theta)\right)a^n = 0, \qquad n = 0,1,2,\dots$$

and the initial condition means that

$$a(t = 0) = a^0 = 1$$

In Figs. 7.5 and 7.6 we compare the behavior of the exact solution with the behavior of this scheme for $\theta = 0$ (forward difference), $\theta = \frac{1}{2}$ (Crank–Nicolson), $\theta = \frac{2}{3}$ (Galerkin), and $\theta = 1$ (backward difference), and for two different values of the time step. It can be seen from Fig. 7.5 that the results from all the schemes exhibit the correct form of behavior when the time step satisfies the relation $\bar{\lambda}\,\Delta t_n = 0.5$ and that the most accurate results are obtained when $\theta = \frac{1}{2}$, that is, with the Crank–Nicolson scheme. This is what we should expect by examining the truncation error in the representation of $da/dt|^{n+\theta}$ in this scheme, which is $O(\Delta t_n^2)$ as compared to $O(\Delta t_n)$ for other values of θ [see Eq. (7.51)].

From Fig. 7.6 we see that increasing the time step to the value given by $\bar{\lambda}\,\Delta t_n = 2.2$ changes markedly the behavior of some of the schemes. With the forward difference method the results show an oscillation with increasing

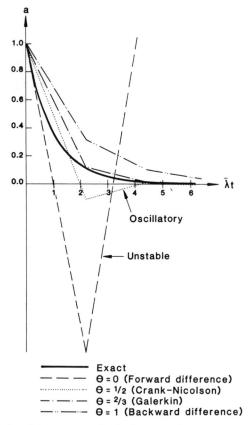

FIGURE 7.6. Result of applying some two-level schemes to the equation $da/dt + \bar{\lambda}a = 0$ with $\bar{\lambda}\Delta t_n = 2.2$.

amplitude, and the method is said to be *unstable* for this value of the time step. The Crank–Nicolson method, although remaining stable, loses its high accuracy and exhibits an oscillation which is not present in the exact solution. The Galerkin and backward difference methods still produce results of the correct form but, as we would expect from Eq. (7.51), the accuracy achieved is not as good as that produced with the smaller value of the time step. We return to a discussion of stability and oscillation properties of time-stepping schemes in the next section.

The three-level scheme of Eq. (7.64) for this problem becomes

$$\left(\frac{1}{2\bar{\lambda}\,\Delta t_n} + \frac{1}{6}\right)a^{n+1} + \frac{2}{3}a^n + \left(-\frac{1}{2\bar{\lambda}\,\Delta t_n} + \frac{1}{6}\right)a^{n-1} = 0$$

and again the given initial condition means that $a^0 = 1$.

To calculate the value of a^1, which we need to enable us to apply the three-level scheme recursively, we can use any of the two-level methods described earlier.

$\bar{\lambda}t$	Exact	Three-Level Method
0	1.0	1.0
0.1	0.9048	0.9
0.2	0.8187	0.8194
0.3	0.7408	0.7362
0.4	0.6703	0.6715
0.5	0.6065	0.6021

In the table the results of implementing the three-level method, starting with a forward difference-type two-level method and with a time step satisfying $\bar{\lambda}\,\Delta t_n = 0.1$, are compared with the exact solution. Good accuracy has been achieved, but the numerical results can be observed to oscillate about the exact solution.

7.5.4. Stability Characteristics for Two-Level Schemes

The general first-order equation (7.32) with the forcing term $\mathbf{f}$ set equal to zero can be solved numerically by using the two-level relationship of Eq. (7.53) in the form

$$\left(\frac{1}{\Delta t_n} + \theta \mathbf{C}^{-1}\mathbf{K}\right)\mathbf{a}^{n+1} = \left(\frac{1}{\Delta t_n} - (1 - \theta)\mathbf{C}^{-1}\mathbf{K}\right)\mathbf{a}^n \qquad (7.65)$$

If $\lambda_1, \lambda_2, \ldots, \lambda_M$ are the eigenvalues and $\alpha_1, \alpha_2, \ldots, \alpha_M$ are the corresponding eigenvectors of the eigenvalue problem (see Section 7.4.2)

$$\mathbf{K}\alpha + \lambda \mathbf{C}\alpha = 0 \qquad (7.66)$$

then we can write in general

$$\mathbf{a}^n = \sum_{m=1}^{M} y_m^n \alpha_m \qquad \mathbf{a}^{n+1} = \sum_{m=1}^{M} y_m^{n+1} \alpha_m \qquad (7.67)$$

and it follows, by substitution into Eq. (7.65), that

$$y_m^{n+1} = \frac{1/\Delta t_n + (1 - \theta)\lambda_m}{1/\Delta t_n - \theta\lambda_m} y_m^n \qquad (7.68)$$

As we have noted previously, if the matrices $\mathbf{K}$ and $\mathbf{C}$ are positive definite, the eigenvalues $\lambda_1, \lambda_2, \ldots, \lambda_M$ of Eq. (7.66) are all negative and real.

The exact solution therefore decays as time increases and, from Eq. (7.67), the numerical solution will certainly exhibit the same behavior if

$$|y_m^{n+1}| < |y_m^n|, \qquad m = 1, 2, \dots, M \tag{7.69}$$

that is, if

$$-1 < \frac{1/\Delta t_n + (1 - \theta)\lambda_m}{1/\Delta t_n - \theta\lambda_m} < 1, \qquad m = 1, 2, \dots, M \tag{7.70}$$

This is the condition that must be satisfied for the *stability* of the time-stepping scheme. In addition, the scheme will be free from oscillation if each mode participation factor y_m^n has the same sign at each time level n, and from Eq. (7.68) it can be seen that this may be accomplished if

$$\frac{y_m^{n+1}}{y_m^n} = \frac{1/\Delta t_n + (1 - \theta)\lambda_m}{1/\Delta t_n - \theta\lambda_m} > 0, \qquad m = 1, 2, \dots, M \tag{7.71}$$

The two-level time-stepping scheme of Eq. (7.36) will therefore be stable and oscillation-free, provided that

$$0 < \frac{1/\Delta t_n + (1 - \theta)\lambda_m}{1/\Delta t_n - \theta\lambda_m} < 1, \qquad m = 1, 2, \dots, M \tag{7.72}$$

The condition of Eq. (7.70) or Eq. (7.71) will, for a given value of θ, normally impose a limit on the maximum size of time step that can be used in a particular problem.

Example 7.5

We will consider the stability and oscillation properties of the two-level schemes applied to the problem of Example 7.4. Here we are interested in solving a single equation, $da/dt + \bar{\lambda}a = 0$, using a scheme of the form

$$\left(\frac{1}{\Delta t_n} + \bar{\lambda}\theta\right)a^{n+1} + \left(-\frac{1}{\Delta t_n} + \bar{\lambda}(1 - \theta)\right)a^n = 0$$

For this case of a single equation, the solution of Eq. (7.66) is trivial, giving $\lambda = -\bar{\lambda}$. The stability condition of Eq. (7.70) then reduces to the requirement that

$$-1 < \frac{1/\Delta t_n - \bar{\lambda}(1 - \theta)}{1/\Delta t_n + \bar{\lambda}\theta} < 1$$

The right-hand inequality is automatically satisfied, and the left-hand inequality requires that

$$(1 - 2\theta)\bar{\lambda} \, \Delta t_n < 2$$

This condition is always satisfied for $\theta \geqslant \frac{1}{2}$, whatever the value of Δt_n, and such schemes are said to be *unconditionally stable*. When $0 \leqslant \theta < \frac{1}{2}$, the stability is conditional, requiring that the time step Δt_n used at any stage of the process should satisfy

$$\bar{\lambda} \, \Delta t_n < \frac{2}{1 - 2\theta}$$

An oscillation-free scheme can be achieved in this case if Eq. (7.72) is satisfied, that is, if

$$\frac{1/\Delta t_n - \bar{\lambda}(1 - \theta)}{1/\Delta t_n + \bar{\lambda}\theta} > 0$$

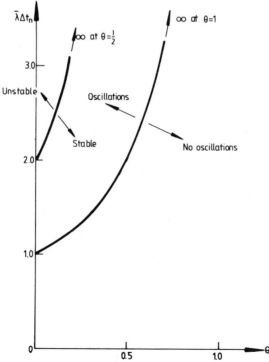

FIGURE 7.7.   Regions of instability and oscillatory behavior for two-level schemes applied to the equation $da/dt + \bar{\lambda}a = 0$.

which is equivalent to requiring that

$$(1 - \theta)\bar{\lambda} \, \Delta t_n < 1$$

This condition is always satisfied for the backward difference scheme $\theta = 1$, whatever the value of the time-step Δt_n, but for all other values of θ the schemes oscillate for sufficiently large time-step values. It follows from this expression that the time step of $\bar{\lambda} \, \Delta t_n = 2.2$ used in the analysis of Example 7.4 is such as to ensure oscillations in the Crank–Nicolson scheme ($\theta = \frac{1}{2}$), but is also such that the Galerkin scheme ($\theta = \frac{2}{3}$) is oscillation-free (see Fig. 7.6). The result is that the Crank–Nicolson scheme, although mathematically more accurate, can in practice give numerical results that are less accurate than those given by other schemes when larger time-step values are used. In Fig. 7.7 we can see the regions of stability and oscillatory behavior in the $(\theta, \bar{\lambda} \, \Delta t_n)$ plane for the two-level schemes in this case.

We have seen in Section 7.4.3 how the use of modal decomposition can convert the equation

$$\mathbf{C}\frac{d\mathbf{a}}{dt} + \mathbf{K}\mathbf{a} = \mathbf{0} \tag{7.73}$$

into a set of decoupled equations [Eq. (7.30)]

$$\frac{dy_l}{dt} - \lambda_l y_l = 0, \qquad l = 1, 2, 3, \ldots, M \tag{7.74}$$

We can therefore gain information on the expected behavior of a two-level scheme applied to Eq. (7.73) by considering the scheme's behavior when applied to each equation of the form of (7.74). This can be achieved by applying the analysis of Example 7.5 to each of the decoupled equations in turn. We can expect an oscillation-free solution if

$$\Delta t_n < \frac{1}{(1 - \theta)|\lambda_l|} \tag{7.75}$$

and a stable solution provided that

$$(1 - 2\theta)|\lambda_l| \, \Delta t_n < 2 \tag{7.76}$$

for each value of λ_l. If we use an unconditionally stable scheme [i.e., we choose θ so that Eq. (7.76) is satisfied automatically] and a time step Δt_n^* which only partially satisfies Eq. (7.75), for example, the time step Δt_n^* may be such that

$$\Delta t_n^* < \frac{1}{(1 - \theta)|\lambda_l|}, \qquad l = 1, 2, \ldots, p \tag{7.77}$$

but

$$\Delta t_n^* > \frac{1}{(1 - \theta)|\lambda_l|}, \qquad l = p + 1, p + 2, \ldots, M \tag{7.78}$$

then the modes with low $|\lambda_l|$ values would be oscillation-free, but oscillations would be present in the high $|\lambda_l|$ value modes. The obvious solution of decreasing the time step to such a level that Eq. (7.75) is satisfied for all λ_l may require excessive computational time, and a frequently adopted alternative is to keep the time step fixed at the value Δt_n^* and to use a numerical device to eliminate the oscillations from the higher modes. Normally such devices consist of averaging successive time steps of the computation in some manner.

Conditionally stable schemes will always result in instability if the time step Δt_n used does not satisfy Eq. (7.76) when λ_l is the maximum eigenvalue in absolute value of the problem of Eq. (7.66). This is a serious limitation on the use of the explicit method of Eq. (7.55), but with this method the necessity of using many small time steps is often compensated by the saving of computational effort produced by the elimination of the matrix inversion procedure.

The stability analysis of this section seems to suggest that the process of estimation of the maximum time-step size which is allowed for a conditionally stable scheme necessitates the solution of the full eigenvalue problem [see Eq. (7.66)],

$$\mathbf{K}\alpha + \lambda \mathbf{C}\alpha = \mathbf{0} \tag{7.79}$$

However, when the space discretization is carried out by the finite element method, it can be shown[3] that the highest global eigenvalue must always be less than the highest local element eigenvalue. These element eigenvalues, obtained from the solution of Eq. (7.79) with the global matrices replaced by reduced element matrices, are easily obtained, and an estimate can then be made of the maximum allowable time step that can be used. It should be noted, however, that such an estimate can prove overconservative, especially when applied to problems in which a nonuniform mesh discretization has been employed.

Example 7.6

Suppose we wish to use the forward difference-type two-level scheme (i.e., $\theta = 0$) for the solution of the equation $\mathbf{C}\, da/dt + \mathbf{K}a = \mathbf{f}$, which results from using linear finite elements in space in the problem described in Example 7.2.

The nonzero components of the typical reduced element matrices $\mathbf{c}^e$ and $\mathbf{k}^e$ are then

$$\mathbf{c}^e = h^e \begin{bmatrix} \frac{1}{3} & \frac{1}{6} \\ \frac{1}{6} & \frac{1}{3} \end{bmatrix}, \qquad \mathbf{k}^e = \frac{1}{h^e} \begin{bmatrix} 1 & -1 \\ -1 & 1 \end{bmatrix}$$

and the element eigenvalue problem [Eq. (7.79)] becomes in this case

$$\det(\mathbf{k}^e + \lambda \mathbf{c}^e) = 0$$

Inserting the above forms for the element matrices requires that

$$\det \begin{vmatrix} 1 + \dfrac{\lambda(h^e)^2}{3} & -1 + \dfrac{\lambda(h^e)^2}{6} \\ -1 + \dfrac{\lambda(h^e)^2}{6} & 1 + \dfrac{\lambda(h^e)^2}{3} \end{vmatrix} = 0$$

and the solution is

$$\lambda_1 = 0, \qquad \lambda_2 = -12/(h^e)^2$$

The stability condition of Eq. (7.76) will thus be satisfied if the time step Δt_n adopted is such that

$$\Delta t_n < \frac{(h^e)^2}{6}$$

for all elements e in space. For the explicit (lumped) scheme we have, using Eq. (7.57), that

$$\mathbf{c}^e = h^e \begin{bmatrix} \frac{1}{2} & 0 \\ 0 & \frac{1}{2} \end{bmatrix},$$

and the element eigenvalues become in this case

$$\lambda_1 = 0, \qquad \lambda_2 = -\frac{4}{(h^e)^2}$$

The stability criterion is thus

$$\Delta t_n < \frac{(h^e)^2}{2}$$

and so for this problem the lumped scheme allows the use of a time step which is three times greater than that allowed for with the unlumped $\mathbf{C}$ matrix.

EXERCISES

7.12. Produce a two-level time-stepping scheme for Eq. (7.32) by a least-squares-type method and minimizing $I = \int_{t_n}^{t_{n+1}} R^2 \, dt$ with respect to $\mathbf{a}^{n+1}$. Show that the scheme involves only symmetric matrices, even if the matrices $\mathbf{C}$ and $\mathbf{K}$ are nonsymmetric.

7.13. Apply the least-squares method to the equation of Example 7.4 and investigate its stability and oscillation characteristics. Compare the performance of the scheme with that of the other methods shown in Fig. 7.5, using $\bar{\lambda} \, \Delta t_n = 0.5$.

7.14. Return to Example 7.3 and solve the equation $\mathbf{C} \, d\boldsymbol{\phi}/dt + k\mathbf{K}\boldsymbol{\phi} = \mathbf{f}$ by an explicit method, using a suitable value for the time step. Perform a few steps of the process and compare the computed temperature variation at the center of the region with that shown in Fig. 7.3.

7.15. By extending the technique used in Eq. (7.57), produce lumped element matrices $\mathbf{c}^e$ when $\alpha = 1$, for the following elements: (a) The one-dimensional Lagrange cubic element with nodes placed at the sampling points of the Gauss–Labatto rule of Eq. (5.27); (b) The four-noded square bilinear element; (c) The three-noded triangular element of Exercise 5.8; (d) The eight-noded serendipity element of Exercise 5.16; (e) The nine-noded Lagrange element of Exercise 5.16.

7.16. The problem of Example 7.3 is to be solved using the simultaneous finite element discretization of the space–time domain shown in the

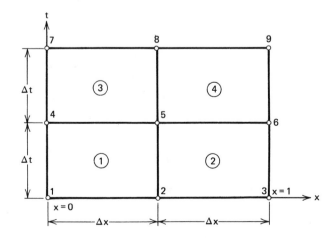

figure. Determine the value of the solution at node 8, using a suitable value for Δt.

7.5.5. Second-Order Equations

We can solve the general second-order equation (7.11)

$$\mathbf{M}\frac{d^2\mathbf{a}}{dt^2} + \mathbf{C}\frac{d\mathbf{a}}{dt} + \mathbf{Ka} = \mathbf{f} \tag{7.80}$$

by extending the methods of the preceding sections.

If we approach the solution of this equation directly, as was done in Section 7.5.1 for first-order equations, we may again write

$$\mathbf{a} \approx \hat{\mathbf{a}} = \sum_{m=1}^{\infty} \mathbf{a}^m N_m(t) \tag{7.81}$$

but we note that now the trial functions $N_m(t)$ have to be at least of degree 2, as a second derivative with respect to time has to be represented. Consider therefore a typical three-noded quadratic element n in time with nodes placed at the points t_{2n}, t_{2n+1}, and t_{2n+2}, as shown in Fig. 7.8. On this element we

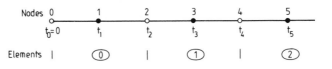

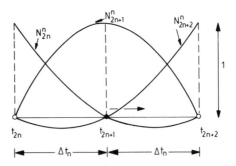

FIGURE 7.8. (*a*) Subdivision of the time domain by means of quadratic finite elements. (*b*) Element shape functions on element n.

have

$$\hat{a} = a^{2n}N_{2n}^n + a^{2n+1}N_{2n+1}^n + a^{2n+2}N_{2n+2}^n \tag{7.82}$$

since all the other trial functions are zero on element n, where

$$N_{2n}^n = -\frac{T(1-T)}{2}, \qquad \frac{dN_{2n}^n}{dt} = \frac{-\frac{1}{2}+T}{\Delta t_n} \qquad \frac{d^2N_{2n}^n}{dt^2} = \frac{1}{\Delta t_n^2}$$

$$N_{2n+1}^n = 1 - T^2, \qquad \frac{dN_{2n+1}^n}{dt} = -\frac{2T}{\Delta t_n} \qquad \frac{d^2N_{2n+1}^n}{dt^2} = -\frac{2}{\Delta t_n^2}$$

$$\tag{7.83}$$

$$N_{2n+2}^n = \frac{T(1+T)}{2}, \qquad \frac{dN_{2n+2}^n}{dt} = \frac{\frac{1}{2}+T}{\Delta t_n} \qquad \frac{d^2N_{2n+2}^n}{dt^2} = \frac{1}{\Delta t_n^2}$$

$$T = \frac{t - t_{2n+1}}{\Delta t_n}, \qquad \Delta t_n = t_{2n+2} - t_{2n+1}, \qquad \Delta t_n = t_{2n+1} - t_{2n},$$

The standard weighted residual technique applied to the second-order equation (7.80) gives

$$\int_0^\infty \left(M\frac{d^2\hat{a}}{dt^2} + C\frac{d\hat{a}}{dt} + K\hat{a} - f \right) W_n \, dt = 0, \qquad n = 0, 1, 2, \ldots \tag{7.84}$$

and, restricting consideration to weighting functions W_n which are only non-zero on element n, the above becomes

$$\int_{t_{2n}}^{t_{2n+2}} \left(M\frac{d^2\hat{a}}{dt^2} + C\frac{d\hat{a}}{dt} + K\hat{a} - f \right) W_n \, dt = 0, \qquad n = 0, 1, 2, \ldots \tag{7.85}$$

Using the results of Eq. (7.83) and performing an analysis similar to that used in deriving Eq. (7.45), we find that we can write this equation as

$$\left[M + \gamma\Delta t_n C + \beta\Delta t_n^2 K \right] a^{2n+2}$$

$$+ \left[-2M + (1 - 2\gamma)\Delta t_n C + \left(\tfrac{1}{2} - 2\beta + \gamma\right)\Delta t_n^2 K \right] a^{2n+1}$$

$$+ \left[M - (1 - \gamma)\Delta t_n C + \left(\tfrac{1}{2} + \beta - \gamma\right)\Delta t_n^2 K \right] a^{2n} = \bar{f}^n\Delta t_n^2 \tag{7.86}$$

in which

$$\gamma = \int_{-1}^{1} \left(T + \tfrac{1}{2}\right) W_n \, dT$$

$$\beta = \frac{\int_{-1}^{1} \tfrac{1}{2}T(T + 1) W_n \, dT}{\int_{-1}^{1} W_n \, dT} \tag{7.87}$$

$$\bar{\mathbf{f}}^n = \frac{\int_{-1}^{1} W_n f(t_{2n+1} + T\Delta t_n) \, dT}{\int_{-1}^{1} W_n \, dT}$$

and where the matrices $\mathbf{M}, \mathbf{C}, \mathbf{K}$ have been assumed constant.

Again, if $\mathbf{f}$ is interpolated in the same manner as $\hat{\mathbf{a}}$, it follows that

$$\bar{\mathbf{f}}^n = \beta \mathbf{f}^{2n+2} + \left(\tfrac{1}{2} - 2\beta + \gamma\right)\mathbf{f}^{2n+1} + \left(\tfrac{1}{2} + \beta - \gamma\right)\mathbf{f}^{2n} \tag{7.88}$$

Equation (7.86) corresponds to the general algorithm first derived by Newmark[4] and is one of the best known recurrence relationships for second-order equations.

Once again various forms of weighting functions can be used, and in Fig. 7.9 we display the values of β and γ corresponding to a series of such weightings. Newmark recommended that the value $\gamma = \tfrac{1}{2}$ should generally be taken, and it can be seen that this corresponds to symmetric weighting functions of all forms. If $\beta = 0$ and the matrices $\mathbf{M}$ and $\mathbf{C}$ are diagonal, then no inversion is necessary to determine $\mathbf{a}^{2n+2}$ and the scheme is explicit.

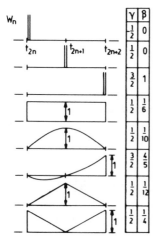

FIGURE 7.9. Weighting functions for some three-level schemes for second-order equations.

However, just as in the case of first-order equations, we shall find here that stability is then conditional, and the time interval Δt_n has to be suitably limited.

It can be observed from Eq. (7.86) that these three-level schemes require two starting values, $\mathbf{a}^0$ and $\mathbf{a}^1$, to begin the computation process, whereas frequently the initial conditions are given in the form

$$\mathbf{a}|_{t=0} = \mathbf{a}^0, \qquad \frac{d\mathbf{a}}{dt}\bigg|_{t=0} = \mathbf{b}^0 \qquad (7.89)$$

In this case, again, it is possible to adopt various starting schemes which use the given conditions to calculate the required value $\mathbf{a}^1$.

7.5.6. Stability of Three-Level Schemes for Second-Order Equations

For simplicity we shall restrict ourselves to considerations of the problem of stability of three-level schemes applied to second-order equations in which the first derivative is absent.[5] As in Section 7.5.4, we consider the free response (i.e., $\mathbf{f} = \mathbf{0}$) of such an equation, and the numerical scheme of Eq. (7.86) then reduces to

$$\left[1 + \beta \Delta t_n^2 \mathbf{M}^{-1}\mathbf{K}\right]\mathbf{a}^{2n+2} + \left[-2 + \left(\tfrac{1}{2} - 2\beta + \gamma\right)\Delta t_n^2 \mathbf{M}^{-1}\mathbf{K}\right]\mathbf{a}^{2n+1}$$

$$+ \left[1 + \left(\tfrac{1}{2} + \beta - \gamma\right)\Delta t_n^2 \mathbf{M}^{-1}\mathbf{K}\right]\mathbf{a}^{2n} = \mathbf{0} \qquad (7.90)$$

If $\omega_1^2, \omega_2^2, \ldots, \omega_M^2$ are the eigenvalues, and $\alpha_1, \alpha_2, \ldots, \alpha_M$ are the corresponding eigenvectors of the eigenvalue problem (see Section 7.4.1),

$$\mathbf{K}\alpha - \omega^2 \mathbf{M}\alpha = \mathbf{0} \qquad (7.91)$$

then we can write in general

$$\mathbf{a}^{2n} = \sum_{m=1}^{M} y_m^{2n}\alpha_m \qquad (7.92)$$

and it follows from Eq. (7.90) that

$$\left[1 + \beta \Delta t_n^2 \omega_m^2\right]y_m^{2n+2} + \left[-2 + \left(\tfrac{1}{2} - 2\beta + \gamma\right)\Delta t_n^2 \omega_m^2\right]y_m^{2n+1}$$

$$+ \left[1 + \left(\tfrac{1}{2} + \beta - \gamma\right)\Delta t_n^2 \omega_m^2\right]y_m^{2n} = 0 \qquad (7.93)$$

Assuming a solution of the form

$$y_m^{2n} = A\mu^{2n} \qquad (7.94)$$

where A is a constant, we find on substitution into Eq. (7.93) that a solution of the assumed form exists if $\mu = \mu_1$ or $\mu = \mu_2$ where μ_1 and μ_2 are the roots of the quadratic equation

$$\left[1 + \beta \Delta t_n^2 \, \omega_m^2\right]\mu^2 + \left[-2 + \left(\tfrac{1}{2} - 2\beta + \gamma\right) \Delta t_n^2 \, \omega_m^2\right]\mu$$
$$+ \left[1 + \left(\tfrac{1}{2} + \beta - \gamma\right) \Delta t_n^2 \, \omega_m^2\right] = 0 \qquad (7.95)$$

It has been noted previously that if the matrices $\mathbf{K}$ and $\mathbf{M}$ are positive definite, then the eigenvalues of Eq. (7.91) are real distinct positive numbers, and the exact solution in this case [Eq. (7.18)] is an undamped persisting oscillation. The numerical solution of Eq. (7.94) exhibits oscillatory behavior if μ_1 and μ_2 are complex, and the oscillation remains bounded if the modulus of these complex values satisfies

$$|\mu_i| \leqslant 1, \qquad i = 1, 2 \qquad (7.96)$$

The numerical solution will thus be stable and of the form of an undamped oscillation as required if

$$|\mu_i| = 1, \qquad i = 1, 2 \qquad (7.97)$$

while the numerical scheme will remain stable, but will be artificially damped if

$$|\mu_i| < 1, \qquad i = 1, 2 \qquad (7.98)$$

The solution of Eq. (7.95) can be written

$$\mu = \frac{(2 - g) \pm \sqrt{(2 - g)^2 - 4(1 + h)}}{2} \qquad (7.99)$$

where

$$g = \frac{\left(\tfrac{1}{2} + \gamma\right)\omega_m^2 \, \Delta t_n^2}{\left(1 + \beta \omega_m^2 \, \Delta t_n^2\right)}$$

$$h = \frac{\left(\tfrac{1}{2} - \gamma\right)\omega_m^2 \, \Delta t_n^2}{\left(1 + \beta \omega_m^2 \, \Delta t_n^2\right)} \qquad (7.100)$$

and these roots are complex if

$$4(1 + h) > (2 - g)^2 \qquad (7.101)$$

that is, if

$$-\omega_m^2 \, \Delta t_n^2 \left[4\beta - \left(\tfrac{1}{2} + \gamma \right)^2 \right] < 4 \tag{7.102}$$

The modulus of the roots can now be evaluated as

$$|\mu| = \sqrt{\mu\mu^*} = \sqrt{1 + h} \tag{7.103}$$

and Eq. (7.96) imposes the requirement that

$$-1 < h \leqslant 0 \tag{7.104}$$

Analyzing the inequalities of Eqs. (7.102) and (7.104), it can be seen that the three-level scheme will be unconditionally stable if

$$\beta > \tfrac{1}{4} \left(\tfrac{1}{2} + \gamma \right)^2$$

$$\gamma \geqslant \tfrac{1}{2} \tag{7.105}$$

$$\tfrac{1}{2} + \gamma + \beta > 0$$

If these conditions are not satisfied, stability can still be achieved provided that the time step Δt_n, which is adopted, is such that the inequality of Eq. (7.102) is satisfied.

For the schemes shown in Fig. 7.9 it is easy to check that only two ($\gamma = \tfrac{3}{2}$, $\beta = 1$ and $\gamma = \tfrac{1}{2}$, $\beta = \tfrac{1}{4}$) are unconditionally stable. All schemes for which $\gamma = \tfrac{1}{2}$ and $\beta \geqslant \tfrac{1}{4}$ are unconditionally stable and show no artificial damping, whereas schemes with $\gamma > \tfrac{1}{2}$ show an appreciable numerical damping, leading to an inaccuracy of the results at large values of Δt_n. This is illustrated in Fig. 7.10, which shows a plot of $|\mu|$ against $\omega \, \Delta t_n / 2\pi$ for various schemes applied to the equation $d^2a/dt^2 + \omega^2 a = 0$ and we note the large damping which is present as Δt_n increases. The damping present at high frequencies is often a desirable feature as it helps reduce spurious oscillations in the higher modes.

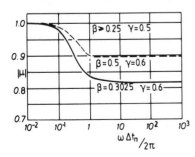

FIGURE 7.10. Variation of $|\mu|$ with $\omega \Delta t_n / 2\pi$ for some three-level schemes applied to the equation $d^2a/dt^2 + \omega^2 a = 0$.

For modes of importance a criterion such as $\omega_m \Delta t_n / 2\pi < 10^{-1}$ is normally adopted and the effect of numerical damping on these modes is then small.

7.5.7. Nonlinear Time-Dependent Problems

A major field of application of numerical techniques is to the solution of nonlinear time-dependent problems where the coefficient matrices in Eq. (7.32) or Eq. (7.80) depend upon the unknown vector $\mathbf{a}$. Now the alternative approach of Section 7.4, in which analytical solution procedures were applied, will prove unsuccessful. The general weighted residual statements of Eq. (7.36) and (7.84) for first- and second-order equations are still applicable, but the resulting two- and three-level schemes of Eqs. (7.41) and (7.86) are no longer valid. For consistency, a full (numerical) integration of the weighted residual statement is necessary, and iteration within each time step is generally required.

In many practical applications the difficulties associated with the integration in time are avoided by use of point collocation procedures; for example, taking $W_n = \delta(T - \theta)$ directly in Eq. (7.39) produces

$$\frac{\mathbf{C}(\mathbf{a}^{n+\theta})}{\Delta t_n}(-\mathbf{a}^n + \mathbf{a}^{n+1}) + \mathbf{K}(\mathbf{a}^{n+\theta})((1-\theta)\mathbf{a}^n + \theta\mathbf{a}^{n+1}) = \mathbf{f}^{n+\theta} \quad (7.106)$$

where $\mathbf{a}^{n+\theta}$ and $\mathbf{f}^{n+\theta}$ can be evaluated by using the approximation of Eq. (7.52), and this equation can be solved iteratively for $\mathbf{a}^{n+1}$.

The stability analyses which have been performed in the preceding sections will also not be valid for nonlinear equations. However, it is possible to obtain some information on the likely performance of the numerical schemes by performing a local linearization of the equation over one time step. For example, in the case of first-order equations we could then consider the stability over the time interval $[t_n, t_{n+1}]$ of the two-level scheme of Eq. (7.65) when applied to the linear equation

$$\mathbf{C}(\mathbf{a}^n)\frac{d\mathbf{a}}{dt} + \mathbf{K}(\mathbf{a}^n)\mathbf{a} = \mathbf{0} \quad (7.107)$$

The maximum time-step size Δt_n for stability and for the onset of oscillations will now be solution dependent and will vary as the solution proceeds.

EXERCISES

7.17. Investigate the stability characteristics of the three-level scheme used in Example 7.4.

7.18. The displacement ϕ of a spring-mass system is governed by the equation $d^2\phi/dt^2 + \phi = 0$ and is subject to the conditions $\phi = 0$, $d\phi/dt = 1$ at

$t = 0$. Compare the exact solution with the results produced by applying the three-level method of Eq. (7.86) with various values of β and γ.

7.19. Show that the general second-order equation

$$\mathbf{M}\frac{d^2\mathbf{a}}{dt^2} + \mathbf{C}\frac{d\mathbf{a}}{dt} + \mathbf{Ka} = \mathbf{f}$$

can be written as

$$\mathbf{C}_1\frac{d\mathbf{c}}{dt} + \mathbf{K}_1\mathbf{c} = \mathbf{f}_1$$

where $\mathbf{c}^T = (\mathbf{a}, \mathbf{b})$, $\mathbf{b} = d\mathbf{a}/dt$, and obtain the form of $\mathbf{C}_1$, $\mathbf{K}_1$, and $\mathbf{f}_1$. Write down an explicit two-level scheme for obtaining successive values of $\mathbf{c}$ and show that this scheme is unconditionally unstable when applied to the single equation $d^2a/dt^2 + \omega^2 a = 0$.

7.20. The propagation of waves on a stretched string fixed at both ends is governed by the equation $\partial^2\phi/\partial x^2 = (1/c^2)\,\partial^2\phi/\partial t^2$, where c is a constant and ϕ is the displacement. Such a string, of unit length and with $c = 1$, represented by three linear finite elements and is initially at rest with displacement $\phi = x(1 - x)$. Indicate how the solution may be obtained using the scheme of Eq. (7.86) with $\beta = 0$, $\gamma = \frac{1}{2}$ and perform a few steps in the process with a suitable value for the time step.

7.21. Investigate the stability properties of the three-level scheme of Eq. (7.86) when applied to the single equation $d^2a/dt^2 + \omega^2 a = 0$ with $\gamma = \frac{1}{2}$ and $\beta = \frac{1}{6}$. If this scheme is used to analyze the problem of Exercise 7.20, determine the maximum allowable value of the time step.

7.22. The displacement of a nonlinear spring-mass system is governed by the equation $d^2\phi/dt^2 + \phi(1 + \phi) = 0$ and is subject to the conditions $\phi = 1$, $d\phi/dt = 0$ at $t = 0$. Obtain a solution using the scheme of Eq. (7.86) with $\gamma = \frac{1}{2}$, $\beta = 0$.

7.23. Investigate the possibility of using four-noded bilinear space–time finite elements to solve the problem of Exercise 7.20.

REFERENCES

[1] The discussion of such techniques will not be made here, and the interested reader is referred to J. H. Wilkinson, *The Algebraic Eigenvalue Problem*, Clarendon, Oxford, 1965.

[2] O. C. Zienkiewicz, *The Finite Element Method*, 3rd ed., McGraw-Hill, New York, 1977.

[3] (a) B. M. Irons, *Applications of a Theorem on Eigenvalues to Finite Element Problems*, Department of Civil Engineering, Swansea, Report CR/132/70, 1970; (b) B. M. Irons and S. Ahmad, *Techniques of Finite Elements*, Ellis Horwood, Chichester, 1979.

[4]N. M. Newmark, "A method for computation of structural dynamics," *Proc. Am. Soc. Civ. Eng.* **85**, EM3, 67–94 (1959).

[5]For a fuller discussion of the stability of three-level schemes for general second-order equations, the reader can consult: (a) L. Fox and E. T. Goodwin, "Some new methods for the numerical integration of ordinary differential equations," *Proc. Camb. Philos. Soc.* **49**, 373–388, 1949; (b) K. J. Bathe and E. L. Wilson, "Stability and accuracy analysis of direct integration methods," *Int. J. Earthquake Eng. Struct. Dynam.* **1**, 283–291 (1973).

SUGGESTED FURTHER READING

S. H. Crandall, *Engineering Analysis*, McGraw-Hill, New York, 1956.

I. Fried, *Numerical Solution of Differential Equations*, Academic, New York, 1979.

A. R. Mitchell and D. F. Griffiths, *The Finite Difference Method in Partial Differential Equations*, Wiley, Chichester, 1980.

G. D. Smith, *Numerical Solution of Partial Differential Equations*, Oxford University Press, Oxford, 1971.

CHAPTER EIGHT ———————————

Generalized Finite Elements, Error Estimates, and Concluding Remarks

8.1. THE GENERALIZED FINITE ELEMENT METHOD

The main theme of this book has been the essential unity of the various approximation processes used in the numerical solution of physical problems governed by suitable differential equations. While the finite difference process of Chapter 1 appeared to present an entirely different type of approximation to the continuous trial function approach of Chapter 2, the local shape functions used in classical finite element forms of Chapter 3 closed the gap. It was indicated there not only that simple finite difference and simple finite element forms at times result in identical approximation equations, but that with suitable local expansions all finite difference expressions can be cast as particular examples of trial function finite element approximations.

Further, the boundary solution processes (touched upon in Chapter 2 and often categorized independently) turn out to be a closely related procedure. The common link of all the approximation processes is the expansion of the unknown function in terms of shape or basis functions and unknown parameters and the determination of such parameters from a set of weighted residual equations. In view of the growing popularity of the finite element method, and its systematic and well-documented procedures, we shall coin the definition of *the generalized finite element method* to embrace all of the different approaches mentioned. This has the advantage of a unified formulation, computer program organization, and theory to cover all approximation processes. Further, trivial arguments concerning the alleged superiority of say finite difference vis-à-vis finite elements become meaningless, as each subclass possesses special merit in particular circumstances.

Indeed it becomes possible to combine different types of approximation in a single, unified computer program, and much has been already achieved in this area. A typical example is the simultaneous use of boundary-type approximations (which are excellent for modeling singularities and infinite domains) and standard, classical finite elements (in regions where the geometric detail is better modeled by small irregular shapes).[1]

The use of global and local shape functions in a simultaneous manner is another one of the possibilities offered. We have observed that, for simple problems, the global shape functions of the type used in Chapter 2 often give good results with fewer parameters than a corresponding local approximation using standard finite element or finite difference forms. In cases where the problem only slightly differs from one in which a global solution is a good approximation, the finite element local form can be used as a hierarchical refinement.[2]

A final example of the advantages accruing due to the unification of concepts is perhaps the recent use of finite difference approximations on an irregular grid. Here the simple difference operators of Chapter 1 are no longer viable, but a generation of local polynomial shape functions based on neighboring nodal points is possible, and collocation or other weighting process will result immediately in a tenable approximation. This unified approach to the approximate numerical solution of real physical problems allows us to address, in a systematic manner, the most important and practical questions of the analysis. These are: (1) How accurate is the solution that has been achieved? (2) How can we achieve a solution of a desired accuracy? It is to answering these questions that the remaining sections of this chapter are addressed.

8.2. THE DISCRETIZATION ERROR IN A NUMERICAL SOLUTION

The errors in approximate, numerical solutions arise due to three primary causes. The first and most important is the *discretization* error, which is due to the incomplete satisfaction of the governing equations and their boundary conditions and is introduced by the trial function approximation. The second, the roundoff error, is due to the fact that only a finite amount of information may be stored at any stage of the calculation process. The third error is due to the approximations involved in the mathematical model to which the numerical solution is applied. With the growing precision of today's computers, the second of these errors can be minimized, while that involved in the mathematical model is clearly beyond the discussion possible in this book (in which we have postulated that the solution of the mathematical model is "exact"). We shall therefore concentrate here purely on the discretization error, which arises due to the approximation processes described.

So far we have limited our discussion of errors to statements [see Eqs. (1.17) and (4.2)], defining the order of discretization error in terms of the typical mesh size. This by itself does not determine the magnitude of the discretization errors, but if a series of solutions on meshes of uniformly decreasing size is available, it allows an approximate estimate of the correct answer to be obtained. However, much more information on the possible errors is desirable if good use is to be made of numerical approximation. Clearly, the ideal situation would be if, for any discretized solution, we could assert with some certainty that the error does not exceed some calculable value and indeed that this value is a reasonable estimate of the real error. This would allow the solution to be pursued by successive refinement until a predetermined precision is reached and, at the same time, increase the confidence of the user in the numerical approximation processes.

Although this objective may at first sound utopian (and indeed some feel that it could never be achieved, as the exact knowledge of error presupposes the knowledge of the exact solution), considerable achievements have already been made, and today practical use of error estimates can be made for realistic discrete computations. Once such an estimate has been made, users can decide whether their answers are satisfactory or whether further refinement of the solution is necessary. Indeed one can expect the knowledge of the distribution of errors to guide the manner in which this refinement is made, and such refinement can then be achieved automatically. Procedures of this kind are known as *adaptive*, and allow the computer to refine the mesh in an efficient manner until sufficient accuracy is achieved. Clearly, many strategies of refinement could be used, some being computationally more economic than others, but all achieving the desired aim. We shall not discuss this aspect further, as the interested reader can refer to the literature,[3] but here we shall briefly consider the crucial question of error estimates.

8.3. A MEASURE OF DISCRETIZATION ERROR

Throughout the major portion of this book we have been concerned with the solution of a general mathematical boundary value problem of finding an unknown function ϕ satisfying a differential equation

$$\mathcal{L}\phi + p = 0 \quad \text{in } \Omega \qquad (8.1a)$$

subject to boundary conditions

$$\mathcal{M}\phi + r = 0 \quad \text{on } \Gamma \qquad (8.1b)$$

where $\mathcal{L}$ and $\mathfrak{M}$ are general linear differential operators and p and r are defined functions.

The approximate solution was obtained by constructing a trial function expansion

$$\hat{\phi} = \sum_{m=1}^{M} a_m N_m \qquad (8.2)$$

and using a weighted residual process and, clearly, the local error E is simply defined by

$$E = \phi - \hat{\phi} \qquad (8.3)$$

where ϕ is the exact solution.

As we shall not be able to evaluate the error at all points, some alternative measures of E can be used. One convenient measure or norm is the so called *energy norm*, defined as

$$\|E\| = \left[\int_\Omega E \mathcal{L} E \, d\Omega \right]^{1/2} \qquad (8.4)$$

This norm is representative of the correctness with which the function is modeled by our approximation and, through application of Green's lemma or integration by parts, gives an idea of the accuracy to which derivatives, such as stresses in elasticity problems, are represented. Further it is possible with this norm to obtain close estimates of the error.

The definition of Eq. (8.4) is closely associated with the domain residual R_Ω which is given by

$$R_\Omega = \mathcal{L}\hat{\phi} + p \qquad (8.5)$$

assuming, for simplicity, that the boundary conditions are exactly satisfied by the expansion of Eq. (8.2). To show this, let us use Eq. (8.3) in Eq. (8.4) and expand to get

$$\|E\|^2 = \int_\Omega \left(\phi \mathcal{L} \phi + \hat{\phi} \mathcal{L} \hat{\phi} - \phi \mathcal{L} \hat{\phi} - \hat{\phi} \mathcal{L} \phi \right) d\Omega \qquad (8.6)$$

If the solution $\hat{\phi}$ is obtained by using the Galerkin method, it can be shown that

$$\int_\Omega \hat{\phi} \mathcal{L} \hat{\phi} \, d\Omega = - \int_\Omega \hat{\phi} p \, d\Omega \qquad (8.7)$$

and, inserting Eqs. (8.7) and (8.1a) into Eq. (8.6), gives

$$\|E\|^2 = -\int_\Omega (\phi - \hat{\phi})(\mathcal{L}\hat{\phi} + p)\,d\Omega \tag{8.8a}$$

With the definitions of Eqs. (8.3) and (8.5), this becomes

$$\|E\|^2 = -\int_\Omega E R_\Omega\,d\Omega \tag{8.8b}$$

We show in the next section how an expression of this form can be used in practical computation to produce an estimate of the discretization error.

8.4. ESTIMATE OF DISCRETIZATION ERROR

The domain residual R_Ω is easily determined for a given finite element approximation $\hat{\phi}$. It is, in general, composed of two parts, and to denote this we write

$$R_\Omega = R_1 + R_2 \tag{8.9}$$

The first component, R_1, is defined in a continuous manner within each element; the second component, R_2, is of the form of a Dirac delta function at element interfaces where often, as we have seen, some of the derivatives present in $\mathcal{L}\hat{\phi}$ are not defined. In such cases an integration by parts can be carried out and, if n is the normal and t the tangential direction to the interface I^e of an element e, then

$$\int_\Omega R_2\,d\Omega = \sum_{I^e \not\subset \Gamma} \int_{I^e} \Theta\,dt \tag{8.10}$$

where Θ represents a discontinuity jump of appropriate derivatives and the summation extends over the element boundaries not contained in Γ. For instance, if we are dealing with the Laplace equation, the jump quantity becomes

$$\Theta = \left.\frac{\partial \phi}{\partial n}\right|_{I^e} \tag{8.11}$$

In elasticity problems, similarly, Θ represents the discontinuity of tractions occurring on element interfaces.

With the knowledge of the residuals, expression (8.8) can be used to estimate the error. From a simple Taylor series expansion about an arbitrary point in the domain we know that, as we refine the mesh so that the measure of the element size h tends to zero, the error in the polynomial approximation of an arbitrary function tends in the limit to be one order higher than the approximating function itself.

Consider an expansion of the form of Eq. (8.2) with a solution $a_1, a_2, \ldots, a_M$. We now make a hierarchical addition of a single shape function N_{M+1} and a parameter a_{M+1}, where N_{M+1} represents the next order of interpolation, for example, if the solution is based on linear elements, N_{M+1} would be the quadratic hierarchical mode. The error, in the vicinity of the additional term, can be approximately written as

$$E \approx a_{M+1} N_{M+1} \tag{8.12}$$

and an approximation to a_{M+1} may be determined by using the weighted residual statement

$$\int_\Omega N_{M+1} \left[\mathcal{L}(\hat{\phi} + a_{M+1} N_{M+1}) + p \right] d\Omega = 0 \tag{8.13}$$

assuming that the parameters $a_1, a_2, \ldots, a_M$ remain unchanged. Here we must take care to interpret the integral in a generalized way, since we have already mentioned that $\mathcal{L} N_m$ may not exist. Then

$$a_{M+1} = - \left[\int_\Omega N_{M+1}(\mathcal{L} N_{M+1}) \, d\Omega \right]^{-1} \int_\Omega N_{M+1} R_\Omega \, d\Omega \tag{8.14}$$

and the reader will observe that the first term on the right-hand side can sometimes be written, after integration by parts, simply as

$$\left[\int_\Omega N_{M+1}(\mathcal{L} N_{M+1}) \, d\Omega \right]^{-1} = K^{-1}_{M+1, M+1} \tag{8.15}$$

which is the inverse of the appropriate matrix coefficient. We now have an expression for the error estimate, using Eq. (8.8b), as

$$\|E\|^2 = - \int_\Omega a_{M+1} N_{M+1} R_\Omega \, d\Omega = K^{-1}_{M+1, M+1} \left[\int_\Omega N_{M+1} R_\Omega \, d\Omega \right]^2 \tag{8.16}$$

As the addition of hierarchical functions may be made for any element, in general such terms could be evaluated over each element, and the individual element contributions could then be summed.

It has been noted that this estimate is only valid in some asymptotic sense as h tends to zero, and unfortunately it is not reliable on finite meshes; for example, N_{M+1} could be orthogonal to the domain residual, and then a zero error estimate would result. Also we have assumed that when the term $a_{M+1} N_{M+1}$ is added to the approximation, the original parameters $a_1, a_2, \ldots,$

a_M remain unchanged. Any interactive effect between the new mode and the original solution has therefore been ignored. Nevertheless Eq. (8.16) is extremely useful as it does *indicate* where the next refinement of hierarchical type is most effectively introduced and as such is extensively used in the process of adaptive refinement.

To derive a more reliable error estimator let us consider the special case of the equation

$$-\frac{d^2\phi}{dx^2} + p = 0 \tag{8.17}$$

and we attempt a solution using linear finite elements.

In the limiting situation discussed above the residual will here be constant on each element, and the hierarchical mode N_{M+1} is a quadratic function which is zero at the nodes, thus eliminating the singular part of the residual. We can now rewrite Eq. (8.16) as

$$\|E\|^2 = K_{M+1,\,M+1}^{-1} R_\Omega^2 \left[\int_\Omega N_{M+1}\, d\Omega \right]^2 \tag{8.18}$$

This can be generalized to the case of nonconstant residuals by evaluating a mean value of R_Ω^2 and arriving at the alternative formula

$$\|E\|^2 = K_{M+1,\,M+1}^{-1} \left[\int_\Omega N_{M+1}\, d\Omega \right]^2 \frac{\int_\Omega R_\Omega^2\, d\Omega}{\int_\Omega d\Omega} \tag{8.19}$$

We have thus ensured, for the linear elements, a nonzero error estimate. In Chapter 4 we also noted that for the problem defined by Eq. (8.17) the hierarchical modes can, by appropriate choice of hierarchic functions, lead to diagonal contributions to the final matrix so that no interactive effects occur and the estimate (8.19) is therefore "asymptotically exact" as h tends to zero.

We leave it as an exercise for the reader to show that, by substituting $N_{M+1}^e = (x/h^e)(1 - x/h^e)$ in Eq. (8.19), the error estimator gives for the element e

$$\|E\|_{[\Omega^e]}^2 = \frac{(h^e)^2}{12} \int_0^{h^e} R_{\Omega^e}^2\, dx \tag{8.20}$$

It remains a nontrivial problem to prove that these estimates are also reliable for more general differential equations when interactive effects occur, and the interested reader should refer elsewhere.[4] In two dimensions the following formula has been applied with some success in practical problems to estimate the error for a general element e when reasonable (obviously not improper) meshes of four-node quadrilateral elements are used:

$$\|E\|^2_{[\Omega^e]} = \frac{(h^e)^2}{24} \int_{\Omega^e} (R_{\Omega^e} - \overline{R}_e)^2 \, d\Omega + \frac{h^e}{24} \int_{I^e} \Theta^2 \, dt \qquad (8.21)$$

The energy norm of the error for the complete mesh follows by summation over all elements. Here $\overline{R}_e$ is the mean value of R_{Ω^e} on the element, and both the regular part of the residual on the domain and the singular part are considered because the hierarchical quadratic mode includes both an interior function confined to the element and an interface mode common to adjacent elements. The mean value $\overline{R}_e$ has here been subtracted, as to some extent the interface and element residuals "self-equilibrate," at least in the sense that

$$\int_{\Omega} R_{\Omega} N_m \, d\Omega = 0 \qquad (8.22)$$

Indeed it has been suggested that only the interface term of expression (8.21) should be considered in practice, as the first term makes a negligible contribution if the problem is sufficiently smooth. This to some extent justifies the heuristic arguments that slope-discontinuities of Eq. (8.10) or the traction jumps in elasticity problems are some measure of the error in the finite element solution.

The error estimates considered here are called a posteriori because they are based on the finite element solution which is substituted into the governing differential equation to identify the residual R_{Ω} and are very much the subject of current research.[3] Extensions are being made to higher order elements where it is apparent that a factor $d^{-\alpha}$ needs to be included in both Eq. (8.20) and Eq. (8.21), where d is the degree of the complete polynomial on the element and α a constant of order 2. It is obvious, however, that the stage has been reached when practical analysis can include a reliable error estimate. In the following examples we first consider a simple problem amenable to hand calculation, and then a problem of some practical significance.

Example 8.1

We will consider the equation

$$-\frac{d^2\phi}{dx^2} + Q = 0, \qquad 0 \leqslant x \leqslant 1$$

with boundary conditions $\phi = 0$ at $x = 0$ and $\phi = 0$ at $x = 1$.

We begin by assuming a constant source term defined by $Q = -1$ and attempt to solve the problem using a single linear finite element. Equation (8.5) gives for the linear element

$$R_\Omega = Q = -1$$

and the error estimate of Eq. (8.20) gives

$$\|E\|^2 = \frac{h^2}{12} \int_0^h R_\Omega^2 \, d\Omega$$

that is,

$$\|E\|^2 = \tfrac{1}{12}$$

To evaluate the accuracy of this estimate we need to know the exact error. The exact solution, found by integrating the governing equation and imposing boundary conditions, is

$$\phi = \tfrac{1}{2}x(1 - x)$$

Since the finite element will have both nodal parameters prescribed as zero, we have immediately from Eq. (8.3) that

$$E = \tfrac{1}{2}x(1 - x)$$

Substituting directly in Eq. (8.8) gives the exact energy norm of the error as

$$\|E\|_{\text{exact}}^2 = -\int_\Omega ER_\Omega \, d\Omega$$

so that

$$\|E\|_{\text{exact}}^2 = \tfrac{1}{12}$$

The error estimator has (as expected for the case of linear elements and constant R_Ω) given the exact value.

We now assume a linear form for the source term, that is, $Q = -x$, and again we consider the error for a single linear element. Equation (8.5) then gives

$$R_\Omega = Q = -x$$

with the error estimate given by

$$\|E\|^2 = \frac{1}{12} \int_0^1 x^2 \, dx$$

that is,

$$\|E\|^2 = \tfrac{1}{36}$$

The reader can verify that the exact solution in this case is

$$E = \tfrac{1}{6}x(1 - x^2)$$

so that the exact energy norm of the error is

$$\|E\|^2_{\text{exact}} = -\int_\Omega \frac{x}{6}(1 - x^2)(-x)\,dx$$

giving

$$\|E\|^2_{\text{exact}} = \tfrac{1}{45}$$

The error estimate has given here an upper estimate of the energy norm of the error.

Now we consider the addition of a single hierarchical quadratic function to the solution of this problem, so that the finite element is now quadratic. The reader should first verify that the finite element solution is

$$\hat{\phi} = \tfrac{1}{4}x(1 - x)$$

The residual given by Eq. (8.5) becomes

$$R_\Omega = -\frac{d^2\hat{\phi}}{dx^2} + Q = \tfrac{1}{2} - x$$

and applying Eq. (8.20) with the suggested d^{-2} factor for the quadratic element produces

$$\|E\|^2 = \frac{1}{12} \times \frac{1}{2^2} \int_0^1 \left(\frac{1}{2} - x\right)^2 dx = 0.0017$$

As above, the exact energy norm of the error is given by

$$\|E\|^2_{\text{exact}} = -\int_0^1 E R_\Omega\,dx$$

$$= -\int_0^1 \left[-\frac{x^3}{6} + \frac{x}{6} - \frac{x}{4}(1 - x)\right]\left(\frac{1}{2} - x\right) dx$$

with the result that $\|E\|^2_{\text{exact}} = 0.0014$.

Again the error estimate is sufficiently accurate for practical purposes.

___

Example 8.2

As a final practical example we consider the finite element analysis of a gravity dam using an adaptive algorithm and a two-dimensional error estimator based

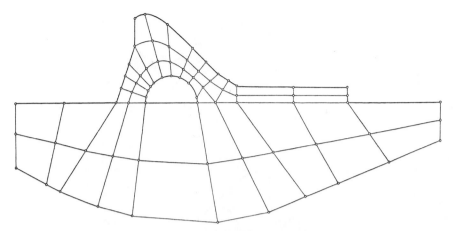

FIGURE 8.1.   Base finite element mesh for hierarchical analysis of a dam.

on the hierarchical projections discussed in this chapter. The mesh of bilinear finite elements used for the initial solution is given in Fig. 8.1. New hierarchical modes were added to these elements in a sequence of solutions which followed. A special adaptive algorithm was used in which only those degrees of freedom that best absorbed the estimated error in the solution were added to the mesh. In the final mesh the polynomial degree therefore varies from element to element, but compatibility is preserved by ensuring that the same polynomial degree exists along any interface between elements.[5]

In Fig. 8.2 the maximum principal stress on the outer surface of the dam and on the inner hollow gallery is plotted at three stages in the process. The predicted energy of the error $\|E\|$ is tabled together with a second estimate which is an upper bound on the error. An estimate of the maximum error in stress was also derived and is plotted on each solution.

A comparison of the three solutions indicates the strong convergence of the finite element solution for the stress to the "exact" solution indicated by the dashed line. This "exact" solution was obtained by analyzing the problem with a highly refined finite element mesh. The simultaneous convergence of the energy of the error and the error estimates to zero is also apparent.

EXERCISES

8.1.   Return to Section 8.3 and prove that $\|E\|^2 = -\int_\Omega \phi R_\Omega \, d\Omega$.

8.2.   Repeat the analysis of Example 8.1 for the differential equation $-d^2\phi/dx^2 + \phi = -Q$ and determine the accuracy of the error estimator of Eq. (8.20).

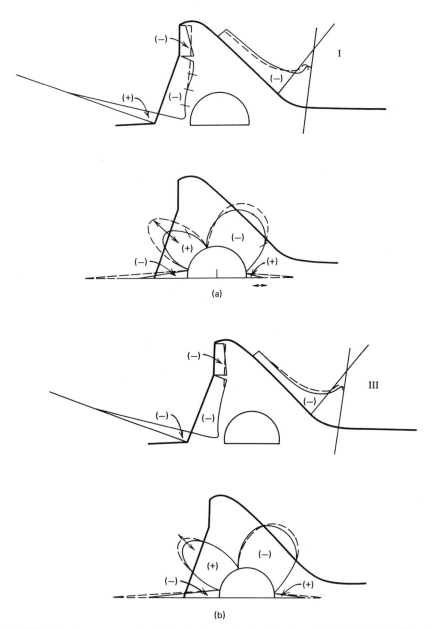

FIGURE 8.2. Convergence of solutions in the analysis of the dam of Fig. 8.1. (*a*) Initial step using linear elements. The percentage errors in the total energy are $\|E\| \leqslant 14.6$ (exact), $\|E\| = 7.3$ (by estimates), and $\|E^*\| = 21.6$ (by corrective upper bound estimates). Number of unknowns = 118. (*b*) Third step. The percentage errors in the total energy are $\|E\| \leqslant 6.0$ (exact), $\|E\| = 3.6$ (by estimates), and $\|E^*\| = 7.8$ (by corrective upper bound estimates). Number of unknowns = 206. (*c*) Fifth step of adaptive refinement process. The percentage errors in the total energy are $\|E\| \leqslant 3.0$ (exact), $\|E\| = 2.9$ (by estimates), and $\|E^*\| = 4.8$ (by corrective upper bound estimates). Number of unknowns = 365.

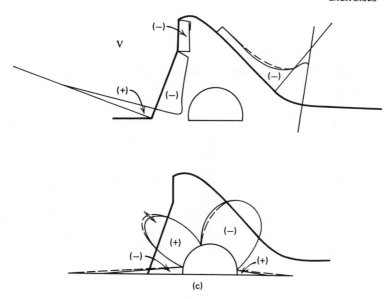

FIGURE 8.2. (*continued*).

8.3. Consider the linear four-node element applied to the two-dimensional Laplace equation

$$\frac{\partial^2\phi}{\partial x^2} + \frac{\partial^2\phi}{\partial y^2} = 0 \quad \text{in } \Omega$$

with $\phi = \bar{\phi}$ on Γ_ϕ and $\partial\phi/\partial n = \bar{q}$ on Γ_q.
(a) Show that the regular part of the residual on the domain is zero. (b) Try to derive the second term in Eq. (8.21) from Eq. (8.16) using the interface quadratic hierarchical shape function indicated in figure *a*. You should obtain the same expression except that the coefficient 24 is replaced by 28.8. (c) Repeat the calculation of (b) above, but now using as N_{M+1} in Eq. (8.16) the cylindrical quadratic function shown in figure *b*. You should now obtain a coefficient of 24.

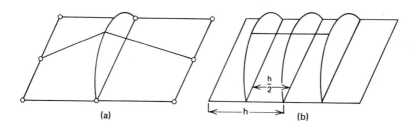

8.5. THE STATE OF THE ART

In this book we have endeavored to introduce to the reader the basic concepts of numerical approximation used in practical solutions of problems governed by differential equations. Many details and elaborations have been left unmentioned and will have to be sought in larger treatises and the current literature. We hope, however, that with the tool kit obtained the reader will be able to see more clearly through the accomplishment and aims of research currently in progress and at the same time to apply intelligently well-established procedures.

The ultimate aim of obtaining the most effective numerical discretization to a new physical problem encountered by the reader and the possibility of devising a procedure which automatically "refines" to obtain a predetermined accuracy will always present difficulties that challenge the imagination. We hope that the material presented here will give sufficient background to enable an intelligent approach to be made.

REFERENCES

[1] O. C. Zienkiewicz, D. W. Kelly, and P. Bettess, Marriage à la mode—The best of both worlds (finite elements and boundary integrals), in *Energy Methods in Finite Element Analysis*, R. Glowinski, E. Y. Rodin, and O. C. Zienkiewicz, Eds., Wiley-Interscience, New York, 1979.

[2] C. D. Mote, Global-local finite element, *Int. J. Num. Mech. Eng.* **3**, 565–574 (1971).

[3] See the reference list of the paper by O. C. Zienkiewicz, D. W. Kelly, J. Gago, and I. Babuška, Hierarchical finite element approaches, error estimates and adaptive refinement, in *The Mathematics of Finite Elements and Applications*, J. Whiteman, Ed., Academic, New York, 1981.

[4] I. Babuška and W. C. Rheinboldt, A posteriori error estimates for the finite element method, *Int. J. Num. Meth. Eng.* **12**, 1597–1615 (1978), and also I. Babuška and W. C. Rheinboldt, Adaptive approaches and reliability estimations in finite element analysis, *Comp. Meth. Appl. Mech. Eng.* **17–18**, 519–540 (1979).

[5] For a full discussion of this problem and for details of the algorithm and the error estimate, the reader should consult D. W. Kelly, J. Gago, O. C. Zienkiewicz and I. Babuška, A posteriori error analysis and adaptive processes in the finite element method, *Int. J. Num. Meth. Eng.* (in press).

SUGGESTED FURTHER READING

J. P. de S. R. Gago, D. W. Kelly, and O. C. Zienkiewicz, A posteriori error analysis and adaptive processes in the finite element method, Department of Civil Engineering, University College Swansea report C/R/364/80, 1980.

J. P. de S. R. Gago, A posteriori error analysis and adaptivity for the finite element method, Ph.D. thesis, University of Wales, 1982.

INDEX